Compliments and best wishes to Bill Voiers and all the staff at Dynastat

[illegible]

Prentice Hall Advanced Reference Series

Computer Science

JAIN AND DUBES *Algorithms for Clustering Data*
MCCONNELL *Internetworking Computer Systems: Interconnecting Networks and Systems*
SHERE *Software Engineering and Management*

Engineering

DENNO *Power System Design and Applications for Alternative Energy Sources*
FERRY, AKERS, AND GREENEICH *Ultra Large Scale Integrated Microelectronics*
JOHNSON *Lectures on Adaptive Parameter Estimation*
MILUTINOVIC *Microprocessor System Design GaAs Technology*
QUACKENBUSH, BARNWELL III, AND CLEMENTS *Objective Measures of Speech Quality*

Science

BINKLEY *The Pineal: Endocrine and Nonendocrine Function*
CAROZZI *Carbonate Rock Depositional Models: A Microfacies Approach*
EISEN *Mathematical Methods and Models in the Biological Sciences: Linear and One-Dimensional Theory*
FRASER *Event Stratigraphy*
MCLENNAN *Introduction to Non-Equilibrium Statistical Mechanics*
PLISCHKE AND BERGERSEN *Equilibrium Statistical Physics*
WARREN *Evaporite Sedimentology*

PRENTICE HALL SIGNAL PROCESSING SERIES

Alan V. Oppenheim, Series Editor

ANDREWS AND HUNT *Digital Image Restoration*
BRIGHAM *The Fast Fourier Transform*
BRIGHAM *The Fast Fourier Transform and its Applications*
BURDIC *Underwater Acoustic System Analysis*
CASTLEMAN *Digital Image Processing*
COWAN AND GRANT *Adaptive Filters*
CROCHIERE AND RABINER *Multirate Digital Signal Processing*
DUDGEON AND MERSEREAU *Multi-Dimensional Digital Signal Processing*
HAMMING *Digital Filters, 2 ed.*
HAYKIN, ET AL. *Array Signal Processing*
JAYANT AND NOLL *Digital Coding of Waveforms*
KAY *Modern Spectral Estimation: Theory and Application*
KINO *Acoustic Waves: Devices, Imaging, and Analog Signal Processing*
LEA, ED. *Trends in Speech Recognition*
LIM *Speech Enhancement*
LIM AND OPPENHEIM *Advanced Topics in Signal Processing*
MARPLE *Digital Spectral Analysis with Applications*
MCCLELLAN AND RADER *Number Theory in Digital Processing*
MENDEL *Lessons in Digital Estimation Theory*
OPPENHEIM, ED. *Applications of Digital Signal Processing*
OPPENHEIM AND SCHAFER *Digital Signal Processing*
OPPENHEIM, WILLSKY, WITH YOUNG *Signals and Systems*
QUACKENBUSH, BARNWELL III, AND CLEMENTS *Objective Measures of Speech Quality*
RABINER AND GOLD *Theory and Application of Digital Signal Processing*
RABINER AND SCHAFER *Digital Processing of Speech Signals*
ROBINSON AND TREITEL *Geophysical Signal Analysis*
STEARNS AND DAVID *Signal Processing Algorithms*
TRIBOLET *Seismic Applications of Homomorphic Signal Processing*
WIDROW AND STEARNS *Adaptive Signal Processing*

Objective Measures of Speech Quality

Schuyler R. Quackenbush

Thomas P. Barnwell III

Mark A. Clements

School of Electrical Engineering
Georgia Institute of Technology
Atlanta

PRENTICE HALL
ENGLEWOOD CLIFFS, NEW JERSEY 07632

Library of Congress Cataloging-in-Publication Data

Quackenbush, Schuyler R.
Objective measures of speech quality / Schuyler R. Quackenbush, Thomas P. Barnwell III, Mark A. Clements.
p. cm.—(Prentice Hall advanced reference series) (Prentice Hall signal processing series)
Bibliography: p.
Includes index.
ISBN 0-13-629056-6
1. Speech processing systems. I. Barnwell, T. P. (Thomas Pinkney). II. Clements, Mark A. III. Title. IV. Series.
TK7882.S65Q33 1988 87-34756
621.38'0413—dc 19 CIP

Editorial and production supervision: Gloria L. Jordan
Manufacturing buyer: M. Gloriande

Prentice Hall Advanced Reference Series
Prentice Hall Signal Processing Series

Printed in the United States of America

10 9 8 7 6 5 4 3 2 1

ISBN 0-13-629056-6

Prentice-Hall International (UK) Limited, *London*
Prentice-Hall of Australia Pty. Limited, *Sydney*
Prentice-Hall Canada Inc., *Toronto*
Prentice-Hall Hispanoamericana, S.A., *Mexico*
Prentice-Hall of India Private Limited, *New Delhi*
Prentice-Hall of Japan, Inc., *Tokyo*
Simon & Schuster Asia Pte. Ltd., *Singapore*
Editora Prentice-Hall do Brasil, Ltda., *Rio de Janeiro*

To Lisa, Aina, and Melinda

Contents

Preface

This book is a result of more than ten years of research performed at the Georgia Institute of Technology between 1975 and 1985. The goal of this research was to develop new objective measures for accurately predicting the perceived quality of speech signals which have been moderately distorted as a result of transmission, coding, or other corruption. In this research we define an objective measure of speech quality to be any measure which can be computed directly from an original (undistorted) speech waveform and a processed (distorted) speech waveform, and which is able to predict the results of a subjective measure of speech quality.

Finding good objective quality measures for coded, distorted, or synthetic speech is one of the most important problems in speech processing. If an objective quality measure could be found which was highly correlated with the results of human preference tests, its utility would be undeniable. For example, it could replace subjective quality measures in the design of speech coding, synthesis and communication systems, in which the measure could be repeatedly applied in the course of the design so as to iteratively optimize the system's parameters. Alternatively, it could be an integral part of a speech system itself, providing the fidelity criteria to be dynamically optimized as part of the coding procedure. Relative to subjective tests, objective tests are less expensive to administer, give more consistent results, and are not subject to the human failings of administrator or subject. And since objective test results are consistent, tests applied at different times and with different personnel and testing facilities could be directly compared. This is not generally the case for subjective test results. Finally, and perhaps most importantly, the form of the objective measure itself can give valuable insight into the nature of the human speech perception process.

The fundamental problem in research into objective quality measures is that the complexities of the speech perception process make such objective measures very difficult to find. Nevertheless, this research has produced a number of objective measures whose performance is significantly better than that of commonly used objective measures.

This research focuses on identifying objective measures which predict speech quality rather than speech intelligibility or speech communicability. The "quality" of a speech signal is defined by the scores it receives on subjective speech quality tests. The particular subjective speech quality measure used throughout this research is the Diagnostic Acceptability Measure (DAM), which was developed and administered at the Dynastat Corporation. This subjective test is appropriate for measuring moderate to high quality speech signals, for which intelligibility tests give poor resolution and communicability tests are prohibitively expensive.

The basic approach used in the research was to generate a very large data base of distorted speech, and then to apply a subjective quality test (the DAM) to each of the distorted speech signals. In all, the data base included 1288 distorted utterances totaling 15456 sentences or 17.2 hours of speech. The distorting systems used to generate the data base were quite diverse, including many speech coding techniques as well as other time and frequency domain distortions. The distorted speech and the associated subjective responses were then used to compare the performance of different objective speech quality measures: an objective measure was tested by statistically determining its ability to predict the subjective quality responses over this very large and diverse set of distortions.

During the course of the research, several thousand distinct objective measures were defined and compared using statistical correlation analysis. The measures tested included most of the objective measures which have been commonly used for designing and testing speech coding systems over the last decade, along with many parametric variations of these measures. Hence, on one level, this research is a carefully controlled comparison of many previously proposed objective measures. The primary result of this part of the research was that most previously proposed measures do not perform very well when applied across a diverse distortion ensemble.

The most significant topic of this book, however, is the development of new objective quality measures. This research has yielded measures which have substantially improved performance relative to that of many commonly used objective measures. Though these new measures are not necessarily the best or the only solution to the problem of objective measurement of speech quality, their improved performance serves to validate our methodology for objective measure design and testing. Furthermore, the fact that these measures provide good quality estimates across an extremely diverse set of speech distortions indicates that they are general in nature rather than specific to a single type of speech degradation. Therefore they should be useful in a broad range of quality testing situations.

This book should be of interest to anyone working in the area of speech communication or speech and hearing research. This would include engineers responsible for the design of speech coders or speech communication systems and designers of speech recognition or analysis/synthesis systems. It also includes scientists investigating speech or psychoacoustic phenomena. The level of presentation in the book assumes that the reader has a good background in digital signal processing, while advanced material and elementary material in different fields are presented in a tutorial manner.

Synopsis

The nine chapters of this book are organized as follows:

Chapter 1 gives an overview of the problem of objective measurement of speech quality and broadly describes our approach to the testing of objective measures. A summary of the major research results is given.

Chapter 2 is a tutorial review of subjective and objective methods for measuring speech quality. It is compact yet comprehensive, and provides for the first time a unified presentation of all major methods of speech quality evaluation which have been documented in the literature.

Chapter 3 discusses the Diagnostic Acceptability Measure, the subjective quality test used in all of the speech quality research reported in this book. The *parametric*, *metametric*, and *isometric* scales of this test are described, as is the method of administering the test. There is also a brief description of how the test was developed.

Chapter 4 describes the method of testing objective measures of speech quality. The first part of the chapter is a detailed description of the three data bases used in the speech quality research: the undistorted speech data, the distorted speech data, and the subjective quality evaluations of each of the distorted speech signals. Next the method for testing objective quality measures using the three data bases is described. Finally, there is a tutorial presentation of the analysis techniques used in objective measure design and evaluation, those being regression analysis, regression diagnostics, regression validation and multidimensional scaling.

Chapters 5 through 9 present the objective measures and document their performance. Chapter 5 covers "basic" objective quality measures, or measures which are in current use and widely discussed in the literature. Chapter 6 covers measures based on models of the ear. Chapter 7 covers measures which yield improved performance by restricting themselves to narrow classes of speech distortions. Chapter 8 presents objective measures for estimating the parametric quality measures of the DAM. These measures attempt to identify the fundamental quantities that determine speech quality. Chapter 9 presents the results of combining all of the parametric objective measures of Chap. 8 into a single "composite" measure of speech quality. This measure turns out to have very good performance.

Finally, since all the research presented here has been previously published in some form, this book is in large part an attempt at coherently integrating these documents into a common framework. If at times the style and manner of presentation seem inconsistent, it is a reflection of the differences in style and authorship of the original documents. It is hoped that this does not substantially affect the readability of the book.

Acknowledgements

We would especially like to recognize the efforts of Bill Voiers, whose subjective test was a cornerstone in this research. Bill not only participated in the subjective testing phase of the research, but also was a major contributor to and editor of the chapter on the Diagnostic Acceptability Measure. We would like to express our deepest appreciation for both his technical contributions, and his support as colleague and friend. We would also like to thank Ira Panzer, of the Dynastat Corporation, for his diligence and enthusiasm in the administration of the subjective tests.

In addition, we would like to thank Aina Barnwell, N. S. Jayant, Adrienne Kavelle, and Peter Kroon for their efforts in reviewing portions of the manuscript. We would also like to thank Deborah Jackson for setting up the computing environment in which this book was edited and typeset, John Hansen for producing the entire set of DAM profile graphs, and Stacy Schultz and Kay Gilstrap for their secretarial support.

Much of the research reported in this book was funded by the National Science Foundation. Much of the supporting research was funded by the Defense Communications Agency.

Schuyler R. Quackenbush
Thomas P. Barnwell III
Mark A. Clements

Chapter 1

Introduction

This monograph is the collected results from a series of studies performed at Georgia Institute of Technology [1.1], [1.2], [1.3], [1.4]. Portions of this work were a collaborative effort between Georgia Tech and the Dynastat Corporation. Work done at the Digital Signal Processing Laboratory at the Georgia Institute of Technology involved the production of distorted and coded speech, the measurement of objective quality measures, and the analysis of the relationship between objective and subjective measures. Work done at the Dynastat Corporation included subjective quality testing and the associated analysis.

The task of investigating the relationships between objective and subjective measures of speech quality follows work by both Georgia Tech [1.5] and the Dynastat Corporation [1.6], [1.7] in related areas.

1.1 Measuring Speech Quality

In the near future, a large fraction of voice communication systems will be entirely digital. This is true for most commercial speech transmission and storage applications, including switched networks for voice communication, mobile radio, cellular telephones, voice mail, and talking machines. This is especially true in the area of defense related systems where security and anti-jamming capabilities are essential. Increased utilization of communication channels, such as satellite links in which a variety of data types must share the channel, make digital speech almost unavoidable. The need to use a channel for as many messages as possible also makes it highly desirable that the speech be coded in a very efficient form.

As everyone is aware, efficiency of representation is often at odds with the fidelity of the resulting speech. The history of speech coding has been an effort at pushing out the envelope of the efficiency-to-quality curve. Of these two quantities in question, the former is highly quantifiable (i.e., bit rate) whereas the latter is highly subject to interpretation. Some commonly used interpretations for quality or fidelity are signal-to-noise ratio, intelligibility, predicted intelligibility, communicability, and subjective preference or scaling evaluations. Although most of these are valuable, given the proper applications, there are a wide range of situations in which the fidelity measures are either highly misleading, or are even completely inappropriate. When competing digital speech coding systems are to be evaluated relative to one another, a method which is repeatable, meaningful, and able to reliably measure the sound quality of the speech is needed.

A possible solution to some of these problems is the use of a highly standardized procedure which uses human listeners to evaluate goodness. Although it has been possible to partially fulfill this need (see Chap. 3, which discusses the Diagnostic Acceptability Measure), such tests are expensive, difficult to administer, and often suspect due to the inherent non-repeatability of human responses. In addition, such tests seldom provide much insight into the factors which may lead to improvements in the evaluated systems. A computable measure of quality which is roughly equivalent in accuracy to that of subjective tests would therefore be highly desirable. Not only could candidate communication systems be evaluated for initial selection, but periodic field testing of all links would be possible.

This monograph presents a set of techniques which can be used for more effective and efficient operational speech quality testing. In general, these *objective fidelity measures* are computed from an "input" or "unprocessed" speech data set, S, and an "output" or "distorted" speech data set, S_Q, as shown in Fig. 1.1. The output speech data set results when the input speech data set is passed through the speech communication system under test. Objective fidelity measures range from very simple to highly complex. A complex measure might incorporate such diverse measures as a spectral distance or other parametric distances between the input and output speech data sets, semantic,

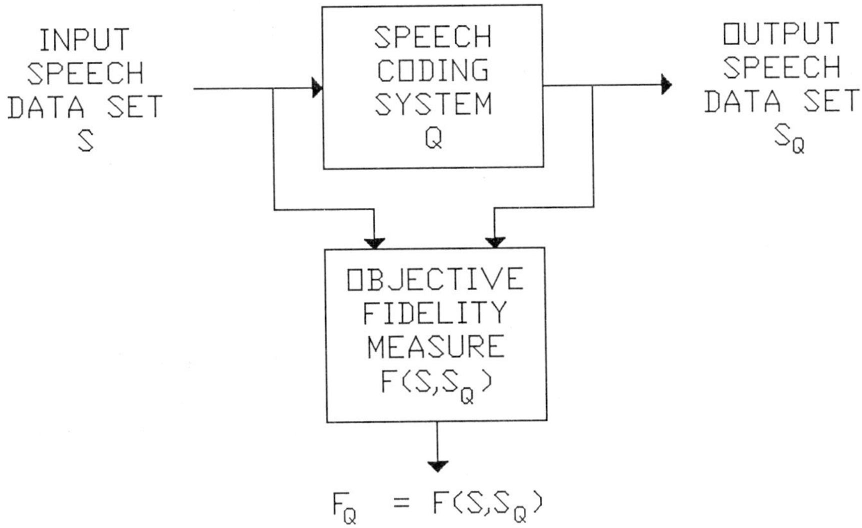

Figure 1.1: System for computing objective fidelity measures.

syntactic, or phonemic information extracted from the input speech data set, or the characteristics of the talker's vocal tract or glottis.

These techniques can be used to identify an objective fidelity measure that accurately predicts the results of human preference tests and which is also compactly computable. Such a measure would be extremely useful since, relative to subjective procedures, it would be less expensive to administer, would give more consistent results, and would not to be subject to the human failings of administrator or subject. It could be used in the design of speech coding systems to iteratively optimize the parameters of the coding system through repeated application of the quality measure—a process which is extremely expensive using subjective tests. Or, if the procedure were analytically tractable, the speech coding system could be designed to explicitly maximize the objective quality measure. In addition, the results of the objective measure applied at different times and at different locations could be compared directly. This is generally not the case for the results of subjective quality tests.

The goal of this research was to find an objective fidelity measure which is both highly correlated with subjective measures over all possible distortions and which is compactly computable. Although the speech perception process is poorly understood, it is apparent that the human listener is an active perceiver, responding to pragmatic, semantic, prosodic, syntactic, and talker related information as well as to phonemic content. In short, he uses his vast knowledge of the language in the speech perception process. The acoustic correlates of the various hierarchically structured elements of the language in the speech signal are simultaneously overlapping and redundant. This means that certain very small distortions which are properly placed with respect to the syntactic structure or the semantic content could cause complete loss of intelligibility, while other more extensive distortions might hardly be perceivable. Hence, it can be argued that objective fidelity measures which do not use high level or language related information could never accurately predict the quality of speech over a wide ensemble of coding systems.

On the other hand, modern speech coding systems typically do not produce distortions which are in any way synchronous with the semantic or syntactic content of the utterance. Hence, the distortions introduced by most speech coding systems represent only a subset of all possible distortions. It is our hypothesis that it is possible to design relatively compact objective measures which correlate well with subjective results over the limited set of distortions introduced by speech coding systems. However, we admit that these measures could never be completely general since they do not reflect the complexities of the speech perception process.

1.2 Testing Objective Quality Measures

Over the years, there have been numerous objective measures suggested and used for the evaluation of speech coding systems. Such measures include signal-to-noise ratios, arithmetic and geometric spectral distance measures, cepstral distance measures, various parametric distance measures, such as pseudo area functions and log area ratios from linear prediction analysis, as well as many more. Although each of these is often successful in predicting fidelity for limited classes of coders, many counterexample systems can be produced for each. (As a simple example, multiplying a signal by -1 yields a SNR of -6 dB)

The task of evaluating the validity of such measures over a nontrivial set of conditions is immense, and it is for this reason that the performance of many commonly used objective measures has not previously been investigated. Since the goal of objective measures is largely to replace the need for subjective procedures, the predictability of the latter from the former would be an appropriate vehicle for evaluation. To this end, we have designed a system for assessing the goodness of objective fidelity measures (i.e., a quality measure for quality measures). A block diagram of this system is shown in Fig. 1.2.

The system involves the creation of three data bases. The first, referred to as the *undistorted data base*, is a set of forty-eight phonetically balanced sentences spoken by three males and one female. This speech is recorded with high fidelity and then digitized by a computer for subsequent processing.

The next is the *distorted data base*, which results from processing the sentences in the undistorted data base in various manners. Two categories of distortions, coding and "controlled," are produced in this manner. Coding distortions result from applying realistic speech coding procedures, representative of most commonly used methods, to the undistorted sentences. Controlled distortions result when distortions which are not attributable to any specific class of coders, but are nonetheless similar to various isolated aspects of coder distortions, are imposed on the waveforms. Eleven classes of coding distortions, including both pitch-excited vocoder and waveform coder techniques, and fourteen classes of controlled distortions, including filtering, additive noise, peak clipping, center clipping, interruption, echo, and frequency variant distortions were produced. The coder distortions included both error free and fixed error rate channel simulations. The speech in this data base had a total duration of roughly eighteen hours. As illustrated in Fig. 1.2, the objective measures require access to both the distorted and undistorted waveforms, since it is through comparison that the degree of distortion is estimated.

The last is the *subjective data base*, which is the result of applying a subjective quality test to all sentences in the distorted data base. The subjective quality test used in this study was the Diagnostic Acceptability Measure (DAM) developed and administered at the Dynastat Corporation. This test has the special feature of providing *parametric* subjective results as well as *isometric* subjective results. The former term refers to measures of isolated features of the speech (e.g., "noisy" or "buzzy"), and the latter refers to global quality assessments (e.g., "acceptability"). This allows the objective measures to be tested for their ability to predict parametric as well as isometric quality assessments. In particular, many of the objective measures studied, including all of the frequency variant measures and the composite measures, could be designed to predict specific parametric results. Such specific predictions tend to be of greater utility to the systems designer than is a single measure of overall quality.

Two aspects of the data base are very significant. The first is its size: 322 distortions are applied to the twelve sentences for each of four speakers for a total of 15456 distorted sentences. The existence of so much data allows a high degree of confidence in the parameters estimated by the regression procedure, even when the number of parameters was rather large. The second is that the subjective tests were conducted by professionals with professional listeners at Dynastat Corporation using the DAM test. This test, when conducted by the individuals involved, is considered by many to achieve the most reliable and repeatable results available anywhere.

A particular candidate objective measure is tested using these data bases as follows. First, the objective quality measure is applied to the undistorted and distorted data bases. Then a statistical analysis is done to determine how well the objective measure predicts the subjective test results. The quantitative results of this analysis are used as a figure-of-merit for comparing the various objective measures.

Several points should be made about this procedure. First, note that the subjective tests are administered only once, regardless of how many objective measures are to be studied. Hence, the most expensive portion of this process, namely the compilation of the subjective data, need only be done once. Also note that the data base may be expanded over a period of time to extend the set of distortions involved. Similarly, subsets of the entire data base may be used if a quality measure over only a limited set of distortions (e.g., waveform coders) is desired.

Second, note that this "quality test for a quality measure" may be used to optimize the free parameters of an objective measure. This optimization may be accomplished either analytically, using techniques such as regression analysis, or heuristically, by applying the test to parametrically different versions of the same objective measure and observing which parameter set yields the best figure-of-merit.

Regression analysis provides a means of both functionally relating the objective results to the subjective results, and quantifying the "goodness" of the relationship. Simple linear regression produces a minimum variance estimate of the subjective results S_d given the objective results O_d, with the estimate computed from

$$\hat{S}_d = \bar{S}_d + \left(\frac{\hat{\rho}\hat{\sigma}_s}{\hat{\sigma}_o}\right)(O_d - \bar{O}_d) \tag{1.1}$$

where d is the distortion index, $\hat{\sigma}_s$ and $\hat{\sigma}_o$ are the estimated standard deviation of the subjective and objective measures, respectively, and the overbar indicates the average value over all d.

In the context of linear regression, two figures-of-merit are employed for each objective measure. The first is an estimate of the coefficient of correlation between the objective measures and the subjective measures, given by

$$\hat{\rho} = \frac{\sum_d(S_d - \bar{S}_d)(O_d - \bar{O}_d)}{[\sum_d(S_d - \bar{S}_d)^2]^{1/2}[\sum_d(O_d - \bar{O}_d)^2]^{1/2}} \tag{1.2}$$

A second figure-of-merit is an estimate of the variance of the error which results when an objective measure is used in place of the subjective test. This is given by

$$\hat{\sigma}_e^2 = \hat{\sigma}_s^2(1 - \hat{\rho}^2) \tag{1.3}$$

where $\hat{\sigma}_e$ is the estimated standard deviation of error. This statistic, which incorporates the variance of the subjective test results as well as the correlation coefficient, is possibly the more appropriate figure-of-merit.

It should be noted that estimates of correlation coefficients computed in this way are only meaningful when comparing objective measures over the same data base, and such estimates, in general, should not be compared when they are estimated from different data bases.

1.3 Research Objectives

The principal objective of the research reported here was to implement and test a large variety of objective measures. This included measures already discussed in the literature, extensions to those measures, and totally new forms of measures. The classes of measures investigated were: global, short-time, and frequency variant signal-to-noise ratio measures; simple spectral distance measures; frequency variant spectral distance measures; time-classified spectral distance measures; other classified measures; parametric distance measures; and composite measures. Two of the terms in the preceding list need clarification. Classified measures are those which classify the speech data into one of several categories and apply a separate measure to each of the categories. Composite objective measures are those which are formed using multiple linear regression on sets of simple measures.

Achieving these objectives involved numerous additional tasks which often led to interesting results in their own right. The most notable of these include: the design and construction of a large data base of distorted speech; the application of subjective quality tests to the distorted data base; and the analysis of subjective test results for each distortion in the data base. Chapter 4 discusses the results of each of these tasks in detail.

1.4 Summary of Major Results

In all, a total of over a million correlation analyses were performed involving more than 2000 different measures from many different classes. One summary of performance is to consider how each of these classes of measures perform relative to the others. Table 1.1 lists the performance of the best measure from each of several classes, as correlated over the entire data base.

Many other important conclusions can be drawn from this extensive set of experiments. A number of outstanding results follow:

1. A very good objective quality measure applicable to waveform coders and additive noise distortions was developed based on frequency variant (banded) short-time signal-to-noise measurements. This measure resulted in an estimated correlation coefficient of 0.93 across all relevant distortions and a $\hat{\sigma}_e$ of 3.3 quality points on a 100 point scale.
2. The best simple measure was found to be a inverted spectral distance measure, which had an estimated correlation coefficient of

Table 1.1: Performance of a sample of objective measures.

Objective Speech Quality Measure	$\lvert \hat{\rho} \rvert$
SNR	0.24*
Segmental SNR	0.77*
Frequency variant seg. SNR	0.93*
LPC-based measures:	
Linear predictor coefficient	0.06
Log predictor coefficient	0.11
Linear reflection coefficient	0.46
Log reflection coefficient	0.11
Linear area ratio	0.24
Log area ratio	0.62
Itakura energy ratio	0.59
Linear spectral distance	0.38
Inverse linear spec. distance	0.63
Log spectral distance	0.60
Nonlinear spectral distance	0.61
Polynomial regression modeling,	
Spectral distance	0.80
Energy ratio	0.64
Time-classified log spec. distance	0.63
Frequency variant log spec. distance,	
LPC-based	0.68
Filter bank	0.72
Frequency variant linear spec. dist.	0.68
Weighted-slope spectral distance	0.74
Composite measures	
Simple and freq. var. measures	0.86
Parametric objective measures	0.82

* SNR measures are only correlated across waveform coder distortions.

0.63 and a $\hat{\sigma}_e = 7.6$. However, this performance was not significantly better than many other spectral distance measures.

3. A simple measure that was nearly as good as the inverted spectral distance was the log area ratio measure, which had an estimated correlation coefficient of 0.62 and a $\hat{\sigma}_e = 7.7$. This is significant, in that the log area ratio requires an order of magnitude less computation than the spectral distance measures.
4. The value of p in the L_p norm in the best spectral distance measures had a consistent, yet relatively small, effect on the performance of the measures.
5. Similarly, frequency weighting in spectral distance measures had a consistent, but small, effect on the performance of the best measures.
6. Weighting the difference between spectra by the short-time energy was found to have little value for any of the measures.
7. The only two parametric measures which did well were the log area ratio measure and the energy ratio measure.
8. The frequency variant spectral distance measures performed with about a 0.1 point improvement in correlation over the simple measures. This was less improvement than was hoped for.

   sloppy
9. Critical band spectral decomposition was found to yield 0.04 points improvement in correlation over LPC-based frequency variant spectral representations.
10. The best measure based on aural models was the weighted slope spectral distance, with an estimated correlation coefficient of 0.74.
11. The best composite measure in the study had an estimated correlation coefficient of 0.84 and $\sigma_e = 4.6$. This measure involved the estimation of thirty-four regression coefficients, which was done as a pilot experiment without the appropriate statistical validity checks. A more conservative measure, done with careful statistical analysis, had an estimated correlation coefficient of 0.82 and $\sigma_e = 5.7$. In both cases, such composite measures typically performed much better than the simple measures on which they are based, and their performance can be considered an estimate of

the limit of the ability of objective measures to predict the results of subjective tests.

12. The use of polynomial regression analysis (3rd order and 6th order) often gave considerable improvement in the performance of both simple and composite objective measures. However, these results must be approached with caution, since better prediction of outlying data points may be the primary reason for improved performance.

13. The statistical reliability of virtually all of the better objective measures was quite high (≥ 0.99), as measured over different sentences in the data base. The reliability of the subjective measures was greater than 0.95.

1.5 Discussion

There are many aspects to this study. On the one hand, a major result is quantitative figures on the ability of a large set of objective measures to predict subjective test results. This permits quantitative comparisons to be made, often for the first time, between many of the commonly used objective quality measures. The research indicates that no single measure (with the noteworthy exception of the frequency variant short-time signal-to-noise ratio for waveform coders) predicts subjective responses well enough so as to entirely replace subjective testing.

On the other hand, the results of this study tell a good deal about the performance of the subjective measures themselves, and offer new data from which to improve the subjective measures. Furthermore, the subjective results can be used to judge the design of the distorted data base. These developments, once again, are quite important, but do not appreciably improve the overall quality testing environment.

The real potential for improvement comes from the use of composite objective measures. As previously stated, this study gives fairly safe estimates of $\hat{\rho} = 0.84$ and $\hat{\sigma}_e = 4.6$ for the best composite measures. However, there are several issues which need to be discussed here. First, the approach used in this study, which was necessitated by the mass of data involved, was essentially a "bulk" approach in which only standard multiple linear regression analysis and coarse, non-data-dependent preclassification was used. In order to design a final "best" measure, the results reported here should be studied to understand the detailed behavior of the composite measures as a function of particular distortions. Only after this is done can pragmatic variations of the composite measure be designed which allow for the specific interaction of the measure with the data. Second, it should be noted that this "best" result was obtained by setting a number of parameters in the composite objective measure so as to obtain maximum correlation across the data base. Thus, this should be considered an upper limit on expected performance.

Another point is that using polynomial regression instead of linear regression often resulted in significantly improved measure performance. Given the large number of data points (typically 1056), 3rd or 6th order polynomial models could be supported from a statistical point of view. However, it is not clear whether a 3rd or 6th order model is appropriate. Exactly how much improvement really can be obtained by polynomial regression is a subject for further study.

Measures based on critical-band filter bank analysis were consistently better than those based on LPC spectral representations. This result is not surprising, considering that the filter band representation more closely models the activity in the human auditory periphery. However, the improvements based on such models will be limited due to the omission of central processing models.

A major point which should be made concerns the reliability of the objective measures. For the number of frames of speech used in this study, the measured reliability was between 0.98 and 0.99 for most "good" measures. This means that whatever an objective measure actually evaluates for a distortion, it is repeatable on different, but equivalent, samplings of the data base. This also suggests that these measures could be utilized with great effectiveness for detecting malfunctions or nonstandard operation of systems in the field.

Some retrospective comment on the contents of the distorted data base is also appropriate. The data base was designed to include numerous frequency variant controlled distortions in order to facilitate the design of frequency variant objective measures. This worked well for time domain measures, but not nearly so well for frequency domain measures. Had this result been known at the outset, relatively more coding distortions would have been included.

A final point, concerning further work in this area, is that the same techniques developed here might also be used to predict the results of other subjective tests. The two most obvious classes of tests are parametric intelligibility tests, such as the DRT, and talker identification tests.

References

[1.1] T. P. Barnwell III and W. D. Voiers, "An Analysis of Objective Measures for User Acceptance of Voice Communication Systems," Final Report, DCA100-78-C-0003, Sep. 1979.

[1.2] T. P. Barnwell III, M. A. Clements, S. R. Quackenbush, and E. P. Farges, "Improved Objective Measures for Speech Quality Testing," Final Report, DCA100-83-C-0027, Sep. 1984.

[1.3] T. P. Barnwell III, "Improved Objective Speech Quality Measures for Low Bit Speech Compression," Final Report, National Science Foundation ECS-8016712, 1985.

[1.4] S.R. Quackenbush, "Objective Measures of Speech Quality," Ph.D. Thesis, Georgia Institute of Technology, 1985.

[1.5] T. P. Barnwell, III, A. M. Bush, R. W. Mersereau, and R. W. Schafer, "Speech Quality Measurement," Final Report, DCA Contract No. RADC-TR-78-122, June 1977.

[1.6] W. D. Voiers et al., "Methods of Predicting User Acceptance of Voice Communications Systems," Final Report, DCA 100-74-C-0056, DCA DCEC, Reston, VA, July 1976.

[1.7] W. D. Voiers, "Diagnostic Acceptability Measure for Speech Communication Systems," *Proc. 1977 IEEE ICASSP*, Hartford, CN, pp. 204–7, May, 1977.

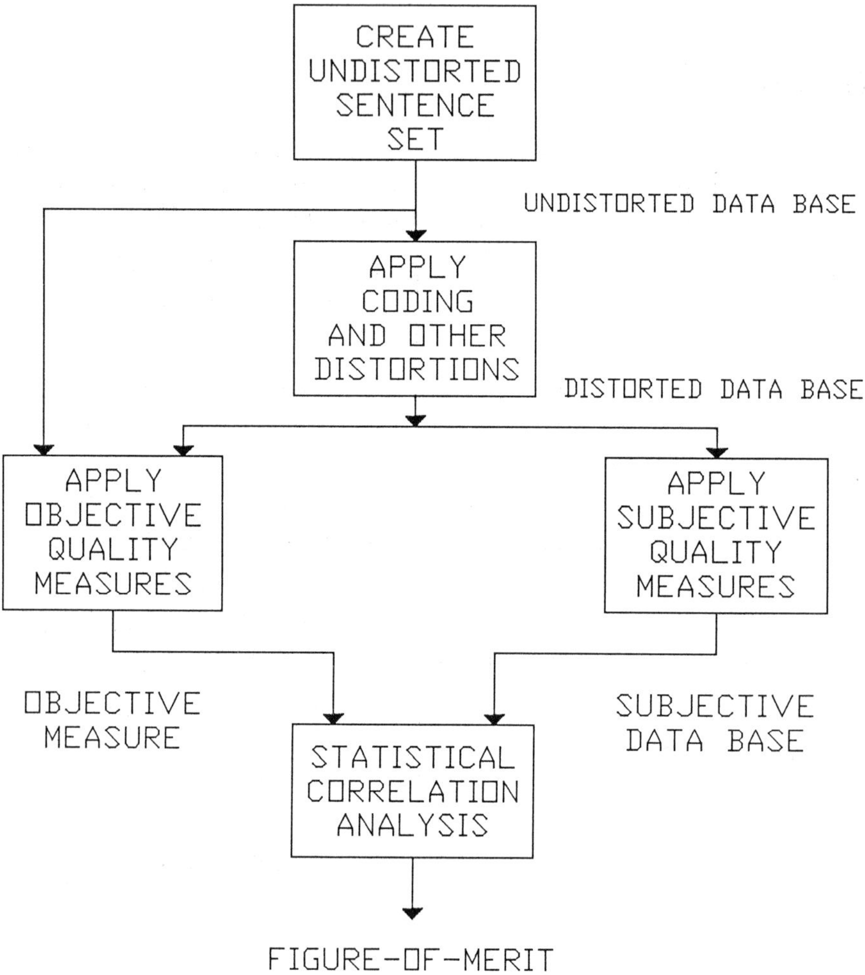

Figure 1.2: System for comparing objective quality measures.

Chapter 2

Review of Speech Quality Measures

Despite the fact that there are a number of papers that present reviews of subjective speech quality measures [2.1], [2.2], [2.3], [2.4], [2.5], none are both comprehensive and up to date. In this chapter, a comprehensive review of both objective and subjective speech quality measurement techniques will be presented.

2.1 Subjective Quality Measures

Obviously, any meaningful definition of *speech quality* must be based on human responses and perception. Objective quality measures, on the other hand, attempt to measure those physical characteristics of the speech signal that are correlated with factors that determine speech quality. Since an objective measure's success is generally evaluated by its ability to predict some subjective quality assessment, the performance of an objective measure cannot be dissociated from the subjective quality measure it estimates. This section is devoted to a discussion of subjective measures because of the important role they play in the development of objective measures.

Subjective measures can be broadly grouped into two categories: *utilitarian* and *analytic* [2.4]. Utilitarian methods have three goals: to be reasonably efficient in test administration and data analysis, to measure speech quality on a unidimensional scale, and to have good reliability in the test method. Reliability is always an important characteristic of any test, as is efficiency. However, the most important characteristic of the utilitarian methods is the reporting of their results on a unidimensional scale. With such a scale, results are summarized by a single number so that the relative quality of speech communication systems can be directly compared. In contrast, analytic methods seek to identify the underlying psychological components that determine perceived quality, and to discover the acoustic correlates of these components. Such methods typically use more than one dimension for reporting their results and are more oriented toward characterizing speech perception than toward measuring perceived quality. Unless explicitly stated, all subjective tests in this section are in the utilitarian category; only the methods discussed at the end of this section are of the analytic type.

2.1.1 Intelligibility Tests

2.1.1.1 Articulation Tests

The earliest highly quantitative subjective measures focused on speech intelligibility, or the degree to which basic speech sounds could be communicated through a speech system. In 1929, Fletcher and Steinberg [2.6] presented a method for assessing the quality of a voice channel in terms of its articulation value. In their test, *articulation* was defined as the percent of transmitted speech units that were correctly observed by the listener. The speech units could be phonemes, monosyllables, words, or sentences. Obviously, semantics affect the comprehension of speech sounds in word articulation as does syntax in sentence articulation. This generally results in scores for both word and sentence articulation which are higher than for phonemes or monosyllables.

Fletcher and Steinberg's methodology forms the basis for two kinds of tests: *intelligibility* tests, which are scored by the percentage of correct identification of the meaning conveyed by the transmitted speech; and *articulation* tests, which are scored by the percentage of correct identification of the sounds, words, or sentences in the transmitted speech. Since there is no meaning associated with nonsense monosyllables, only articulation can be associated with this type of test stimulus.

One problem with articulation tests is the difficulty of generating lists of monosyllables, words, or sentences in which all items are equally difficult to perceive. A large number of such lists are required to minimize memory effects in the subjective test. Since nonsense monosyllables have no associated meaning, they are least susceptible to memory effects and are therefore used most often. In Fletcher and Steinberg's original work, they constructed random lists of sixty-six nonsense monosyllables in the form of consonant-vowel-consonant (/C-V-C/). These lists were constrainted to have each leading and trailing consonant and each vowel used only once, with no syllables having forbidden leading or trailing consonants. In the test procedure, a crew of four men and four women was used to read the lists. Each member read a syllable list and four of the remaining crew members recorded their observations. The articulation of the system under test was equal to the average of the articulation of each talker-listener pair.

Egan [2.7] improved on Fletcher's test method by proposing two modifications. First, he required that the relative frequency of the speech sounds used in the articulation syllable lists approximately correspond to their distribution in normal conversational speech. This served to increase the procedure's validity as a test of the ability of systems to transmit intelligible speech. Second, Egan required that the distribution of difficulty of the test items (i.e., syllables, words, sentences) in each list be sufficiently broad so as to differentiate between a broad class of systems under test. This basically involves two conflicting criteria: first, that the range of difficulty of items should be large enough so that no system receives nearly zero or nearly one-hundred percent articulation; and second, that the range of difficulty of some items should be small enough so as to differentiate between nearly equivalent systems.

Egan also noted the effect of using syllables, words, or sentences as speech units for the articulation test. The use of nonsense syllables most accurately measures the ability of the speech system to reproduce individual phonemes. Nonsense syllables, however, do have the disadvantage that listeners must record responses as phonemes rather than as words. This requires a degree of listener training which can be eliminated through the use of short words instead of nonsense syllables. The use of sentences is less accurate than the use of either syllables or words since contextual information, as specified by syntactic and

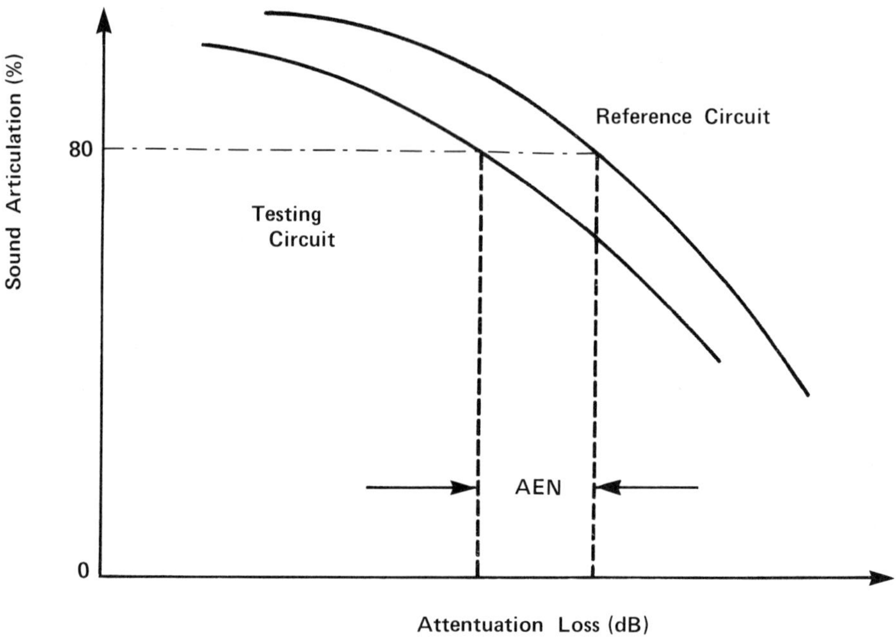

Figure 2.1: Measurement of AEN (from [2.1], ©1984 IEEE).

semantic constraints, assists in phoneme comprehension and therefore complicates the interpretation of articulation test results.

Articulation tests can be modified so that scores have a comparable origin through the introduction of a reference signal [2.1], [2.8]. Such methods involve testing the syllable articulation of each system, including the reference, with various levels of attenuation inserted in series with the system. As the series attenuation is increased, the system's articulation is reduced in a controlled manner. The score of this modified test is expressed in terms of ***articulation equivalent loss*** (AEN), which is defined to be the difference between the two attenuation values that give reference and test systems eighty percent articulation. An example of articulation value as a function of attenuation is plotted for a reference and test signal and AEN is indicated in Fig. 2.1.

2.1.1.2 Rhyme Tests

Fairbanks [2.9] modified the articulation test by constraining the permissible listener responses to be a set of "rhyming" words. In Fairbanks' test, all items in the test list were monosyllables of the form /C-V-C/. On the listener's response sheet, the trailing vowel-consonant was specified, and the listener chose the leading consonant based on his or her interpretation of the test speech. In this way, each word tests the intelligibility of a single consonantal phoneme. Tests were run using lists of fifty monosyllabic words, each with a unique stem (i.e., the word minus the initial consonant). The stimulus words were chosen so that each associated stem corresponds to approximately ten possible choices from an average adult's vocabulary. Stimulus words not in common use were avoided. In the test, words were spoken at five second intervals with articulation results averaged over all listeners.

House, in the Modified Rhyme Test (MRT) [2.10], made administrative improvements to the Fairbanks Rhyme Test by restricting the listener responses to be from a finite set of rhyming words. For each stimulus word, the listener response sheet had a closed set of six alternatives from which to choose the observation. As in the Fairbanks Test, there were fifty items in a word list, so each response sheet consisted of fifty sets of six rhyming words. A new stimulus word list could easily be constructed by randomly selecting one word from each set of six words on the response sheet, and new response sheets could be constructed by randomizing both the order of sets and the order within sets. This has the advantage of making the test easy to automate in both administration and scoring. Furthermore, because the entire set of possible choices was available on the response sheet, the learning time required for listeners to become familiar with the message set was usually eliminated, which made the test more suitable for untrained listeners. Finally, the use of a known, finite message set effectively eliminated the variation in response attributable to word-frequency effects. In conducting the test, the stimulus speech was recorded, using constant vocal effort and preceding each stimulus word with a carrier phrase of the form "Number 10 is ——." Stimulus words (with the carrier phrase) were spoken at five second intervals.

Voiers [2.11] refined the concept of rhyme tests for intelligibility in his Diagnostic Rhyme Test (DRT). His contribution was to make each of the possible listener responses differ not only in the leading consonant (so that they would be rhyming words), but also to restrict them to differ in just one distinctive feature of the leading phoneme. (See Jakobson [2.12] for a discussion of the distinctive feature set used in the design of the DRT.) Construction of response sets was simplified by limiting the listener's responses to two possible choices, one of which is the stimulus word. The distinctive features considered in the DRT are voicing, nasality, sustension, sibilation, graveness, and compactness.

The classification of consonant phonemes based on these features is shown in Table 2.1. In terms of the feature opposition scales presented by Jakobson, sustension corresponds to continuant–interrupted and sibilation to strident–mellow, where strident describes a phoneme with an irregular waveform and mellow describes one with a regular waveform. Graveness and compactness are part of the grave–acute and compact–diffuse scales, in which grave describes a phoneme whose spectrum is dominated by low frequencies and acute indicates that the spectrum is dominated by high frequencies, while compact describes a phoneme with a dominant mid-frequency concentration, and diffuse describes a phoneme with separated spectral peaks. Each of the DRT stimulus words tests for the intelligible transmission of a specific phonetic feature. The advantage of this technique is that the DRT gives not only an overall intelligibility rating for a speech system, but also a "diagnostic" rating consisting of intelligibility scores in each of the distinctive feature categories.

When administering the DRT, it is best if recordings of the test materials are used so that speech level and other speaker variations can be controlled as carefully as possible. More than one individual should be used in the recording, with the stimulus words spoken at approximately 1.5 second intervals. In each test session, at least eight to ten listeners should be used with all listener responses pooled together.

The most straightforward scoring of the DRT yields seven scores, one for each of the six consonant distinctive feature categories and one mean score. All scores are corrected for random guessing effects as follows:

$$S_j = \frac{100(R_j - W_j)}{T_j} \tag{2.1}$$

where S is the "true" percent-correct responses, R is the number of correct responses, W is the number of incorrect responses, and subscript j denotes the distinctive feature class. The mean score is a weighted average of the diagnostic scores, which is equivalent to grouping all responses into a single class. Figure 2.2 shows a diagnostic profile made from the scores of the DRT for three communication systems.

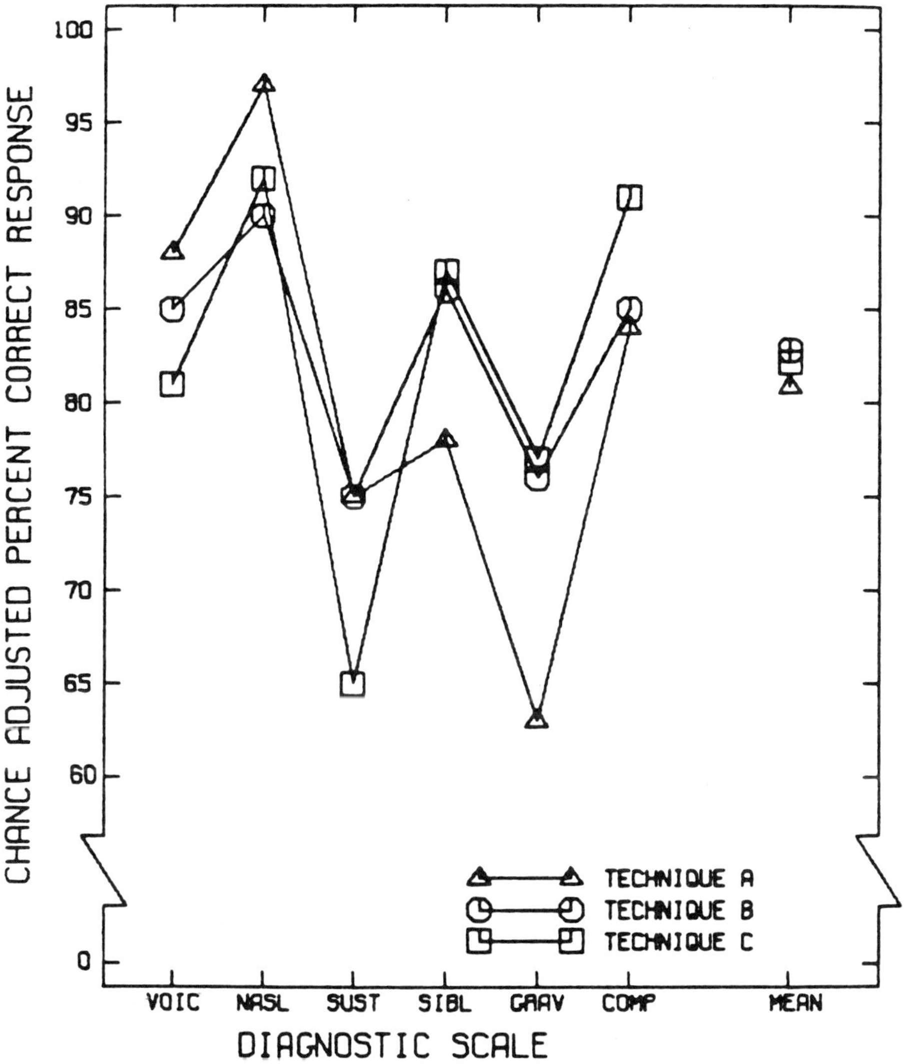

Figure 2.2: DRT diagnostic profile (from [2.11]).

Table 2.1: Distinctive features used in the DRT (from [2.11]).

		m	n	v	∂	z	ž	dž	b	d	g	w
Voicing		+	+	+	+	+	+	+	+	+	+	+
Nasality		+	+	−	−	−	−	−	−	−	−	−
Sustension		−	−	+	+	+	+	−	−	−	−	+
Sibilation		−	−	−	−	+	+	+	−	−	−	−
Graveness		+	−	+	−	−	0	0	+	−	0	+
Compactness		−	−	−	−	−	+	+	−	−	+	−
Vowel-like*		−	−	−	−	−	−	−	−	−	−	+
	r	l	j	f	θ	s	š	tš	p	t	k	h
Voicing	+	+	+	−	−	−	−	−	−	−	−	−
Nasality	−	−	−	−	−	−	−	−	−	−	−	−
Sustension	+	+	+	+	+	+	+	−	−	−	−	+
Sibilation	−	−	−	−	−	+	+	+	−	−	−	−
Graveness	−	0	0	+	−	−	0	0	+	−	0	0
Compactness	−	0	+	−	−	−	+	+	−	−	+	+
Vowel-like*	+	+	+	−	−	−	−	−	−	−	−	−

* The DRT does not test for the apprehensibility of the opposition, vowel-like – nonvowel-like. However, test words are chosen so as not to confound this attribute with the six attributes for which discriminability is tested.

2.1.1.3 Speech Interference Test

Nakatani [2.13] proposed a speech quality test based on articulation principles, but with application to speech quality measurement. The method is based on the fact that small differences in quality may affect the ease with which speech can be understood even though the speech itself is highly intelligible. In this procedure, interfering speech is added to the speech under test to reduce its ease of comprehension. The quality of the speech being tested is related to the level of interfering speech, which produces 50 percent articulation.

Nakatani's test is administered as a sentence articulation test, but it uses nonsense sentences so as to minimize the effect of semantics. In order to produce an absolute rather than relative quality scale, the articulation of both the speech signal under test and a reference speech

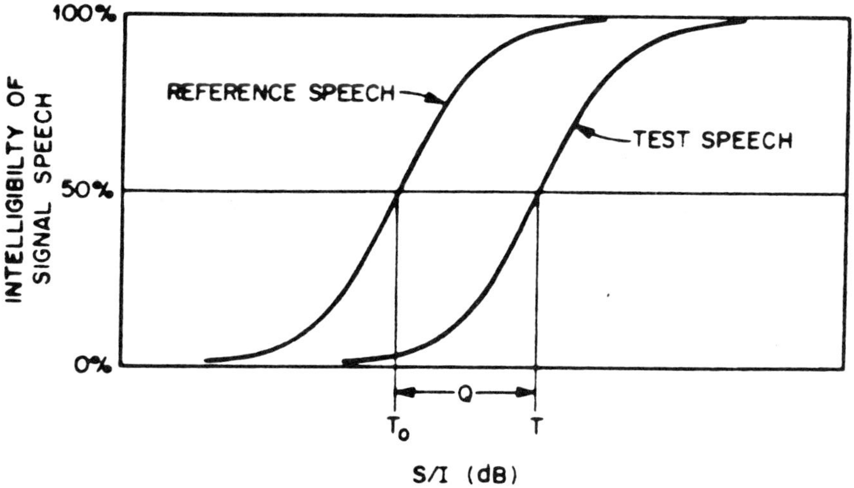

Figure 2.3: Measurement of Q (from [2.18]).

signal are measured. A fixed level of interfering speech is added to both the speech under test and to the reference speech before they are presented to the listeners. The articulation test is repeated for different levels of test and reference speech until the 50 percent articulation level for both systems is bracketed. For both the test and reference signals, the interfering speech is of the same type (high fidelity, lowpass filtered, vocoder, etc.) as the speech to which it is added. The test results identify two thresholds, T_0 and T, which are the speech-to-noise levels associated with 50 percent articulation for the reference and test signals, respectively. This is illustrated in Fig. 2.3. The quality of the test speech, denoted as Q and measured in dB, is equal to the difference between the thresholds T_0 and T.

2.1.2 Quality Tests

Speech intelligibility tests are only appropriate for communication systems that produce moderate to severely degraded speech since, by their very nature, they are unable to distinguish between speech signals that are highly intelligible. Highly intelligible systems may still differ in other perceivable attributes such as pleasantness, naturalness, or the recognizability of the talker. Subjective tests that do discriminate between highly intelligible speech systems are more appropriately termed speech quality tests.

2.1.2.1 Paired-Comparison Methods

An early speech quality test was the *isopreference* method for evaluating speech transmission systems, developed by Munson and Karlin [2.5]. In this method, test signals, each having a different speech level and corrupted by different levels of additive noise, are passed through the transmission system under test. In Munson's experiments, the transmission system most often consisted of a bandlimiting filter, which was meant to simulate one characteristic of a typical analog speech transmission system. Subjects listen to pairs of test signals and specify which one they would prefer to use for a telephone call. The parameters of the test are the loudness of the speech and the power of the noise. (Loudness is measured as the long-term rms sound pressure level in dB, re 0.0002 microbar.) Test results are reported by means of "isopreference contours" in the two-dimensional parameter space of speech level versus noise level, as shown in Fig. 2.4. Any two points on the isopreference contours specify pairs of parameter values that yield a nearly equal split of preferences for the two test signals. Using this method, contours explicitly indicate the optimum speech transmission level for a system's expected background noise level.

In the two-dimensional isopreference space, systems that have higher preference levels are judged better than those with lower levels, but how much better is not clearly specified. To make the test more useful as a tool for comparing system performance, Munson presents two methods for converting the contours to a unidimensional scale. One is the *Transmission Preference Level* (TPL) scale in which a system is scored in terms of the noise level of a reference signal that is judged to be equal in quality to the test signal at its optimum level. The other is the *Transmission Preference Units* (TPU) scale, in which differences in scores are related to the proportion of listeners who would prefer the system having the higher score. Rothauser [2.14] proposed a modified isopreference method in which the loudness of the test speech is fixed at its optimum level, as specified by the isopreference contours. This reduces the number of parameters in the test, simplifies the testing procedure, and makes the test more reliable.

Although the isopreference test attempts to account for the per-

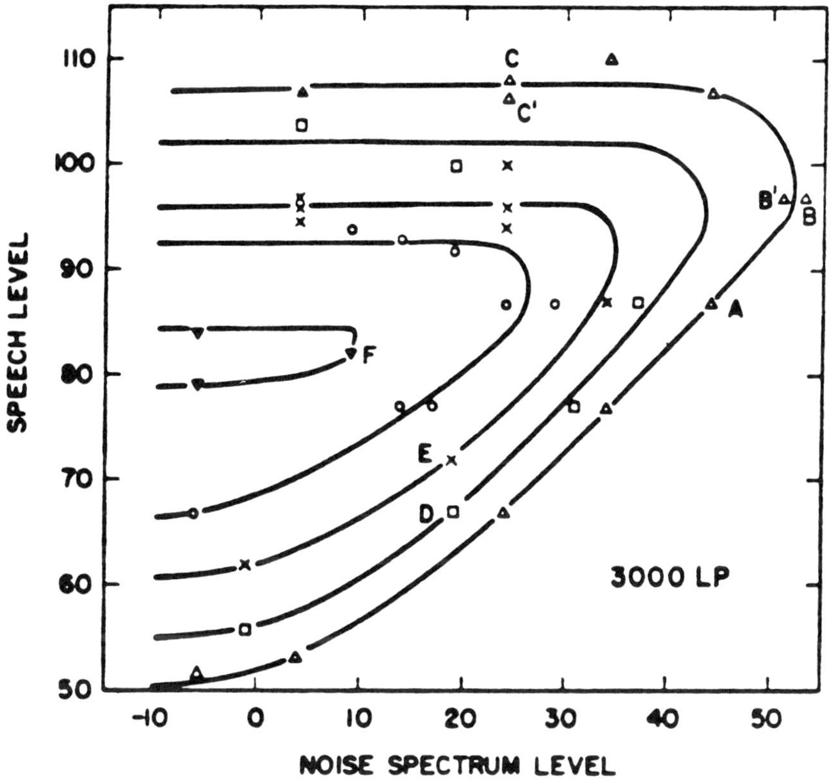

Figure 2.4: Isopreference contours (from [2.5]).

ceptual impact of speech loudness on speech quality, the test results, when reported in terms of isopreference contours, make comparisons of systems difficult. Conversion to one of the unidimensional transmission preference scales makes test comparisons possible, although these scales are largely experimental. Valid comparisons of systems effectively require that their isopreference contours be merged into a single space, which may not be possible or appropriate for all distortions.

Hecker and Williams [2.4] modified the isopreference method to get what is termed a *relative preference* test. It uses five types of distorted speech as reference signals instead of the speech and noise used by Munson and it does not require that isopreference conditions be explicitly determined. The test is conducted in two phases. In the first phase, listeners order the five reference signals by preference, and, in the

Table 2.2: Reference signals for modified isopreference method.

	Signal Description
A	Speech transmitted optimally. Bandwidth 50-10,000 Hz. SNR 45 dB.
B	Speech bandpass filtered (800-3000 Hz).
C	Speech lowpass filtered (3000 Hz) and combined with lowpass filtered white noise (500 Hz). Peak SNR 10 dB.
D	Speech combined with reverberant echo. Delay of first echo 150 ms.
E	Speech peak-clipped, then bandpass filtered (300-2000 Hz).

second phase, listeners rank their preference of the test signal relative to the reference signals. The ordering is done by pairwise comparisons of first, the reference signals, and second, of test and reference signals. Table 2.2 lists the five reference signals used. The preference rating for the reference signals determines an absolute rating scale, which is defined to range from 0 to 1. The position of a given reference signal on the scale is equal to the fraction of reference signals which are judged to have lower preference. Similarly, the position of a test signal on the scale is equal to the fraction of reference signals which are judged to have lower preference.

A disadvantage of this relative preference method is that five reference signals may not be adequate for more than a relatively narrow class of distorting systems. The five signals define only an ordinal quality scale and might easily lead to a scale that has either too small a range or too coarse a granularity. This could result in most of the test signals receiving the same rank relative to the reference signals and therefore the same quality rating [2.2], [2.15]. Another disadvantage is lack of reproducibility: the identical reference signals are required to construct a similar quality scale, and without these signals comparisons of system quality from different applications of the test are not valid.

Several researchers have attempted to simplify and improve the isopreference testing method of Munson and Karlin [2.16], [2.17], [2.18], [2.19]. In the method due to Rothauser [2.16] (which was adopted by the IEEE Subcommittee on Subjective Measurements [2.3]) the test signal is speech produced by the system under test and the reference signal is speech degraded by a continuously variable level of noise. No noise is added to the test speech and the loudness of all speech is fixed at a level that is comfortable for the average listener. Each test signal is evaluated in a separate run in which it is paired with various levels of reference degradation. During each run, the signal pairs are presented twice with each signal having a duration of five seconds and with a three second pause between pairs. After each pair is presented, listeners choose which of the two they would prefer as a source of information. Data from a test run is analyzed by ordering the preference results according to the increasing speech-to-noise level of the reference signal, and then constructing a histogram of the number of listeners that prefer the reference signal to the test signal. The histogram should approximate a cumulative normal distribution, with the mean of the distribution specifying the isopreferent condition and an associated reference signal-to-noise ratio. The quality of the test system is expressed as the signal-to-noise level associated with the isopreferent condition.

One advantage of this preference measurement technique is that one can reasonably expect the relative preference of the reference signal to be monotonically related to the reference noise level, so that preference ordering of the reference signals is unnecessary. Also, the granularity and range of the reference scale is not an issue since a continuously variable noise level yields reference signals of any desired quality level. The noise used in producing the reference signals is either additive or multiplicative noise. The latter reference signal is specified by Schroeder [2.17] as

$$r_\alpha(k) = \frac{s(k) + \alpha n(k)}{(1 + \alpha^2)} \qquad (2.2)$$

where r is the reference, s is the speech, n is the noise, and k is the time sample index. This formulation produces an error that is correlated to the speech energy and is therefore more suitable than additive noise for evaluating certain classes of digital speech coders, such as adaptive waveform coders. It also has the advantage that the speech-to-noise level ($1/\alpha^2$) can be exactly controlled by the parameter α.

2.1.2.2 Direct Methods

A major shortcoming of all paired-comparison preference tests is that the reference signals can only represent a limited scope of speech distortions. Since listeners are required to judge the test speech in terms of the reference speech, listeners may be forced into a smaller perceptual descriptor space than might be desired. This problem related to a limited set of quality descriptors can be addressed by the use of direct or absolute judgment quality tests, in which judgments of overall quality are directly solicited from the listener without the need for reference comparisons.

The most widely used direct method of subjective quality evaluation is the category judgment method, which results in a mean opinion score (MOS) [2.1], [2.3], [2.15], [2.20], [2.21], [2.22], [2.23], [2.24], [2.25]. In this method, listeners rate the speech under test on the following five-point scale:

Rating	Speech Quality	Level of Distortion
5	Excellent	Imperceptible
4	Good	Just Perceptible, but not annoying
3	Fair	Perceptible and slightly annoying
2	Poor	Annoying, but not objectionable
1	Unsatisfactory	Very annoying and objectionable

in which listeners' impressions are assigned a numerical value. Though five categories may seem too few, Miller's [2.26] studies of the limits of human information processing indicate that as few as five but no more than nine categories should be used. The measured quality of the speech under test is equal to the average value of the scores received from all listeners. If desired, data from all listeners can be adjusted to zero mean and unit variance prior to averaging to remove any individual bias.

The category judgment test is typically administered in two phases: training and evaluation. In the training phase, subjects hear a set of reference speech signals that exemplify the high, low, and middle judgment categories. This process, known as "anchoring," is meant to give all listeners the same subjective range and origin in their quality ratings. In the evaluation phase, subjects listen to the speech under test and rate each sample in terms of the quality categories. A standard set of reference signals must be used if the test results are to be compared with the results of other test sessions.

An advantage of a mean opinion score test is that listeners are free to assign their own perceptual meaning to "good," (that is, what kinds of distortions are just perceptible but not annoying). This makes the test applicable to a wide variety of speech distortions. At the same time, this freedom poses a serious disadvantage in that individual listeners' "goodness" scales may vary greatly [2.15]. This variation can result in a significant bias in a listener's judgments, even after the training phase. The bias often stems from a listener's preference for a particular speech distortion being much lower or higher than the population norm. Careful listener selection and training can be used to minimize the judgment bias and combat this problem. However, the need for similarity among listeners must be balanced with the need to represent the full range of preferences in the population which will ultimately use the communication system. Alternatively, bias can be minimized by using a given listening crew to evaluate only perceptually similar speech distortions, so that each individual's bias remains relatively constant throughout the test. Although somewhat restrictive, this scheme is certainly adequate for quality comparisons arising from testing variations of a single type of speech coder.

As in the articulation method, reference signals can be used to help normalize the mean opinion score so that systems tested at different times and places can be compared more easily [2.1], [2.19], [2.21]. A reference signal of speech degraded by varying amounts of multiplicative white noise is most often used since its distortion is similar in quality to adaptive and companded waveform coder noise [2.17]. Equation (2.2) specifies exactly this type of reference signal in which $(1/\alpha^2)$ denotes the speech-to-speech-correlated-noise ratio, referred to as Q in this test method. Reference signals with a range of different Q values are evaluated by the listening crew along with the speech under test. A plot of MOS as a function of the Q of the reference signals is then constructed. This plot is used to transform MOS to an equivalent Q, or speech-to-speech-correlated-noise ratio for the test speech. This can be used as a basis for comparisons of system quality.

2.1.2.3 Indirect Judgment Tests

Indirect judgment tests address the problem of preference differences among individual listeners by attempting to assess speech quality by measuring the underlying parameters that determine perceived quality. The early work in the area of parametric descriptors of speech quality was done by Voiers [2.27], McGee [2.28], and McDermott [2.29], and is described in Sec. 2.1.3. The method was developed by Voiers in his Paired Acceptability Rating Method (PARM) and Quality Acceptance Rating Test (QUART) [2.15], and in the Diagnostic Acceptability Measure (DAM) [2.30], [2.36]. The DAM is the subjective measure used for the bulk of the research reported in this book, and is discussed in detail in Chap. 3.

In an indirect judgment test, listeners are asked to judge the test speech on several perceptually specific quality scales rather than on a scale of overall quality (or acceptability, as in Voiers' terminology [2.30]). Examples of descriptors for such scales are "clicking or ticking," "babbling or gurgling," and "fluttering or twittering." These scales, called *parametric quality scales*, provide an indirect measurement of speech acceptability in that they measure factors that determine, but are not identical to, acceptability. The motivation for the indirect measurement method is that listeners generally agree on the degree to which a specific speech degradation is present in the distorted speech signal, but vary in their preference of that specific degradation [2.15], [2.29]. Therefore indirect judgment tests only solicit judgments of the degree to which a speech signal exemplifies a parametric quality scale. Scales are chosen to complement rather than replicate each other in an attempt to give a multidimensional characterization of the quality of the speech under test. Estimates of overall system quality or system acceptability are functionally derived from the parametric quality scores (and possibly other direct judgment quality scores as well). The weights used in combining the parametric scores must be derived from calibration data for which accurate direct judgments of acceptability are available.

2.1.2.4 Communicability Tests

Communicability, as measured by task performance, can be used as a measure of subjective speech quality. This concept has many variations, although the common idea is for the subjects to perform a specified task that requires communicating over the system under test. One method, proposed by the British Telephone Administration, [2.31] requires a pair of subjects to carry on a two-way conversation over the speech communication system as a means of accomplishing the task. An example of such a task is for the subject at one end of the speech link to describe a simple line drawing to the subject at the other end. The latter is encouraged to engage in clarifying discussion, though the objective is to reproduce the drawing in as short a time as possible. Immediately after completion of the task the subjects are asked to rate the difficulty of their task in terms of conversational effort, based on the following five-point scale:

Rating	Description
5	Complete relaxation possible; no effort required.
4	Attention necessary; no appreciable effort required.
3	Moderate effort required.
2	Considerable effort required.
1	No meaning understood with any feasible effort.

As with mean opinion scores, the measured quality of the system under test is equal to the average rating on this scale. Another means of reporting system quality is to record the length of time required for the subjects to complete the task. The ratio of the time required using the test system to the time required using a high-quality system is called *message rate efficiency*, and was found by Richards [2.31] to fall below 95 percent for those systems requiring more than moderate conversational effort.

Another method of measuring system quality, also proposed by Richards [2.8], [2.31] is similar to sentence articulation, but is based on subjective responses rather than numerical recognition scores. In this test, listeners hear sentences spoken over a communication system and are asked to note which sentences they understood immediately and without significant listening effort. The sentences should be chosen so that all can be immediately understood in a high quality communication system, but still represent a rather broad range of comprehensibility in a marginal communication system. The measured quality of the system is proportional to the percentage of sentences that are immediately comprehended by the listeners.

Yet another test, proposed by Hall, [2.32], is based on speaker discrimination. In this test, a listener is presented with a pair of sentences and is asked whether the same speaker uttered both sentences. To vary the difficulty of the task, both sentences in the pair are masked with a specified level of additive noise. Fifteen utterances are used for the stimulus set, with three sentences from each of five speakers. Sentence pairs are formed by randomly choosing from the set of fifteen, and four levels of white noise are used to mask the pairs. Although the test was set up to run in a laboratory, the resulting test utterances also could be produced by a communication system. The varying noise levels then would be replaced by the various distortions introduced by different communication systems. In this test, the measured quality of the speech is proportional to the percentage of correct listener responses.

2.1.2.5 Task Loading Tests

Nakatani, [2.33], Barnwell [2.34], and Pisoni [2.35] have investigated word recall as a "loading task" for use in speech quality measurement. The method is based on the observation that word recall varies with speech perception effort [2.33], so that as the quality of a communication system deteriorates, the listener is able to recall fewer words. Speech that requires great effort to understand may be ineffective for communication, with users rejecting the system because they find it fatiguing or annoying.

In such tests, listeners hear a set of words produced by a speech coder or communication system and, after a short pause, are asked to recall as many of the words as possible. In Pisoni's test, for example, a list of twenty-five monosyllabic words was presented to listeners at one second to five second intervals. After all words are presented, listeners are given ninety seconds to recall as many words as possible. The measured quality of a speech system is then equal to the percentage of words the listeners were able to recall correctly.

The advantage of both communicability and task performance methods is that they address the fundamental issues of what makes a communication system acceptable to the user—issues such as ease of communication and ease of speaker recognition. However, they all suffer from lack of reproducibility, since it is difficult to ensure that the abilities of the evaluators and the difficulty of the tasks to be performed are well matched in different test situations. In addition, the resolving power of these tests is typically low. This means that either a great many subjects must be used or the task must be very severe, making it unpleasant for the subjects.

2.1.3 Analytic Measures

An entirely different approach to measuring subjective speech quality is represented by several analytic methods. These methods attempt to characterize the underlying psychological factors that determine perceived speech quality and to discover acoustic features that are correlated with these factors. Analytic procedures provide insight into the basic mechanism of speech quality perception and can lead to practical tests of speech quality of both the unidimensional (utilitarian) and multidimensional (analytic) type [2.36].

McGee [2.28] investigated the perceptual space for the quality of filtered speech (lowpass, highpass, or bandpass filtering distortions). In his experiment, a sample of speech from a single male speaker was subjected to fourteen different filtering distortions. Two tests were conducted. The first was a paired-comparison test in which 108 listeners rated all possible pairs from a set of fifteen speech signals, fourteen of which were distorted signals and one an undistorted signal. Listeners rated the pairs on a 22-point preference scale that indicated how much better or worse the first signal of the pair was perceived to be. In the second test, listeners rated individual speech signals on thirty semantic-differential scales with endpoint descriptors such as "rich–poor," "muffled–clear," or "rough–smooth." The semantic differential data was reduced using factor analysis, and two primary factors representing the significant perceptual dimensions of speech quality for this data were identified. The two factors accounted for 95 percent of the variance in the preference quality scales, with the data suggesting that these factors could be labeled "intelligibility" and "naturalness."

McDermott [2.29] also conducted paired-comparison tests, but with listeners rating pairs on similarity as well as on preference. The reference speech was from two male and two female speakers and was subjected to twenty-two distortions ranging from waveform distortions such as peak clipping, rectification, and additive noise to frequency distortions such as bandlimiting. Listeners rated all possible test signal pairs on a ten point scale and indicated which member of the pair they preferred.

Multidimensional scaling [2.37], [2.38] was used to analyze the data. The scaling of preference data takes special account of individual differences in preference ratings and represents an individual's "ideal preference" as a point in a multidimensional perceptual space. The distorted signals also are represented as points in this space, so that the relationship between individual preference and acoustic distortion, as represented by the test signals, can be examined. Multidimensional scaling also was used with the similarity data, but without the individual preference points. In this analysis, the similarity of test signals was represented by the distance between points in a multidimensional perceptual space, in which similar signals correspond to points near each other, and dissimilar signals correspond to points distant from each other. The scaling of both types of data produced very similar configurations of points in three dimensions for the test signals. In the preference scaling, the listener preference points were located near each other on one of the three dimensions but had considerable scatter on the other two dimensions, indicating that individuals did indeed have a varied perspective of what an "ideal" speech signal would be.

Multidimensional scaling procedures provide a geometric description of the data, but do not identify the dimensions of the perceptual space. However, the data itself can suggest labels for the axes of the space. In both analyses, the data were scaled into three dimensions, which McDermott labeled clarity, signal-background distortion, and subjective loudness.

Voiers also has done analytical work [2.27] investigating the perceptual factors that permit speaker differentiation. Although most closely related to the problem of speaker identification, this work also is relevant to speech quality assessment in that some speech processing systems, most notably vocoders, are known to produce speech whose quality varies with the speaker. In Voiers' experiment, thirty-two listeners described their perceptions of sixteen voices using forty-nine semantic-differential scales. Factor analysis was used to determine the dimensionality of the perceived differences between speakers, with four factors accounting for nearly 90 percent of the variance in mean speaker ratings. These factors were labeled clarity, roughness, magnitude, and animation.

2.2 Objective Quality Measures

The ideal objective speech quality measure would be able to assess the quality of distorted or altered speech by simply observing a small portion of the speech in question, with no access to the original speech. The results of such an assessment should be indistinguishable from those obtained from human observers. In order to perform such a task, the objective measure would need to have knowledge of all the levels of human processing of speech: psychoacoustics, acoustic-phonetics, morphology, prosodics, syntax, semantics, linguistics, and pragmatics. The scientific study of most of these areas is still in a rather early stage of development, and cannot in practice be employed. Although compromises can be made, such as allowing access to the original speech and limiting the types of distortions, the goal of replicating human performance is still maintained. And it is from this point of view that the success or failure of an objective measure must be judged.

At present, virtually all objective speech quality measures model only the lowest possible level of speech perception; and even here, the approximations are extremely crude. Generally, nothing more is considered than comparing the original and distorted signals over intervals so short that the speech characteristics do not appreciably change within an interval. Since most consonants and many vowels are characterized by their dynamics, these intervals are usually shorter in duration than the phonemes which they comprise. Accordingly, such measures really only quantify certain physical differences between the distorted and original speech waveforms, measured at a microscopic level. Hence such measures cannot operate at a level of perception beyond simple psychoacoustics.

When observed from this point of view, most currently used objective measures are inherently limited. However, if the problem of interest is specifically limited to the objective evaluation of the distortion generated by speech coding systems, then it is important to note that most modern speech coding systems make no explicit use of syntactic or semantic information. Hence speech coders typically do not produce distortions that are specifically related to high-level perception processes. It is therefore not unreasonable to hypothesize that relatively simple objective measures can be used to predict the subjective quality of at least some speech communication systems.

There is some consensus in the literature as to the methods that should be used in the design of objective measures of speech quality [2.30], [2.39], [2.40], [2.41], [2.42], [2.43], [2.44], [2.45]. In assessing the quality of a digital speech communication system objectively, it is currently necessary to make direct comparisons between the reference (i.e., original) and the distorted speech waveforms. Whereas subjective measures such as the MOS rely on a subject's internalized models of "good" speech, objective measures have no such stored model and must instead rely on the original speech signal to provide a reference of quality. However, there are several notable exceptions to this approach, such as the intelligibility test described in the following section.

Since there is a need for direct comparison of waveforms, synchronization of the reference and distorted speech is extremely important. If not synchronized, comparisons might reflect primarily the changing statistics of the speech waveform and have little to do with the distortions introduced by the speech communication system. This is most severe in time-domain comparisons, as improper alignment can make fine resolution operations, such as waveform subtraction, totally meaningless.

A common method of implementing an objective measure is to break the speech into short segments, or frames, of 10 to 30 ms duration and to compute a distortion measure for each frame. Since the vocal tract remains approximately stationary over the time span of one frame, one can make an estimate of the vocal tract system function which, in turn, can be used to estimate the envelope of the speech magnitude spectrum. It is widely felt that distortions in the envelope of the short-time speech magnitude spectrum are the main determinants of speech quality. Whether calculated in the time or frequency domain, objective measures can usually be broken down into two parts: one that computes a distortion for each speech frame, and another that combines the frame distortions into a single, global measure of speech distortion.

Finally, one might want to impose certain mathematical restrictions on the objective measures, such as

(a)	$d(x,y) > 0$ for $x \neq y$	positive definiteness
	$d(x,x) = 0$	
(b)	$d(x,y) = d(y,x)$	symmetry
(c)	$d(x,y) \leq d(x,z) + d(z,y)$	triangle inequality

where x, y are reference and distorted speech signals, respectively, and $d(x, y)$ is an objective measure. A measure that satisfies all three properties is a metric. Few speech processing tasks require measures that are metrics. If a measure satisfies properties (a) and (b), it can appropriately be called a distance measure. Although symmetry is a desirable property, there is no general requirement that objective quality measures be symmetric. Indeed, since one always can arrange for the first argument to be the reference and the second to be the distorted speech, there is little need for symmetry. Even (a), positive definiteness, which often is required of objective measures, is not always appropriate since there exist good measures—log spectral distance for example—for which negative values are possible.

2.2.1 Articulation Index

The first widely used objective measure of speech quality, or more specifically speech intelligibility, was the Articulation Index, or AI. It was proposed by French and Steinberg in 1947 [2.46] and subsequently was developed by other researchers [2.47], [2.48], [2.49], [2.50], [2.51]. It was designed to test the performance of analog speech communication systems and therefore is valid only for systems that can be modeled as linear systems with additive noise. Although the Articulation Index measures only one aspect of speech quality, intelligibility, it is remarkably accurate and, because of this, will be presented here in detail.

The Articulation Index models the intelligibility of speech by assuming that the overall loss of intelligibility is equal to the component losses in contiguous frequency bands spanning the speech spectrum. It also assumes that the distortion in one frequency band is independent of the distortion in all other bands, and that all distortion results from either signal attenuation or additive noise. In the AI test, the *articulation* in a frequency band is equal to the fraction of the original speech energy in the band which, when reproduced by the communication system, is still theoretically perceivable to the listener. The Articulation Index is then simply the average of the articulation in each of the frequency bands.

French and Steinberg [2.46] have asserted that the appropriate frequency scale for use in calculating the AI is one that is close to the mel scale [2.52]. They divide the frequency scale into twenty contiguous

Table 2.3: Articulation index filter bands.

Band	Frequency Limits (Hz)	Band	Frequency Limits (Hz)
1	250–375	11	1930–2140
2	375–505	12	2140–2355
3	505–645	13	2355–2600
4	645–795	14	2600–2900
5	795–955	15	2900–3255
6	955–1130	16	3255–3680
7	1130–1315	17	3680–4200
8	1315–1515	18	4200–4860
9	1515–1720	19	4860–5720
10	1720–1930	20	5720–7000

bands, ranging from 250 Hz to 7000 Hz as shown in Table 2.3, with the width of each adjusted so that all bands contribute equally to speech intelligibility. They found that frequencies below 250 Hz or above 7000 Hz do not significantly contribute to speech intelligibility.

In the calculation of the Articulation Index, speech that is above the ear's threshold of hearing and below the region of overload is considered perceivable to the listener. Speech levels that *overload* the ear are those levels above which the stimuli are unpleasant. If the entire dynamic range of the speech from the communication system falls between threshold and overload, then, in the absence of noise, the communication system would have an Articulation Index of 1.0. Figure 2.5 graphically illustrates this concept. In the practical case in which noise is introduced by the communication system, the long-term noise spectrum is measured, in the absence of speech, and plotted on the graph of Fig. 2.5. The final Articulation Index of the system is then equal to the fraction of the dynamic range of the speech, as represented by the shaded center area, which is above threshold, below overload, and above the masking noise spectrum.

One of the best explanations of how to compute the Articulation Index is given by Kryter [2.50]. A summary explanation is given by Beranek [2.48], which will be outlined here. In the measurement procedure, Beranek defines *orthotelephonic gain* as

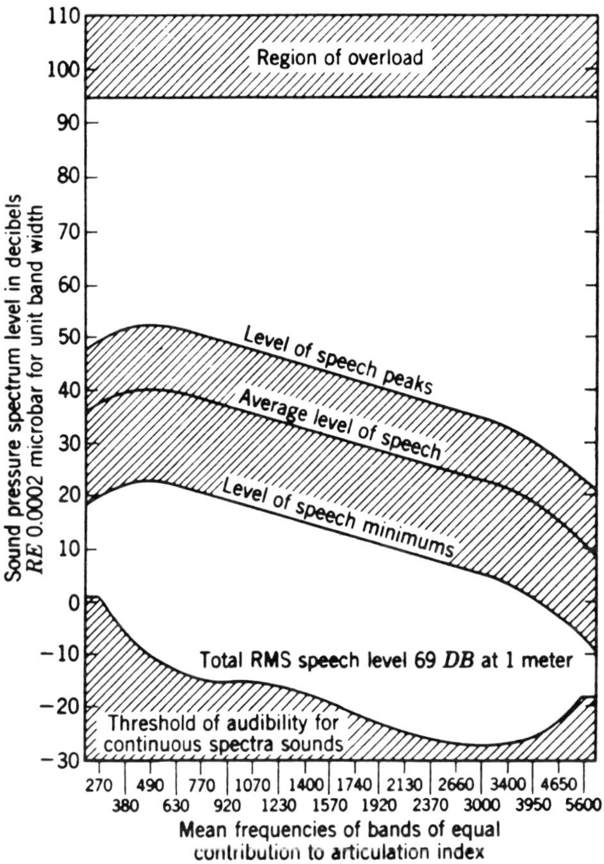

Figure 2.5: Graph for calculation of AI (from [2.48], ©1947 IRE, now IEEE).

$$\text{Orthotelephonic gain } = 20\log_{10}\frac{p_2}{p_1} \tag{2.3}$$

where p_1 is the long-time average spectrum of the original speech and p_2 is the long-time average spectrum of the speech reproduced by the communication system under test.

The steps required to compute the AI are outlined as follows:

1. Compute the orthotelephonic gain in each of the twenty frequency bands. Adjust the plot in Fig. 2.5 so that the "average level

of speech" in each frequency band is moved up or down by an amount equal to the orthotelephonic gain in that band.

2. Measure the long-time average spectrum of the noise produced by the communication system in each of the twenty frequency bands. The noise is measured as the output of the system when the input is silence. The noise spectrum is plotted on the graph in Fig. 2.5.

3. For each band, determine the fraction, F_i, of the resultant shaded center area (average level of speech +18 dB and −12 dB) that remains below overload (+95 dB), above threshold, and above the noise spectrum. This can be computed as:

$$F_i = \frac{\min(\mathrm{SP},\mathrm{OV}) - \max(\mathrm{SM},\mathrm{TH},\mathrm{N})}{30} \tag{2.4}$$

where SP is the level of speech peaks, OV is the region of overload, SM is the level of speech minima, TH is the threshold of audibility, and N is the level of noise.

4. The Articulation Index is equal to the average of all F_i, for i equals 1 to 20.

5. Once the AI is measured, an estimate of word articulation can be obtained from empirically derived graphs that Beranek provides.

The most important aspects of the Articulation Index are that it provides a reasonably accurate estimate of intelligibility and that it has been validated through many applications. However, its use must be constrained to fit the assumptions upon which it is based. The Articulation Index, as one of the oldest objective measures of speech quality, was designed to test the performance of analog speech communication systems. As applied to modern digital speech coders, the most important assumptions to satisfy are, first, that the effects of loss of intelligibility within frequency bands are independent, and second, that the communication system can be modeled as a linear system with additive noise. For many digital speech coders, both of these assumptions are violated. In particular, any coder whose output can be modeled as speech plus speech-correlated noise will not fit the AI model since the the whole concept of assessing the intelligibility impact of the noise through superposition would be invalid. Furthermore, nonlinear coders "smear" noise across the entire frequency spectrum, violating the assumption of independence between bands. Therefore one must be careful in using the Articulation Index with digital speech communication systems.

2.2.2 Speech Transmission Index

A much more recent extension of the Articulation Index concept, developed by Steeneken and Houtgast [2.53], overcomes some of the limitations of the classical AI and is generally more applicable to the measurement of intelligibility in digital speech communication systems. The new method, which produces the Speech Transmission Index, or STI, uses synthetic signals to test the communication channel.

With a linear communication system that does not produce harmonic distortion, the STI is computed from the effective signal-to-noise ratio in individual frequency bands. Unlike the AI, the STI separates the speech spectrum into seven one-octave bands. The unique aspect of the STI is its method of calculating the effective signal-to-noise ratio. A block diagram of the STI procedure is shown in Fig. 2.6. The synthetic test signal used by the STI is generated in two steps. First, a white Gaussian noise signal is passed through a spectrum shaping filter with a frequency response that approximates the long-term spectral estimate of speech. Then, the signal is intensity modulated using a raised cosine wave. (Intensity modulation is simply amplitude modulation using the positive square root of the raised cosine.) The output of the system is split into octave bands, and the envelope of the signal in each band is squared. The effective SNR is calculated from the value of the modulation index, m_k, of the output of each octave band. Calculation of the modulation index is schematically illustrated in Fig. 2.7 and is explicitly calculated as

$$m_k = \frac{T_k}{(N_k + T_k)} \tag{2.5}$$

where T_k is the long-time average power of the test signal in band k and N_k is the long-time average power of the noise in band k.

Upward noise masking, affecting band k based on noise in band $k-1$, can be modeled as a decrease in the modulation index in band

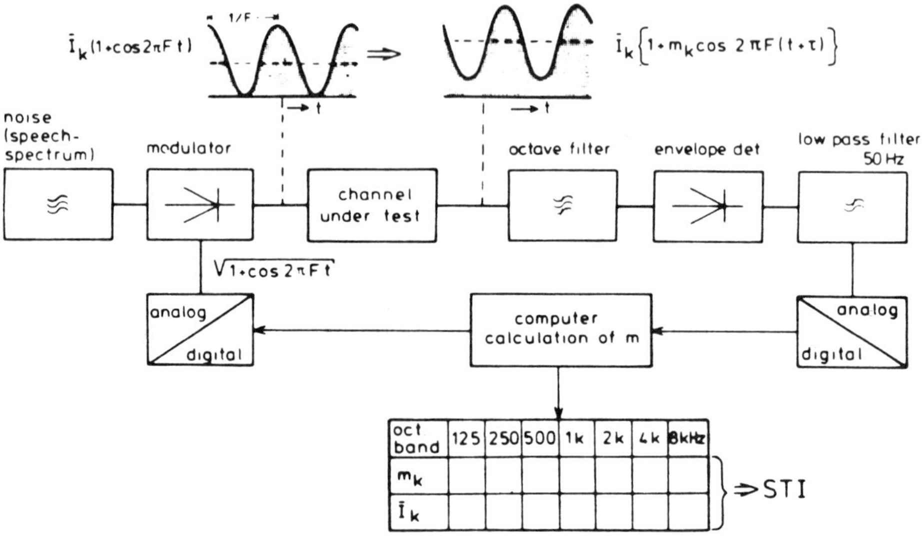

Figure 2.6: Calculating the STI (from [2.53]).

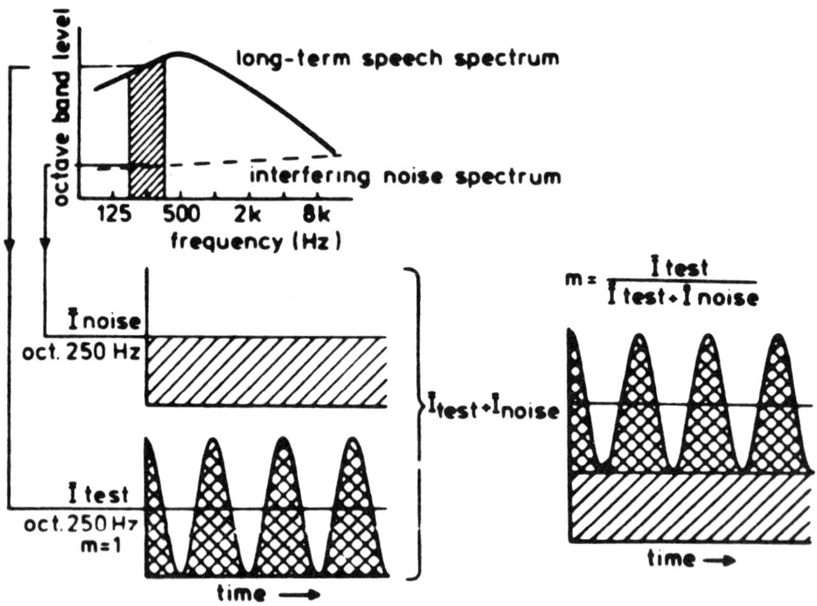

Figure 2.7: Calculating the modulation index (from [2.53]).

k

$$m'_k = (\text{ACF})(m_k) \tag{2.6}$$

where m'_k is the adjusted modulation index and ACF is the auditory correction factor. The auditory correction factor is determined empirically. The effective signal-to-noise ratio in band k, SNR_k, can be calculated as

$$\text{SNR}_k = 10\log_{10}\frac{T_k}{N_k} \tag{2.7}$$

$$= 10\log_{10}\frac{m'_k}{(1-m'_k)} \tag{2.8}$$

where T_k is the long-time average power of the test signal in band k, N_k is the long-time average power of the noise in band k and m'_k is the adjusted intensity modulation index for band k.

The STI for a communication system is computed as a weighted sum of the effective SNR in each band

$$\text{STI} = 100\sum_k(W_k\text{TI}_k)\ \text{percent} \tag{2.9}$$

and

$$\text{TI}_k = (\text{SNR}_k - S)/R,\ \text{limited to } 0 < \text{TI} < 1 \tag{2.10}$$

where W_k is the weight for band k, TI_k is the transmission index, and S and R are empirically determined weights.

The weights W_k have been determined by regression analysis on a set of test data and were adjusted so as to yield a STI that is maximally correlated with subjective intelligibility scores of phonetically balanced word lists. The STI, therefore, is effectively a measure of the intelligibility of phonetically balanced word lists. Steeneken has shown that if second and third order terms are added to the model of Eq. (2.9), improved prediction of intelligibility results.

The STI can be extended to cover nonlinear communication systems that produce distortion harmonics. This is done by exciting the frequency bands not under test (the bands other than band k) with speech-like pseudo-random noise. The signal in these other bands will, in general, produce distorting harmonics in band k which reduced the effective SNR in band k. Measurements are made as before. The appropriate level of the noise in the bands other than band k has been determined empirically so as to maximize the correlation of STI and subjective intelligibility for this class of communication systems.

The Speech Transmission Index is an improvement over the Articulation Index principally in two areas. First, it is more amenable to automatic computation because it uses a synthetic excitation for the test, and second, it has specific extensions to cover some cases of nonlinear distortion, for which the AI is not applicable. However, one of its shortcomings that Kryter [2.51] indicates is that it splits the speech spectrum into only seven one-octave bands, as opposed to the twenty critical bands of the AI. Finally, as with the AI, the STI is still not applicable to certain types of speech communication systems. For one, it does not account for frequency shift or frequency multiplication in the system under test. For example, it is not appropriate for pitch-excited vocoders, since the test signal does not have a pitch-like structure, but approximates only the long-term spectral envelope of speech. Additionally, all systems based on linear prediction that use an all-pole model of the vocal tract will generally respond erratically to the nonlinear distortion part of the STI measurement. This probably is due to the difficulty the all-pole model has in approximating the zeros in the test signal spectrum. Finally, speech coders that are specific to a single speaker or that encode frame templates, such as vector coders, fall entirely outside the scope of the STI approach.

Objective measures that are intended to estimate speech quality, as opposed to speech intelligibility, are not as well understood or as well validated as the articulation index. The remainder of this section will discuss objective measures that are specifically intended to estimate the more general and more complex case of speech quality.

2.2.3 Signal-to-Noise Ratio Measures

An objective measure that is applicable to analog or waveform coding systems is the signal-to-noise ratio, or SNR. Several variations of SNR have been proposed: classical SNR, segmental SNR, frequency weighted segmental SNR, and granular segmental SNR [2.22], [2.23], [2.24], [2.30], [2.34], [2.43], [2.54], [2.55], [2.56], [2.57], [2.58], [2.59]. Signal-to-noise measures are only appropriate for distorting systems that reproduce a facsimile of the input waveform such that the original and distorted signals can be time aligned and noise can be accurately calculated. It may be necessary to correct phase errors in the distorted signal or to interpolate between samples in a sampled data system to achieve correct time alignment. The classical signal-to-noise ratio is measured as

$$\mathrm{SNR} = 10\log_{10}\frac{\sum_n x_\phi^2(n)}{\sum_n (x_\phi(n) - x_d(n))^2} \tag{2.11}$$

where $x_\phi(n)$ is the undistorted input speech signal, $x_d(n)$ is the distorted speech reproduced by the communication system, and n is the sample index. A related measure uses a pure tone instead of speech as the x_ϕ signal [2.59]. In this approach, the SNR for several tones spread across the voice bandwidth is calculated and used to characterize the system under test. This gives some indication of the quality of stationary, non-adaptive systems but is obviously not adequate for adaptive systems that exploit the local stationarity of speech.

2.2.4 Segmental Signal-to-Noise Ratio Measures

It often has been demonstrated [2.23], [2.29], [2.34], [2.56] that classical SNR is a poor estimator of subjective speech quality for a broad range of speech distortions and therefore is of little interest as a general objective measure of speech quality. However, if the same measurement is taken over short segments of the speech waveform and summed over all segments in that waveform, the result, in a wider range of cases, is an extremely good estimator of speech quality [2.30], [2.54]. This objective measure, called *segmental* SNR, can be formulated as

$$\mathrm{SEG\ SNR} = 10/M \sum_{m=0}^{M-1} \log_{10} \sum_{n=Nm}^{Nm+N-1} \left(\frac{x_\phi^2(n)}{[x_d(n) - x_\phi(n)]^2} \right) \tag{2.12}$$

where N is the segment length and M is the number of segments in the speech signal. The segments are produced by windowing the speech, with window length typically chosen to be 15 to 20 ms. Segmentation of the signal-to-noise computation permits the objective measure to assign equal weight to both loud and soft portions of the utterance.

This definition of segmental SNR poses a problem if there are intervals of silence in the speech utterance, as can normally be expected. In segments in which the original speech is nearly zero, any amount of noise will give rise to a large negative signal-to-noise ratio for that segment, which could appreciably bias the overall measure of segmental SNR. A way around this problem involves identifying the silent frames and excluding them from the sum in Eq. (2.12). This can be approximated by computing the energy in each frame of the reference speech, and including the SNR of that frame only if the frame's energy is above a specified threshold.

A very successful extension of segmental SNR is the frequency weighted segmental SNR measure, of which several variations have been proposed [2.56], [2.60]. Tribolet formulates the measure as

$$\text{FW SNR} = \frac{10}{M} \sum_{m=0}^{M-1} \left(\frac{\sum_j W(m,j) 10 \log_{10}[\sigma^2_{\phi,j}/\sigma^2_{n,j}]}{\sum_j W(m,j)} \right) \tag{2.13}$$

where the m is the segment index, j is the frequency band index, $W(m,j)$ is a frequency weight, $\sigma^2_{\phi,j}$ is the undistorted speech variance in band j, and $\sigma^2_{n,j}$ is the noise variance in band j. The speech signal is filtered into B frequency bands, which are usually spaced proportional to the ear's critical bands [2.61], [2.62]. Barnwell [2.30] has a slightly different formulation for this measure, in which the summations over segments and frequency bands are interchanged. This permits the weights to be adjusted for optimal correlation to subjective estimates of speech quality. This frequency weighted segmental SNR is

$$\text{FW SNR}(\alpha, \gamma) = b_0 + \sum_{j=1}^{B} b_j \left(\frac{\sum_m E(m)^\alpha 10 \{\log_{10}[\sigma^2_{\phi,j}/\sigma^2_{n,j}]\}^\gamma}{\sum_m E(m)^\alpha} \right) \tag{2.14}$$

where the b_j's are frequency weights determined by regression analysis, $E(m)$ is the power in segment m of the original waveform and, as before, $\sigma^2_{\phi,j}$ is the undistorted speech variance in band j, and $\sigma^2_{n,j}$ is the noise variance in band j. The parameter α varies the energy weighting, while the parameter γ produces a nonlinearity on the SNR of the speech segment. By varying the parameters α and γ, a whole family of related objective measures can be produced.

Another variation on SNR is *granular segmental* SNR [2.23], [2.24], which is only appropriate for the evaluation of delta modulation or differential waveform coders. The measure is formulated as

$$\text{GRAN SNR} = 1/M \sum_{m=0}^{M-1} 10\log_{10}\left(\frac{\sum_{n=mN}^{mN+N-1} G[x_\phi^2(n)]}{\sum_{n=mN}^{mN+N-1} G[x_\phi(n) - x_d(n)]^2}\right) \tag{2.15}$$

where m is the segment index, N is the segment length, $x_\phi(n)$ is the original speech, $x_d(n)$ is the coded speech, and n is the sample index. Granularity is measured by the function G(), which restricts $x_\phi(n)$ and $x_d(n)$ to be defined only when the speech coder is not overloaded—that is, when the quantization error is less than one-half the quantizer step size.

A complementary measure, also proposed by McDermott, is the *percent of samples overloaded.* In this measure, also appropriate only for differential waveform coders, $x_d(n)$ is the coder output waveform, $e(n) = x_d(n) - x_\phi(n)$ is the coder error, and $D(n)$ is the coder step size at sample n. The percent of samples overloaded, P, is

$$P = 100\left(1 - \frac{\sum_{n=1}^{N} S[x_d(n)]/x_d(n)}{N}\right) \text{ percent} \tag{2.16}$$

where

$$S[x_d(n)] = \begin{cases} x_d(n) & \text{if } e(n) \leq D(n)/2 \\ 0 & \text{if } e(n) > D(n)/2 \end{cases}$$

The two previous measures also can be formulated as segmental and frequency weighted segmental measures.

2.2.5 Log Likelihood Ratio Measures

The Itakura distance measure, or log likelihood ratio [2.41], [2.63], [2.64], [2.65], [2.66], [2.67], [2.68], is based on the dissimilarity between all-pole models of the reference and distortion speech waveforms. The measure assumes that over intervals of roughly 15 ms speech can be represented by a pth order all-pole model

$$x_\phi(n) = \sum_{i=1}^{p} a_\phi(i)x_\phi(n-i) + Gu(n) \tag{2.17}$$

where $x_\phi(n)$ is the sampled speech signal, $a_\phi(i)$ are the coefficients of the all-pole filter, G is the filter gain and $u(n)$ is a unit variance white filter excitation. The measure often is presented in terms of the autocorrelation method of linear prediction analysis, in which the speech waveform is windowed to form frames 15 to 30 ms in length. (See [2.69].) The log likelihood ratio compares two windowed speech waveforms, x_ϕ and x_d, and can be defined as

$$d(\vec{a}_d, \vec{a}_\phi, m) = \log\left(\frac{\vec{a}_d \mathbf{R}_\phi \vec{a}_d^{\,\mathrm{T}}}{\vec{a}_\phi \mathbf{R}_\phi \vec{a}_\phi^{\,\mathrm{T}}}\right) \tag{2.18}$$

where m is the frame index indicating the position of the window; $\vec{a}_\phi$ is the LPC coefficient vector $(1,\ -a_\phi(1),\ -a_\phi(2),\ \ldots,\ -a_\phi(p))$ for the original speech $x_\phi(n)$; and $\vec{a}_d$ is the LPC coefficient vector $(1,\ -a_d(1),$ $-a_d(2),\ \ldots,\ -a_d(p))$ for the distorted speech $x_d(n)$.

The transpose of $\vec{a}$ is indicated as $\vec{a}^{\,\mathrm{T}}$, and $\mathbf{R}_\phi$ is the autocorrelation matrix for $x_\phi(n)$. The elements of $\mathbf{R}_\phi$ are defined as

$$r_\phi(|i-j|) = \sum_{n=1}^{N-|i-j|} x_\phi(n) x_\phi(n+|i-j|) \text{ for } |i-j| = 0, 1, ..., p \tag{2.19}$$

where N is the length of the frame used in the LPC analysis.

A less efficient formulation of the log likelihood ratio, although one that has a more enlightening interpretation, is to consider the measure as a filtering operation. The inverse filter of sequence $x_\phi(n)$, denoted as $A_\phi(z)$, is defined as

$$A_\phi(z) = 1 - \sum_{i=1}^{p} a_\phi(i) z^{-i} \tag{2.20}$$

so that $A_\phi(z)$ is the inverse of the all-pole filter that models the spectral envelope of $x_\phi(n)$. Similarly, $A_d(z)$ is the inverse filter for the distorted speech sample $x_d(n)$. In computing the measure for a given frame, the undistorted speech is inverse filtered both by $A_d(z)$ and $A_\phi(z)$, as shown in Fig. 2.8. The output of filter $A_d(z)$ is $e_d(n)$,

$$e_d(n) = x_\phi(n) - \sum_{i=1}^{p} a_d(i) x_\phi(n-i) \tag{2.21}$$

which is the prediction error or residual for sequence $x_\phi(n)$. The output of filter $A_\phi(z)$ due to $x_\phi(n)$ is the residual $e_\phi(n)$. Let the power in the

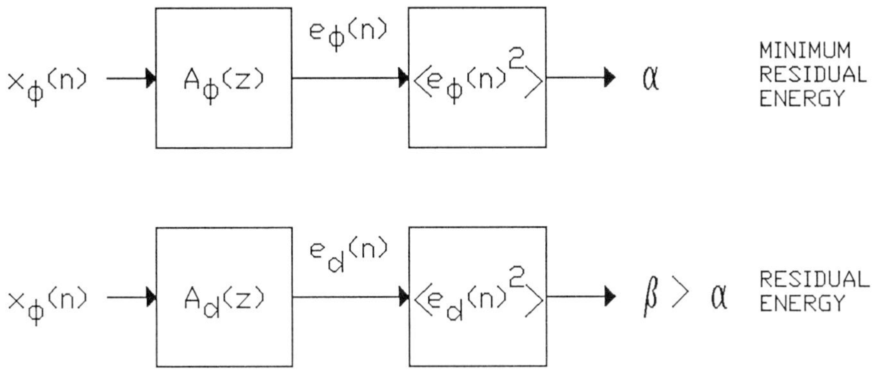

Figure 2.8: Interpretation of log likelihood ratio.

residual signals for a given speech frame be β for filter $A_d(z)$ and α for filter $A_\phi(z)$. It can be shown that $\beta > \alpha$ for all filters $A_d(z) \neq A_\phi(z)$, and because of this, $e_\phi(n)$ is called the minimum residual. In terms of these residuals, the log likelihood ratio for a given frame can be written as

$$d(\vec{a}_d, \vec{a}_\phi, m) = \log(\beta/\alpha) \tag{2.22}$$

which is equivalent to Eq. (2.18). Using Parseval's relation, one can rewrite Eq. (2.22) as

$$d(\vec{a}_d, \vec{a}_\phi, m) = \log\left(1 + \int_{-\pi}^{+\pi} \left| \frac{A_\phi(e^{j\omega}) - A_d(e^{j\omega})}{A_\phi(e^{j\omega})} \right|^2 \frac{d\omega}{2\pi}\right) \tag{2.23}$$

which provides a frequency domain interpretation of the log likelihood ratio [2.41]. This shows that the spectral differences between the reference and distorted speech are most heavily weighted when $1/|A_d(e^{j\omega})|$ is large, which is generally near the formant peaks of the speech.

Since only the distorted speech waveform is inverse filtered, the log likelihood ratio is not a symmetric distance measure so that, in general, $d(x_\phi, x_d) \neq d(x_d, x_\phi)$. In situations that require a symmetric measure, a combined likelihood measure can be formed as

$$d'(\vec{a}_d, \vec{a}_\phi, m) = 1/2\,[d(\vec{a}_d, \vec{a}_\phi, m) + d(\vec{a}_\phi, \vec{a}_d, m)] \tag{2.24}$$

Up to this point in the discussion, the log likelihood ratio has been used only to measure the distortion in a single frame of speech. In order to measure the distortion of an entire utterance, the log likelihood ratio for all frames must be summed to form a resultant global measure of distortion, denoted as D. Although many formulations could be envisioned, one proposed by Barnwell [2.30] is

$$D = \frac{\sum_{m-1}^{M} W(m) d(\vec{a}_\phi, \vec{a}_d, m)}{\sum_{m=1}^{M} W(m)} \tag{2.25}$$

where m is the speech frame index and $W(m)$ is a weight on the entire speech frame.

A slightly different form of the measure due to Itakura and Saito is defined as [2.41]

$$d_{IS}(x, y) = \int_{-\pi}^{+\pi} \left[e^{V(\theta)} - V(\theta) - 1 \right] \frac{d\theta}{2\pi} \tag{2.26}$$

where

$$V(\theta) = \log \left(\frac{{\sigma_\phi}^2}{|A_\phi(e^{j\theta})|^2} \right) - \log \left(\frac{\sigma_d^2}{|A_d(e^{j\theta})|^2} \right)$$

While both the log likelihood ratio and the Itakura-Saito measures are non-negative, a desirable property of the latter is that it is equal to zero if $({\sigma_\phi}^2/|A_\phi(e^{j\theta})|^2)$ and $(\sigma_d^2/|A_d(e^{j\theta})|^2)$ are identical. If the two spectra are normalized such that their gains are equal, the relationship between the log likelihood ratio in Eq. (2.22) and the Itakura-Saito measure defined in Eq. (2.26) is

$$d_{IS}(x_\phi, x_d) = \log(\beta/\alpha) - 1 \tag{2.27}$$

If a symmetric measure is desired, two nonsymmetric Itakura-Saito measures can be combined to form the *cosh* measure

$$d'_{IS} = 1/2 \ \left[d_{IS}(x_\phi, x_d) + d_{IS}(x_d, x_\phi) \right] \tag{2.28}$$

This be expressed directly in terms of $V(\theta)$ making the motivation for the name more obvious

$$d'_{IS} = \int_{-\pi}^{+\pi} (\cosh[V(\theta)] - 1) \frac{d\theta}{2\pi} \tag{2.29}$$

Chu [2.67] has proposed a frequency-weighted Itakura-Saito measure

$$d_{IS}(x, y) = \int_{-\pi}^{+\pi} (e^{V(\theta)} - V(\theta) - 1) \ |W(e^{j\theta})|^2 \frac{d\theta}{2\pi} \tag{2.30}$$

where $W(e^{j\theta})$ is a nonnegative factor that weights low frequencies more heavily than high frequencies, because of decreasing perceptual frequency resolution with increasing frequency.

An important consideration in using the log likelihood ratio as a measure of speech distortion is the assumption that both speech signals, $x_\phi(n)$ and $x_d(n)$, can be described using an all-pole model. This assumption may not be valid if $x_d(n)$ is the product of a speech coder or speech communication system that significantly alters the statistics of the original speech [2.65].

2.2.6 LPC Parameter Measures

Objective measures can be based on parameterizations of the linear prediction vocal tract models of the reference and distorted speech. The parameters used in such measures can be the predictor coefficients themselves, or transformations of the predictor coefficients, such as PARCOR or area ratio coefficients. Each set of parameters provides a different way of quantifying the differences between the vocal tract models of the original and distorted speech.

For measures of this type, the reference and distorted speech waveforms are usually divided into analysis frames of 15 to 30 ms duration. A linear prediction analysis is done for each frame of speech, and the distance measure is computed from the results of the analysis.

Barnwell [2.30], [2.34] has proposed parameter distance measures of the form

$$d(Q,p,m) = \left(\frac{1}{N}\sum_{i=1}^{N}|Q(i,m,\phi) - Q(i,m,d)|^p\right)^{1/p} \tag{2.31}$$

where $d(Q,p,m)$ is the distance for analysis frame m, p is the power in the norm, and N is the order of the LPC analysis. $Q(i,m,\phi)$ and $Q(i,m,d)$ are the ith parameters of the corresponding frames of the original and distorted speech, respectively. The distance for each frame is summed for all frames in the test utterance as follows:

$$D(p) = \frac{\sum_{m=1}^{M} W(m)d(Q,p,m)}{\sum_{m=1}^{M} W(m)} \tag{2.32}$$

where $D(p)$ is the resultant distance measure, M is the total number of frames, and $W(m)$ is a weight associated with the distance measure for each frame. The weighting could, for example, be the energy in the reference analysis frame. Barnwell has investigated this measure for Q equal to the predictor coefficients raised to a power, the magnitude of the PARCOR's or reflection coefficients raised to a power, the area ratios, and the log of the area ratios. Of all of the LPC parameter distances, as will be shown in Chap. 5, the log area ratio measure has the highest correlation with subjective quality. Meister [2.70] and Viswanathan [2.71] also have investigated the performance of log area ratio parameter distance measures.

2.2.7 Cepstral Distance Measures

Linear prediction coefficients also can be used to compute a distance measure based on cepstral coefficients. Unlike a cepstrum computed directly from the speech waveform, one computed from the predictor coefficients provides an estimate of the smoothed speech spectrum. This can be expressed as

$$\log\left(\frac{1}{A(z)}\right) = \sum_{k=1}^{\infty} c(k) z^{-k} \tag{2.33}$$

where $A(z)$ is the LPC model polynomial, $c(k)$ denotes the cepstral coefficients and z can be set equal to $e^{j\omega}$. The coefficients can be computed recursively from the predictor coefficients using the following expression

$$nc(n) - na(n) = \sum_{k=1}^{n-1} (n-k)c(n-k)a(k) \text{ for } n = 1, 2, 3, \ldots \tag{2.34}$$

where $a(0) = 1$ and $a(k) = 0$ for $k > p$. In this expression, $a(k)$ are the LPC predictor coefficients and p is the order of the LPC analysis [2.72].

A measure based on cepstral coefficients [2.25], [2.41] can be computed as

$$d(c_\phi, c_d, 2, m) = \left[[c_\phi(0) - c_d(0)]^2 + 2\sum_{k=1}^{L} [c_\phi(k) - c_d(k)]^2 \right]^{1/2} \tag{2.35}$$

where d is the L_2 distance for frame m and $c_\phi(k)$ and $c_d(k)$ are the cepstral coefficients for the original and distorted speech, respectively.

Although the sequence of cepstral coefficients is infinite, Eq. (2.35) only sums over the first $L+1$ coefficient differences (since $c(k) = c(-k)$). In practice, the value of L can be as small as the order of the LPC analysis and still yield a value for $d(2, m)$ that is very highly correlated with the untruncated cepstral distance measure [2.41]. Note that, by Parseval's relation, the cepstral distance measure in Eq. (2.35) is equivalent to a log spectral distance of cepstrally smoothed spectra. As with the other parametric distance measures, $d(c_\phi, c_d, 2, m)$ is summed over all frames using Eq. (2.32).

2.2.8 Spectral Distance Measures

A final type of objective measure is based on the smoothed spectrum of the reference and distorted speech waveforms. The smoothed spectrum can be obtained by several methods: from LPC analysis, as the spectrum of the vocal tract function; from cepstral analysis, as the exponentiation of the transform of the low-time windowed cepstrum; from the DFT; and from filter bank analysis. The reference and distorted speech waveforms are divided into analysis frames, and the spectra are computed for all frames in the speech utterance.

Spectral distance measures are perhaps the most widely investigated of all speech distortion measures [2.25], [2.40], [2.41], [2.57], [2.60], [2.66], [2.73], [2.74], [2.75]. The distance between two frames of speech can be computed as

$$d(p, m) = \left| \frac{\sum_{k=1}^{N} W\{V_\phi(m,k), V_d(m,k), k\} |F(m,k)|^p}{\sum_{k=1}^{N} W\{V_\phi(m,k), V_d(m,k), k\}} \right|^{1/p} \tag{2.36}$$

where $V_\phi(m,k)$ and $V_d(m,k)$ are the estimated magnitude spectra of frame m of the original and distorted speech, respectively, k is an index over N frequency samples, and p is the power in the L_p norm. W is a weighting factor and is a function of the two spectra. The weighting factor permits parts of the frequency spectrum to receive different emphasis. Also, in the L_p norm, the larger the value of p, the more emphasis that is put on large spectral distances.

The term $F(m,k)$ is a measure of spectral distance for frame m of the original and distorted speech. It can assume the nonlinear form

$$F(m,k) = |V_d(m,k)|^\delta - |V_\phi(m,k)|^\delta \tag{2.37}$$

where the parameter δ specifies the nonlinearity ($\delta = 1$ produces a linear spectral distance). Or it can be a log spectral distance

$$F(m,k) = 20\left[\log_{10} V_d(m,k) - \log_{10} V_\phi(m,k)\right] \tag{2.38}$$

Finally, it could be a ratio of the two spectra

$$F(m,k) = \left[\frac{V_d(m,k)}{V_\phi(m,k)}\right]^\delta \tag{2.39}$$

where, as before, δ specifies a nonlinearity in F. Clearly, many other distance functions also are possible.

The frame distance, $d(p,m)$, is summed for all frames in the utterance to yield $D(p)$, which is the final spectral distance for the reference and distorted speech

$$D(p) = \frac{\sum_{m=1}^{M} W(m) d(p,m)}{\sum_{m=1}^{M} W(m)} \tag{2.40}$$

where m is the frame index and M is the number of frames in the utterance. $W(m)$ is a weight associated with the entire frame of speech, such as the energy in the reference speech frame raised to a power, and is different from $W(V_\phi(m,k), V_d(m,k), k)$.

Dropping the frame index and using a continuous variable instead of k, the smoothed magnitude spectrum, $V(e^{j\omega})$, is estimated by the linear prediction model as

$$V^2(e^{j\omega}) = \frac{G^2}{|A(e^{j\omega})|^2} \tag{2.41}$$

where G is the LPC gain coefficient. Setting G equal to 1 produces a normalized spectrum with mean value equal to 0 on the log scale [2.72]. The inverse filter system function, $A(e^{j\omega})$, can be computed from

$$A(e^{j\omega}) = 1 - \sum_{i=1}^{p} a(i) e^{-j\omega i} \tag{2.42}$$

where the $a(i)$'s are the linear prediction coefficients and p is the order of the prediction analysis.

The speech spectrum also could be estimated using the discrete Fourier transform. An appropriately tapered window should be used prior to transforming in order to minimize the effects of spectral leakage due to window sidelobes, and several transforms of short speech segments must be averaged to yield a smoothed spectrum with low error variance. The total waveform duration represented in the spectrum should be 15 to 30 ms.

A very successful variation on the preceding measures is the frequency variant spectral distance measure [2.30], [2.60]. In this measure, the spectral distance is computed separately in each of several frequency bands. These frequency bands usually are nonoverlapping, with bandwidths proportional to the ear's critical bands. Multiple linear regression is used to determine the weights for each band in summing the individual spectral distances into a global distance measure, and this frequency weighting produces maximum correlation between objective and subjective quality scores. The following expression is used to compute the spectral distance in a given band for a given frame

$$d(p,m,k) = |D(m,k)|^p \tag{2.43}$$

where m is the frame index, k is the band index, and, as before, p is the power in the norm. In Eq. (2.43), the speech spectra V_ϕ and V_d are normalized prior to computing $D(m,k)$ so that the magnitude spectrum in each frequency band has an area equal to 1. Since spectral tilt is a perceptually mild distortion, the normalization procedure is intended to remove any constant spectral difference in $D(k)$ and hence reduce the influence of spectral tilt on the resultant spectral distance measure. The distances for each frame are combined into a spectral distance for each band for the entire utterance as follows:

$$D(p,k) = \frac{\sum_{m=1}^{M} W(m) d(p,m,k)}{\sum_{m=1}^{M} W(m)} \tag{2.44}$$

where $D(p,k)$ is the overall spectral distance in band k, M is the number of frames in the utterance, and as in Eq. (2.42), $W(m)$ is a weighting applied to the distances in frame m. A single distance measure is obtained by forming a weighted sum of $D(p,k)$

$$D(p) = b_0 + \sum_{k=1}^{K} b_k D(p,k) \tag{2.45}$$

where weights b_k are chosen by means of regression analysis so as to maximize the correlation between objective and subjective quality estimates.

A concept that can be applied to any measure, but which has been specifically applied to the spectral distance measure, is that of time classification prior to computing a distance measure [2.73]. This method, proposed by Breitkopf, uses an objective classifier to divide the speech frames of an utterance into different phonemic categories. The underlying assumption in this approach is that the same distortion applied to different phonemic classes may be perceived very differently. This is used to advantage by computing a separate distortion measure for each phonemic class. Breitkopf uses four classes in his spectral distance measure: *vocalic*, *nasal*, *fricative*, and *silence*. The spectral distance is computed from the spectral envelope derived from linear prediction analysis, with the speech analyzed in 32 ms frames. The frames are objectively classified using the frame energy and the frame linear prediction gain.

After all speech frames are classified, a spectral distance measure is computed for each class. The measure is identical to the one specified in Eqs. (2.36), (2.40), and (2.42) except that the distance has an additional parameter to specify phonemic category, indicated by c in the per-frame distance $D(c,p,m)$ and the global distance $D(c,p)$. The global spectral distance for the four classes is combined as follows:

$$D(p) = \sum_{j=1}^{4} b_j D(j,p) \tag{2.46}$$

where the b_j's are weights on the spectral distance of each class.

2.2.9 Weighted Slope Spectral Distance Measure

The speech spectrum also can be estimated using a filter bank. Klatt [2.74], [2.76] has proposed a measure of phonetic distance based on this approach that also works well as a speech quality measure. He uses thirty-six overlapping filters of progressively larger bandwidths to estimate the smoothed short-time speech spectrum every 12 ms. The filter bandwidths approximate critical bands so as to give equal perceptual weight to each band [2.61]. Rather than measuring the absolute spectral distance per band, Klatt measures a weighted difference between the spectral slopes in each band. The motivation for this technique is that a distance measure should be sensitive to differences in spectral peak locations and insensitive to differences in the heights of those peaks. Furthermore, it also should be relatively insensitive to any differences in the neighborhood of spectral valleys or the spectral tilt.

The measure is computed by first finding the spectral slope in each band. If $V_\phi(k)$ and $V_d(k)$ are the reference and distorted spectra measured in decibels, $S_\phi(k)$ and $S_d(k)$ are the slopes of those spectra

$$\begin{aligned} S_\phi(k) &= V_\phi(k+1) - V_\phi(k) \\ S_d(k) &= V_d(k+1) - V_d(k) \end{aligned} \tag{2.47}$$

where k is the band index. Next, a weight for each band must be calculated. The magnitude of the weight is related, first, to whether the band is near a spectral peak or valley, and second, to whether that peak is the largest peak in the spectrum. Klatt computes the weights for each spectrum separately and then averages the two sets of weights to yield $W_a(k)$. The weight, $W(k)$, for either the reference or distorted speech spectrum is computed as

$$W(k) = \frac{K_{max}}{[K_{max} + V_{max} - V_\phi(k)]} \frac{K_{locmax}}{[K_{locmax} + V_{locmax}(k) - V_\phi(k)]} \tag{2.48}$$

where k is the band index. The weighting function is composed of two terms. The first term weights most heavily those bands whose spectral magnitude is closest to the maximum spectral magnitude among all bands. In this term V_{max} is the largest log spectral magnitude among all bands. The second term increases the weight for bands which are close to a local spectral maximum. In this term $V_{locmax}(k)$ is the value of the peak which is nearest to band k. Constants K_{max} and K_{locmax} can be adjusted so as to maximize measure performance. For a highly distorted speech signal, in which the formant structure may be severely altered, it may be preferable to compute $W_a(k)$ based only on the reference speech. The per-frame spectral distance measure is then

$$d(m) = K_{spl}(K_\phi - K_d) + \sum_{k=1}^{36} W_a(k)\{S_\phi(k) - S_d(k)\}^2 \tag{2.49}$$

where K_ϕ and K_d are related to the overall sound pressure level of the reference and distorted utterances and where K_{spl} is a parameter that can be varied. The frame distance, $d(m)$, is summed over all frames in an utterance to produce a measure of overall spectral distance, D, as follows:

$$D = (1/M) \sum_{m=1}^{M} d(m) \tag{2.50}$$

where m is the frame index and M is the number of frames in the utterance.

An advantage of the weighted spectral slope distance measure is that one does not have to explicitly identify and align speech formants prior to computing a spectral difference or decide how to normalize the spectra so as to remove differences in spectral tilt. The critical band representation of the speech spectra also has the advantage of implicitly weighting the spectral differences. This is due to the varying bandwidths in the filters used to estimate the short-time spectrum, and yields a perceptually meaningful frequency weighting.

2.3 Summary

This chapter has presented a review of subjective and objective measures of speech quality. We have attempted to cite both the advantages and disadvantages of each measure, but have avoided any precise statements on the relative performance of the objective measures. This is for good reason: there is little ground for direct comparison of different researchers' work.

The criterion of performance most often applied to objective speech quality measures is the correlation between objective and subjective quality estimates, as measured by the magnitude of the correlation coefficient. To permit valid comparisons of the performance of two objective measures, one must ascertain that the performance of each is measured under approximately equal conditions. Two criteria should be satisfied in making this judgment. First, the two measures must assess the same perceived deficiencies in a speech signal. It would be inappropriate, for example, to compare an objective measure of intelligibility with an objective measure of the naturalness of speech. To make this criterion more restrictive, one could insist that both measures must estimate not only the same perceptual qualities of speech, but also the same subjective measure of those qualities.

A second criterion is that the two objective measures must estimate speech quality over a similar class of speech distortions, and similar levels of those distortions. One should not compare the performance of a measure assessing the quality of waveform coder distortions with one assessing the quality of linear prediction distortions. Again, this can be made more restrictive by insisting that both objective measures estimate speech quality across the identical set of speech distortions.

The research presented in the remainder of this book satisfies all aspects of the preceding criteria. Specifically, all objective measures estimate the same subjective test scores for the same set of distorted speech signals. Therefore valid comparisons of objective measure performance can be made by comparing the magnitude of the estimated correlation coefficient associated with each objective measure.

References

[2.1] N. Kitawaki, M. Honda, and K. Itoh, "Speech Quality Assessment Methods for Speech-Coding Systems," *IEEE Communications Magazine, vol. 22*, no. 10, pp. 26–33, Oct. 1984.

[2.2] E. H. Rothauser, "A Comparison of Preference Measurement Methods," *J. Acoust. Soc. Amer., vol. 49*, no. 4, pp. 399–410, 1971.

[2.3] "IEEE Recommended Practice for Speech Quality Measurements," *IEEE Trans. on Audio and Electroacoust.*, pp. 227–46, Sep. 1969.

[2.4] M. H. L. Hecker and C. E. Williams, "Choice of Reference Conditions for Speech Preference Tests," *J. Acoust. Soc. Amer., vol. 39*, no. 5, pp. 946–52, Nov. 1966.

[2.5] W. A. Munson and J. E. Karlin, "Isopreference Method for Evaluating Speech-Transmission Circuits," *J. Acoust. Soc. Amer., vol. 34*, no. 6, pp. 762–74, Jun. 1962.

[2.6] H. Fletcher and J. C. Steinberg, "Articulation Testing Methods," *Bell Sys. Tech. Journal, vol. 8*, pp. 806–54, 1929.

[2.7] J. P. Egan, "Articulation Testing Methods," *Laryngoscope, vol. 58*, no. 9, pp. 955–91, 1948.

[2.8] J. Swaffield and D. L. Richards, "Rating of Speech Links and Performance of Telephone Networks," *Proc. IEE, vol. 106*, no. 26, pp. 65–76, Mar. 1959.

[2.9] G. Fairbanks, "Test of Phonemic Differentiation: The Rhyme Test," *J. Acous. Soc. Amer., vol. 30*, no. 7, pp. 596–600, Jul. 1958.

[2.10] A. S. House, C. E. Williams, M. H. L. Hecker, and K. D. Kryter, "Articulation Testing Methods: Consonantal Differentiation with a Closed-Response Set," *J. Acous. Soc. Amer., vol. 37*, no. 1, pp. 158–66, Jan. 1965.

[2.11] W. D. Voiers, "Diagnostic Evaluation of Speech Intelligibility," in M. E. Hawley (Ed.), *Benchmark Papers in Acoustics.* Stroudsburg, Pa.: Dowden, Hutchenson and Ross, 1977.

[2.12] R. Jakobson, C. G. M. Fant, and M. Halle, *Preliminaries to Speech Analysis: The Distinctive Features and Their Correlates.* Cambridge, Mass.: M.I.T. Press, 1967.

[2.13] L. H. Nakatani, "Measuring the Ease of Comprehending Speech," *Bell Lab. Tech. Memo.* Jun. 1971.

[2.14] E. H. Rothauser, "Modified Isopreference Method for Audio Quality measurements," *J. Acoust. Soc. Amer., vol. 35*, 1899(A), 1963.

[2.15] W. D. Voiers, "Methods of Predicting User Acceptance of Voice Communication Systems," Final Report, DCA100-74-C-0056, Jul. 1976.

[2.16] E. H. Rothauser, G. E. Urbanek, and W. P. Pachl, "Isopreference Method for Speech Evaluation," *J. Acoust. Soc. Amer., vol. 44*, no. 2, pp. 408–18, Jul. 1967.

[2.17] M. R. Schroeder, "Reference Signals for Signal Quality Studies," *J. Acoust. Soc. Amer., vol. 44*, no. 6, pp. 1735–36, 1968.

[2.18] L. H. Nakatani and K. D. Dukes, "A Sensitive Test of Speech Communication Quality," *J. Acoust. Soc. Amer., vol. 53*, no. 4, pp. 1083–92, 1973.

[2.19] M. Nakatsui and P. Mermelstein, "Subjective Speech-to-Noise Ratio as a Measure of Speech Quality for Digital Waveform

Coders," *J. Acoust. Soc. Amer., vol. 72*, no. 4, pp. 1136–44, Oct. 1982.

[2.20] C. B. Grether and R. B. Stroh, "Subjective Evaluation of Differential Pulse Code Modulation using the Speech 'Goodness' Rating Scale," *Record, 1972 Conference on Speech Communication and Processing*, pp. 175–8, Feb. 1972.

[2.21] D. J. Goodman, B. J. McDermott, and L. H. Nakatani, "Subjective Evaluation of PCM Coded Speech," *Bell Sys. Tech. Journal, vol. 55*, no. 8, pp. 1087–109, Oct. 1976.

[2.22] D. J. Goodman, C. Scagliola, R. E. Crochiere, L. R. Rabiner, J. Goodman, "Objective and Subjective Performance of Tandem Connections of Waveform Coders with an LPC Vocoder," *Bell Sys. Tech. Journal, vol. 58*, no. 3, pp. 601–29, Mar. 1979.

[2.23] B. J. McDermott, C. Scaglia, and D. J. Goodman, "Perceptual and Objective Evaluation of Speech Processed by Adaptive Differential PCM," *Proc. 1978 IEEE ICASSP*, Tulsa, pp. 581–5, Apr. 1978.

[2.24] B. J. McDermott, C. Scagliola, and D. J. Goodman, "Perceptual and Objective Evaluation of Speech Processed by Adaptive Differential PCM," *Bell Sys. Tech. Journal, vol. 57*, no. 5, pp. 1597–619, May 1978.

[2.25] N. Kitawaki, K. Itoh, and K. Kakei, "Speech Quality of PCM System in Digital Telephone System," *Electronics and Communication in Japan, vol. 65-A*, no. 8, pp. 1–8, 1982.

[2.26] G. A. Miller, "The Magical Number Seven, Plus or Minus Two," In S. Ducker (Ed.), *Time-Compressed Speech* (vol. 1). Metuchen, N. J.: The Scarecrow Press, 1974.

[2.27] W. D. Voiers, "Perceptual Bases of Speaker Identity," *J. Acoust. Soc. Amer., vol. 36*, no. 6, pp. 1065–73, Jun. 1964.

[2.28] V. E. McGee, "Semantic Components of the Quality of Processed Speech," *J. Speech and Hearing Research, vol. 7*, no. 4, pp. 310–23, Dec. 1964.

[2.29] B. J. McDermott, "Multidimensional Analysis of Circuit Quality Judgments," *J. Acoust. Soc. Amer., vol. 45*, no. 3, pp. 774–81, 1969.

[2.30] T. P. Barnwell III and W. D. Voiers, "An Analysis of Objective Measures for User Acceptance of Voice Communication Systems," Final Report, DCA100-78-C-0003, Sep. 1979.

[2.31] D. L. Richards and J. Swaffield, "Assessment of Speech Communication Links," *Proc. IEE, vol. 106*, no. 26, pp. 77–89, Mar. 1959.

[2.32] J. L. Hall, "Auditory Discrimination in the Presence of Masking Noise," *IEEE Trans. Acoust., Speech, Signal Processing*, pp. 100–2, Feb. 1976.

[2.33] L. H. Nakatani, "Evaluation of Speech Quality Using Immediate Recall of Digit Strings," *Bell Lab. Tech. Memo.*, Jan. 1970.

[2.34] T. P. Barnwell III, A. M. Bush, R. M. Mersereau, and R. W. Schafer, "Speech Quality Measurement," Final Report, RADC-TR-78-122, May 1978.

[2.35] D. B. Pisoni, "Perceptual Evaluation of Voice Response Systems: Intelligibility, Recognition and Understanding," *Proc., Workshop on Standardization for Speech I/O Technology*, pp. 185–207, Mar. 1982.

[2.36] W. D. Voiers, "Diagnostic Acceptability Measure for Speech Communication Systems," *Proc. 1977 IEEE ICASSP*, pp. 204–7, May 1977.

[2.37] J. B. Kruskal, "Multidimensional Scaling by Optimizing Goodness of Fit to a Numerical Hypothesis," *Psychometrika, vol. 29*, pp. 1–27, 1964.

[2.38] J. B. Kruskal, "Nonmetric Multidimensional Scaling: A Numerical Method," *Psychometrika, vol. 29*, pp. 115–29, 1964.

[2.39] J. Makhoul and R. Viswanathan, "A Framework for the Objective Evaluation of Vocoder Speech Quality," *Proc. 1976 IEEE ICASSP*, Philadelphia, pp. 103–6, Apr. 1976.

[2.40] R. Viswanathan, J. Makhoul, and W. Russell, "Towards Perceptually Consistent Measures of Spectral Distance," *Proc. 1976 IEEE ICASSP*, Philadelphia, Pa. pp. 485–8, Apr. 1976.

[2.41] A. Gray, Jr. and J. D. Markel, "Distance Measures for Speech Processing," *IEEE Trans. Acoust., Speech and Signal Processing, vol. ASSP-24*, no. 5, pp. 380–91, Oct. 1976.

[2.42] T. P. Barnwell III, "Objective Measures for Speech Quality Testing," *J. Acoust. Soc. Amer., vol. 66*, no. 6, pp. 1658–63, Dec. 1979.

[2.43] T. P. Barnwell III, "Correlation Analysis of Subjective and Objective Measures for Speech Quality," *Proc. 1980 IEEE ICASSP*, Denver, pp. 706–9, Apr. 1980.

[2.44] R. Billi and C. Scagliola, "Artificial Signals and Identification Methods to Evaluate the Quality of Speech Coders," *IEEE Trans. Comm., vol. COM-30*, no. 2. pp. 325–35, Feb. 1982.

[2.45] K. Itoh, N. Kitawaki, and K. Kakehi, "Objective Quality Measures for Speech Waveform Coding Systems," *Review Electrical Communication Laboratories, vol. 32*, no. 2, pp. 220–8, 1984.

[2.46] N. R. French and J. C. Steinberg, "Factors Governing the Intelligibility of Speech Sounds," *J. Acous. Soc. Amer., vol. 19*, no. 1, pp. 90–119, Jan. 1947.

[2.47] J. L. Flanagan, *Speech Analysis, Synthesis and Perception* (2nd ed.). New York: Springer-Verlag, 1972.

[2.48] L. L. Beranek, "The Design of Speech Communication Systems," *IRE Proceedings, vol. 35*, no. 9, pp. 880–90, 1947.

[2.49] L. L. Beranek, *Acoustics*. New York: McGraw-Hill, 1954.

[2.50] K. D. Kryter, "Methods for the Calculation and Use of the Articulation Index," *J. Acoust. Soc. Amer., vol. 34*, no. 11, pp. 1689–97, Nov. 1962.

[2.51] K. D. Kryter, "Validation of the Articulation Index," *J. Acoust. Soc. Amer., vol. 34*, no. 11, pp. 1698–702, Nov. 1962.

[2.52] S. S. Stevens, J. Volkman, and E.B. Newman, "A Scale for the Measurement of the Psychological Magnitude of Pitch," *J. Acoust. Soc. Amer., vol. 8*, pp. 185–90, Jan. 1937.

[2.53] H. J. M. Steeneken and T. Houtgast, "A Physical Method for Measuring Speech-Transmission Quality," *J. Acoust. Soc. Amer., vol. 67*, no. 1, pp. 318–26, Jan. 1980.

[2.54] P. W. Noll, "Adaptive Quantization in Speech Coding Systems," IEEE Int. Zurich Seminar, Oct. 1974.

[2.55] R. E. Crochiere, L. R. Rabiner, N. S. Jayant, and J. M. Tribolet, "A Study of Objective Measures for Speech Waveform Coders," *Proceedings of 1978 Zurich Seminar on Digital Communications*, pp. H1.1–7, Mar. 1978.

[2.56] J. M. Tribolet, P. Noll, B. J. McDermott, and R. E. Crochiere, "A Study of Complexity and Quality of Speech Waveform Coders," *Proc. 1978 IEEE ICASSP*, Tulsa, pp. 586–90, Apr. 1978.

[2.57] T. P. Barnwell, III and A. M. Bush, "Statistical Correlation between Objective and Subjective Measures for Speech Quality," *Proc. 1978 IEEE ICASSP*, pp. 595–6, Apr. 1978.

[2.58] P. Mermelstein, "Evaluation of a Segmental SNR Measure as an Indicator of the Quality of ADPCM Coded Speech," *J. Acoust. Soc. Amer., vol. 66*, no. 6, pp. 1664–7, Dec. 1979.

[2.59] R. V. Cox, J. H. Snyder, R. E. Crochiere, D. E. Block, and D. J. Johnston, "Testing of Wideband Digital Coders," *Proc. 1984 IEEE ICASSP*, pp. 19.3.1–4, Apr. 1984.

[2.60] T. P. Barnwell III, "A Comparison of Parametrically Different Objective Speech Quality Measures Using Correlation Analysis with Subjective Quality Results," *Proc. 1980 IEEE ICASSP*, Denver, pp. 710–3, Apr. 1980.

[2.61] E. Zwicker, "Subdivision of the Audible Frequency Range into Critical Bands," *J. Acoust. Soc. Amer., vol. 33*, no. 4, p. 248, Feb. 1961.

[2.62] B. Scharf, "Critical Bands," in J.V. Tobias (Ed.), *Foundations of Modern Auditory Theory.* New York: Academic Press, pp. 157–202, 1970.

[2.63] F. Itakura, "Minimum Prediction Residual Principle Applied to Speech Recognition," *IEEE Trans. Acoust., Speech and Signal Processing, vol. ASSP-23*, no. 1, pp. 67–72, Feb. 1975.

[2.64] J. M. Tribolet, L. R. Rabiner, and M. M. Sondhi, "Statistical Properties of an LPC Distance Measure," *IEEE Trans. Acoust., Speech, Signal Processing, vol. ASSP-27*, no. 5, pp. 550–8, Oct. 1979.

[2.65] R. E. Crochiere, J. E. Tribolet, and L. R. Rabiner, "An Interpretation of the Log Likelihood Ratio as a Measure of Waveform Coder Performance," *IEEE Trans. Acoust., Speech, Signal Processing, vol. ASSP-28*, no. 3, Jun. 1980.

[2.66] R. M. Gray, A. Buzo, A. H. Gray, Jr., and Y. Matsuyama, "Distortion Measures for Speech Processing," *IEEE Trans. Acoust., Speech, Signal Processing, vol. ASSP-28*, no. 4, pp. 367–76, Aug. 1980.

[2.67] P. L. Chu and D. G. Messerschmitt, "A Weighted Itakura-Saito Spectral Distance Measure," *IEEE Trans. Acoust., Speech, Signal Processing, vol. ASSP-30*, no. 4, pp. 545–60, Aug. 1982.

[2.68] B. H. Juang, "On Using the Itakura-Saito Measure for Speech Coder Performance Evaluation," *AT&T Bell Laboratories Tech. Jour., vol. 63*, no. 8, pp. 1477–98, Oct. 1984.

[2.69] L. R. Rabiner and R. W. Schafer, *Digital Processing of Speech Signals*. Englewood Cliffs, N. J.: Prentice-Hall, 1978.

[2.70] S. Meister, "Quality Comparison Measure for Linear Predictive Systems," *Proc. IEEE 1976 ICASSP*, pp. 107–9, Apr. 1976.

[2.71] R. Viswanathan, W. Russell, and J. Makhoul, "Objective Speech Quality Evaluation of Narrowband LPC Vocoders," *Proc. 1978 IEEE ICASSP*, Tulsa, pp. 591–4, Apr. 1978.

[2.72] J. D. Markel and A. H. Gray, *Linear Prediction of Speech*. New York: Springer-Verlag, 1976.

[2.73] P. Breitkopf and T. P. Barnwell III, "Segmental Preclassification for Improved Objective Speech Quality Measures," *Proc. 1981 IEEE ICASSP*, Atlanta, pp. 1101–4, Mar. 1981.

[2.74] D. H. Klatt, "Prediction of Perceived Phonetic Distance from Critical-Band Spectra: A First Step," *Proc. 1982 IEEE ICASSP*, Paris, pp. 1278–81, May 1982.

[2.75] V. R. Viswanathan, W. H. Russell, and A. W. F. Huggins, "Objective Speech Quality Evaluation of Medium and Narrowband Real-Time Speech Coders," *Proc. 1983 IEEE ICASSP*, Boston, pp. 543–6, Apr. 1983.

[2.76] D. H. Klatt, "A Digital Filter Bank for Spectral Matching," *Proc. 1976 IEEE ICASSP*, pp. 573–6, Apr. 1976.

Chapter 3

The Diagnostic Acceptability Measure

3.1 Background

The Diagnostic Acceptability Measure or DAM is a relatively new quality measure, first described by Voiers [3.1] in 1977. This chapter will discuss the DAM, how it was developed and how it is administered and scored. It is meant to be a comprehensive presentation, since there are currently few publications [3.1], [3.2], which provide a detailed description of this important subjective test.

A great difficulty in subjective speech quality evaluation is deciding exactly what quantities to measure when making the quality assessments. Speech intelligibility is unquestionably an important factor in determining the quality of voice communication systems, but it is clearly not sufficient in all cases. For example, in high quality systems where easy and effective communication might be the overall performance goal, it also is desirable for the speech to sound natural and for the user to be able to recognize the speaker. The DAM is able to assess a broad range of speech degradations, and therefore is valid for a wide range of speech qualities.

The Diagnostic Acceptability Measure is unlike other subjective measures of speech quality in that it incorporates a multidimensional approach to its evaluation. The DAM evaluates a speech signal on sixteen separate scales, all of which have a range of 0 to 100 points. These scales can be divided into three categories: signal quality, background quality, and total quality (see Table 3.1). The three total quality scales, I, P and A, represent primarily the conventional, direct approach of evaluating speech quality, in which the speech is rated relative to some general perceptual quality such as "intelligibility," "pleasantness," or "acceptability." The signal and background quality scales represent the indirect, or "parametric," evaluation method, which is based on listeners' judgments of the speech relative to various specific characteristics of the perceived speech degradation. The profile of these parametric quality ratings provides a fine-grained portrait of how the communication system distorts the speech signal. The composite acceptability scale summarizes the information of these DAM scales into a single unidimensional scale, and is simply a weighted average of the signal, background, I, P and A scales.

The multiplicity of scales is an important feature of the DAM. Accuracy in listener responses is obtained by permitting a distortion to be categorized as affecting either the speech signal or the background and, furthermore, by rating the distortion on several distinct scales within the signal and background quality groups. This fine-grained structure of the DAM parametric quality scales ensures that listeners are not required to compromise their subjective impressions, and thus diminish their accuracy, by forcing the impressions to align with a much more limited set of quality descriptors. Thus the DAM provides for a multidimensional description of subjective quality which is able to accurately describe the complex differences between modern digital speech coding systems.

Although the DAM provides excellent characterization of the distortions in a speech signal, it would be of little use if it were not, in addition, a consistent and reliable measure of speech quality. From a statistical point of view, the consistency and stability of a test can be measured by its index of reliability, which is equal to the correlation between the outcomes of two independent runs of a test. The index of reliability of the DAM is 0.96 [3.3].

The reliability statistic can also be interpreted as an estimate of the random error associated with DAM quality scores. The ratio of the variance of the random error, ${\sigma_e}^2$, to the variance of the quality scores,

Table 3.1: Diagnostic Acceptability Measure Quality Scales.

PARAMETRIC SCALES			
Name		Descriptors	Example
Signal	SF	fluttering, bubbling	AM speech
	SH	distant, thin	highpass speech
	SD	rasping, crackling	peak clipped speech
	SL	muffled, smothered	lowpass speech
	SI	irregular, interrupted	interrupted speech
	SN	nasal, whining	bandpass speech
	TSQ	total signal quality	
Back-ground	BN	hissing, rushing	Gaussian noise
	BB	buzzing, humming	60 Hz hum
	BF	chirping, bubbling	narrowband sys. errors
	BR	rumbling, thumping	low freq. noise
	TBQ	total background quality	

METAMETRIC SCALES	
I	intelligibility
P	pleasantness

ISOMETRIC SCALES	
A	acceptability
CA	composite acceptability

${\sigma_y}^2$, can be expressed as

$$({\sigma_e}^2/{\sigma_y}^2) = 1 - R^2 \tag{3.1}$$

where R is the index of reliability [3.4]. The ratio in Eq. 3.1 is therefore equal to the fraction of the quality score variance which is due to random errors in the quality testing method, and is approximately eight percent. The random error associated with a quality score is a limiting factor in the accuracy of a test since a reported score will only be an estimate of the true quality score, being distributed about the true, but unknown, score with a standard deviation equal to σ_e. The DAM composite acceptability scale's standard deviation of error, σ_e, is approximately three points on a scale of 0 to 100 [3.3].

Another criterion of the usefulness of a test is provided by the level of acceptance it has gained among those who have a need to test speech quality. Voiers has produced two tests which are widely used for evaluating speech intelligibility and speech quality. These are the Diagnostic Rhyme Test (DRT) and the DAM. Both the DRT and the DAM are used by the U.S. Department of Defense for the evaluation of communication systems. The DRT is the NATO standard for evaluating intelligibility in communication systems. It is also, along with the Modified Rhyme Test, included in the forthcoming version of the American National Standards Institute (ANSI) standard for evaluating intelligibility in communication systems. In addition to wide use by government agencies, the DRT and DAM are used by many companies involved in speech communication systems research and development.

3.2 Design of the DAM

Any subjective testing situation is prone to error associated with human variability. Therefore, in discussing the characteristics of the DAM, it is important to consider the sources of error involved in the measurement process and to attempt to minimize their impact.

3.2.1 Control of measurement error

There are four major sources of systematic or random variation in subjective quality testing: differences between listeners (interlistener), differences within listeners (intralistener), differences between speakers (interspeaker), and differences within speakers (intraspeaker).

Error due to interlistener variation Listener factors, both systematic and random, are potential sources of error in any psychoacoustical experiment. Their impact on test results is likely to be most significant when listeners' ratings depend on their personal preferences in speech quality. Though costly, the most direct means of reducing random interlistener variation is to increase the number of listeners in the test, thereby decreasing the variance of the sample mean. Systematic errors may arise from differences in the individual's perspective on the quality rating task, such that one listener may consistently rate *all* speech coding systems to be of lower quality than another listener.

Systematic variations can be reduced by estimating the bias in an individual's responses and correcting for this bias.

Error due to intralistener variation Significant error may result from random and systematic variation in the responses of a given listener. Random variation can be reduced through repetition of the experiment (reducing the variance of each listener's mean response). Systematic error may result from fatigue, learning, or other experimental factors. Particularly important is error which results from shifts in the listener's adaptation level, or subjective reference point. Both good experimental design incorporating anchors to track adaptation level, and randomization of the order in which speech is presented to the listener for evaluation can be used to minimize intralistener systematic error.

Error due to interspeaker variation Speaker differences, particularly as they interact with system characteristics, are also potential sources of error in the measurement of speech acceptability. For example, the intelligibility of vocoded speech can be affected by the pitch of the speaker [3.5], [3.6]. In this case the error is composed of both random and systematic components. However, since any number of speakers may use a communication system, this source of error may best be reduced by using several speakers in the subjective test, each with varying speech characteristics.

Error due to intraspeaker variation Many investigators have observed that the intelligibility of speech varies with a number of factors, such as a speaker's vocal effort [3.7]. It follows that speech acceptability must also be subject to intraspeaker variation. Whether systematic or not, effects such as stress and enunciation can be controlled during the course of recording a speaker's voice and hence their variation reduced.

3.2.2 Isometric and Parametric Evaluation

An investigation by McDermott [3.8] provides significant insight into a possible methodology for speech acceptability evaluation. In this investigation, preference data and similarity data were collected from a large number of listeners for a set of twenty-one speech transmission systems. The study demonstrated that it is possible to predict user preference or acceptance from judgments made relative to other subjective characteristics, a number of which involved judgments of similarity. A significant aspect of these results was that similarity data which were unadjusted for listener idiosyncrasies could be used to predict preference judgments which were adjusted for listener idiosyncrasies. This finding suggests a means of circumventing what is perhaps the greatest obstacle to the development of valid, practical methods of acceptability evaluation: the fact that listeners tend to agree on what they hear far more than they agree on how much they like what they hear. Therefore measurements of what individuals perceive to be the distinguishing features of a distorted speech signal can serve as a valid basis for the prediction of user acceptability, independently of the values placed on these features by the individual listeners.

Given these results, and the previous discussion on control of measurement error, it appears that the problem of measuring system acceptability can be solved in either of two ways: isometric evaluation, or parametric evaluation.

Isometric approach to acceptability evaluation The first, and more common method of acceptability evaluation is the *isometric* approach, in which opinions of the overall quality of a speech signal are directly solicited from the listener. However, this approach assumes that the sample of listeners used is representative of the population of interest. If this assumption is not valid, predictions of system acceptability based on isometric evaluations will not be accurate.

Parametric approach to acceptability evaluation A second method of acceptability evaluation is the *parametric* approach, in which the listeners perception of the degraded speech, rather than his or her preference for that degradation, is used as a basis for predicting system acceptability. The validity of the parametric approach rests on two assumptions:

1. That the listener sample and the population of interest both have an equal capacity to discriminate a set of perceived characteristics in the speech signal under test, regardless of their preferences for the perceived characteristics.

2. That some correlation exists, either on the aggregate or the individual level, between the perceived characteristics of a speech signal and its degree of acceptance by the population of interest.

The key element in this approach is the ability to isolate the effect of listener preference from predictions of system acceptability. Given the validity of these assumptions, a listener whose personal preference does not match that of the population of interest can still provide useful information about the degree to which a speech signal is described by a set of perceived characteristics. Such information can, in turn, be used to predict the acceptance reactions of the population of interest.

The isometric evaluation of a speech communication system is intuitively appealing, since the resulting score can be related directly to overall quality and hence to the expected degree of user acceptance. However, the problem with this approach is that listener ratings are subject to extremely large interindividual variation, which serves to increase the error in isometric quality estimates. As was previously mentioned, this variation is due not only to random errors in judgment, but also to bias arising from an individual's perspective on the rating task [3.9]. Bias in an individual's quality ratings may stem from differences in the scale and origin of each listener's personal criteria for rating speech acceptability. Careful listener training and the use of "anchors" and "probes" in the data can help minimize the bias, while averaging over a relatively large and varied selection of listeners can minimize any random error component.

By comparison, the parametric approach, though not as easy to implement or interpret, tends to yield more accurate scores than the isometric approach. Voiers has hypothesized that this is primarily because listener "taste" does not enter as heavily into the assessment of parametric speech quality, thereby diminishing a major source of interindividual variation [3.2]. The parametric method requires a listener to rate the degree to which a specific characteristic, such as "rasping and crackling," is present in the speech, and does not ask for preference judgments of that characteristic. The parametric scales of the DAM do not, in general, have a practical interpretation by themselves, although they do provide a rich and detailed characterization of a distorted speech signal. Rather, their chief importance lies in the fact that they are part of a system for producing accurate estimates of the DAM composite acceptability score.

3.2.3 The Response Scales

Because of its higher accuracy, the DAM uses the parametric approach to speech quality assessment. Implicit in this approach is the necessity of developing a set of perceptual characteristics which are relevant to the prediction of speech acceptability. Also necessary is the determination of quantitative relationships between listener ratings of these characteristics and evaluations of speech acceptability. Based on experimental evaluation of a large amount of test speech using a large pool of potential quality descriptors, and on factor analysis of ratings relative to these descriptors, Voiers developed a set of *listener response scales* or *rating scales*, shown on the DAM listener response sheet, Fig. 3.1. These scales are designed to characterize the perceptual nature of diverse forms of speech degradation, and are distinct from the DAM *reporting scales* shown in Table 3.1. Of the response scales, ten scales are used to rate the speech signal, seven to rate the background, and three to rate overall quality. Voiers' analysis identified these twenty scales as, first, most different from each other and hence providing the greatest dimensionality in the subjective response, and, second, best able to serve as a basis for highly reliable predictions of speech acceptability.

Further use of factor analysis revealed that in various combinations the response scales could be used to produce ten nearly independent perceptual qualities, six pertaining to the signal and four pertaining to the background. These are the DAM reporting scales of Table 3.1. Each of the DAM reporting scales is derived from a weighted combination of one or two of the rating scales, as listed in Table 3.2. Therefore there are two sets of scales in the DAM, the final, reporting scales (Table 3.1), and the listener response scales (Table 3.2), with the former being functionally dependent on the latter. That the final DAM parametric quality scales are able to describe all significant perceptual aspects of speech quality is indicated by the fact that almost all of the variation in the composite acceptability scale can be accounted for by the scores of the ten parametric scales [3.10].

The relationship between the twenty rating scales and the DAM's four total quality scales was established using regression analysis over

Make a slash at the appropriate point on each scale to indicate the degree to which this transmission sample is characterized by the indicated quality.

THE SPEECH SIGNAL

10 20 30 40 50 60 70 80 90 100
negligible FLUTTERING extreme
TWITTERING PULSATING

10 20 30 40 50 60 70 80 90 100
negligible MUFFLED extreme
SMOTHERED LOW

10 20 30 40 50 60 70 80 90 100
negligible DISTANT extreme
SMALL COMPACT

10 20 30 40 50 60 70 80 90 100
negligible RASPING extreme
SCRAPING GRATING

10 20 30 40 50 60 70 80 90 100
negligible THIN extreme
TINNY HIGH

10 20 30 40 50 60 70 80 90 100
negligible UNNATURAL extreme
MECHANICAL LIFELESS

10 20 30 40 50 60 70 80 90 100
negligible BABBLING extreme
CHORTLING SLOBBERING

10 20 30 40 50 60 70 80 90 100
negligible IRREGULAR extreme
SPASMOTIC FITFULL

10 20 30 40 50 60 70 80 90 100
negligible INTERRUPTED extreme
INTERMITTENT CHOPPED

THE BACKGROUND

10 20 30 40 50 60 70 80 90 100
negligible HISSING extreme
SIMMERING FIZZING

10 20 30 40 50 60 70 80 90 100
negligible CHIRPING extreme
CHEEPING CLICKING

10 20 30 40 50 60 70 80 90 100
negligible ROARING extreme
RUSHING GUSHING

10 20 30 40 50 60 70 80 90 100
negligible CRACKLING extreme
SCRATCHING STATICY

10 20 30 40 50 60 70 80 90 100
negligible BUZZING extreme
HUMMING WHIRRING

10 20 30 40 50 60 70 80 90 100
negligible RUMBLING extreme
THUMPING THUDDING

10 20 30 40 50 60 70 80 90 100
negligible BUBBLING extreme
GURGLING PERCOLATING

THE TOTAL EFFECT

10 20 30 40 50 60 70 80 90 100
negligible INTELLIGIBLE extreme
UNDERSTANDABLE MEANINGFUL

10 20 30 40 50 60 70 80 90 100
negligible PLEASANT extreme
RICH MELLOW

10 20 30 40 50 60 70 80 90 100
negligible ACCEPTABLE extreme

Figure 3.1: DAM System rating form (From [3.3]).

more than 200 test systems. In an initial analysis, the "total effect" listener response scales were the dependent variables and the remaining speech and background rating scales were the independent variables. The resulting estimates of the intelligibility, pleasantness and acceptability "total effect" scores were used as initial estimates of the I, P, and A scores of the reporting scales. The composite acceptability score was initially estimated as an average of parametric, I, P and A scores. These regression equations were further refined using additional test

Table 3.2: Derivation of DAM Scales from Rating Scales (From [3.8]).

DAM Perceptual Quality Scale	Based on Rating Scale
SF	1, 7
SH	3, 5
SD	4, 14
SL	2
SI	8, 10
SN	9
BN	11, 13
BB	15
BF	12, 17
BR	16

data to produce expressions which related the signal and background rating scores to the final parametric and acceptability scores, the rating intelligibility to final intelligibility, the rating pleasantness to final pleasantness, and the final DAM parametric and total quality scores to the final composite acceptability score. This analysis, done only once as part of the design of the Diagnostic Acceptability Measure, is particularly critical since it determines one of the measure's most influential scales, composite acceptability.

3.3 Description of Quality Scales

The DAM quality scales, listed in Table 3.1, consist of twelve parametric scales and four metametric or isometric scales. Each scale ranges from 0 to 100, with a score of 80 interpreted as undistorted speech. Scales have a mnemonically useful code, such as SF for "signal fluttering and bubbling," which suggests the type of distortion which is most exemplary of that scale. The parametric scales are separated into six ratings of signal quality and four of background quality, and yield estimates of the specific character of signal or background degradation. Two additional scales evaluate total signal quality (TSQ) and total background quality (TBQ). Three scales evaluate the character of the speech in terms of intelligibility (I), pleasantness (P) and acceptability (A). The last, and most important scale is composite acceptability (CA), which is a single number specifying overall acceptability and which effectively summarizes the information provided by all the other scales. The six signal and four background scales, along with their associated total quality scales, are "parametric" in that each measures one aspect, or dimension, of composite acceptability. The intelligibility and pleasantness scales are "metametric" in that they measure characteristics similar to, but not the same as, acceptability. The acceptability and composite acceptability scales are "isometric" since they are direct measures of acceptability.

Although the composite acceptability scale of the DAM yields the score of greatest practical interest, an individual parametric scale can be interpreted by itself as estimating the level of acceptability a speech signal would have if it were deficient in only that parametric quality. This is simply an expression of the fact that the DAM parametric reporting scales are designed to be perceptually independent. Because of this independence, the values of the parametric scores indicate the relative degree to which the ten parametric qualities enhance or diminish the overall quality of the speech system under test.

3.4 Test Materials and Procedures

3.4.1 Speech Materials

The speech materials used with the DAM consist of twelve sentences from each of six speakers. Each sentence is six syllables in length and has its phonetic makeup controlled in the following manner: it is constrained to have at least one vowel from each of the three categories (Front, Mid, Back) listed in Table 3.3, and at least one consonant from each of the three categories (Sibilant, Stop, Fricative) listed in Table 3.4. Three of the speakers are male and three are female and a different set of twelve sentences are used by each of the speakers.

3.4.2 Evaluation Procedures

The Diagnostic Acceptability Measure is typically used to evaluate speech from six to twenty-four coder or speech communication sys-

Table 3.3: Vowel categories (from [3.9]).

	FRONT	MID	BACK
HIGH	team - /i/ tip - /I/		tool - /u/ took - /U/ tone - /o^u/
MID		ton - /ʌ / bird - /ɝ/	
LOW	ten - /ϵ/ tap - /æ/		talk - /ɔ/ top - /a/

Table 3.4: Consonant categories (from [3.9])

SIBILANTS	STOPS	FRICATIVES
zip - /z/ sit - /s/ chat - /č/ shot - /š/ jot - /ǰ/	pat - /p/ top - /t/ bat - /b/ dot - /d/ get - /g/ kit - /k/	vat - /v/ for - /f/ thin - /θ/ that - /ð/

tems in a single testing session. For each system, twelve sentences of speech with total duration of approximately one minute is required. Sentences are presented to a crew of twelve listeners at a rate of one sentence every four seconds and systems are presented in random ordering. Prior to each listening session, listeners rate two "anchors" and four "probes." The high and low anchors, which are examples of high and low quality speech at specific DAM scores, give the listeners a frame of reference so that their evaluations can be more consistent relative to previous sessions, while the probes are used to estimate any coincidental fixed errors which may affect that particular session.

3.4.3 Listener Selection and Calibration

In addition to the corrective action resulting from anchors and probes, the DAM uses carefully trained listening crews. Candidate listeners' ratings should be consistent over time and have a moderately high correlation to the listening crew's historical average ratings over an extremely diverse set of more than 200 training utterances. Listeners who are able to do this are then calibrated to determine their own subjective origin and scale relative to the historical average listener ratings. The calibration is done using a simple regression analysis of the listener ratings onto the historical average ratings. The magnitude of the coefficient of correlation obtained from the regression analysis is used to weight the individual's data in subsequent tests. The slope and intercept, representative of the individual's subjective origin and scale, are used to correct that individual's scores.

3.5 Calculation of DAM Scores

The raw data from the DAM consists of the listener scores on the response form of Fig. 3.1. So that all scores have the same sense of larger values indicating higher quality, the scales for the signal and background data are inverted for each listener as follows:

$$R'_{ij} = A_j - R_{ij} \tag{3.2}$$

where R'_{ij} is the inverted score for the ith system and the jth rating scale, R_{ij} is the raw score, and A_j is the historical average rating of the high anchor on the jth rating scale. Scores are further transformed using

$$R''_{ij} = (b_j)(R'_{ij}) + c_j \tag{3.3}$$

where b_j and c_j are selected such that R''_{ij} approximates the acceptability rating that system i would receive if its degradation could be completely measured in terms of the characteristics of scale j. The values of R''_{ij} are used by themselves or are combined, depending on j, (as specified in Table 3.2) to form the raw scores S_{ij} of the DAM quality scales. These raw scores are corrected for the listener's subjective origin and scale using

$$S'_{ijk} = b_{jk}(S_{ij}) + c_{jk} \tag{3.4}$$

where b_{jk} is a scale factor which adjusts listener k to have the same range as the historical average listener for quality scale j, and c_{jk} is the difference in subjective origin between listener k and the historical average listener. The DAM score S_{ij} for system i and scale j is a weighted average of S'_{ijk}, and is computed as

$$S_{ij} = \frac{\sum_{k=1}^{N} r_{jk} S'_{ijk}}{\sum_{k=1}^{N} r_{jk}} \tag{3.5}$$

where N is the number of listeners in the test (usually twelve) and r_{jk} is the coefficient of correlation computed while calibrating listener k. It is equal to the correlation between the ratings of listener k on scale j and the historical average listener ratings on scale j as measured over the set of 200 test system conditions. The effect of this sum is to give the greatest weight to those listeners whose responses are most highly correlated with those of the historical average listener.

A final, minor adjustment of all averaged adjusted scores is made to minimize the effect of any transient coincidental influences which may affect the entire listening crew in the course of a test session. The adjustment is computed using

$$\bar{S}_{ij} = S_{ij} - (0.5)(P_j - P_j^{(h)}) \tag{3.6}$$

where $\bar{S}_i j$ is the *fully adjusted* average rating of condition i on scale j, P_i is the average rating of the four probes on scale j and $P_j^{(h)}$ is the historical average of the identical listening crew's ratings of the four probes, and 0.5 is the estimated coefficient of reliability (session to session) of the probe average.

The fully adjusted perceptual quality scores $\bar{S}_{ij}$ give a detailed portrait of the way in which a system distorts the speech signal. Their information can be summarized into two higher-level criteria of system performance: total signal quality (TSQ) and total background quality (TBQ). These measures are computed from the respective signal or background quality scores as follows:

$$\text{TSQ} = (\sum_{j=1}^{6} b_j^S \bar{S}_j) + C_1^S (\prod_{j=1}^{6} \bar{S}_j)^{1/3} + C_2^S (\prod_{j=1}^{6} \bar{S}_j)^{1/6} + C_3^S \tag{3.7}$$

$$\text{TBQ} = (\sum_{j=7}^{10} b_j^B \bar{S}_j) + C_1^B (\prod_{j=7}^{10} \bar{S}_j)^{1/3} + C_2^B (\prod_{j=7}^{10} \bar{S}_j)^{1/6} + C_3^B \tag{3.8}$$

where the parametric signal scales are indexed 1 to 6, and the parametric background scales are indexed 7 to 10. The b_j's are weights on these scales, and the C's are chosen to normalize the TSQ and TBQ scales relative to the acceptability scale. In this way, TSQ or TBQ ratings are approximately the acceptability rating that the system would receive if it were deficient in only signal or background characteristics, respectively.

A measure of acceptability is computed from TSQ and TBQ

$$A = \sum_{j=1}^{10} b_j^A \bar{S}_j + C_1^A(\text{TSQ})(\text{TBQ}) + C_2^A \tag{3.9}$$

where the b_j^A, C_1^A, and C_2^A are regression coefficients which are computed in a regression analysis involving the set of over 200 test systems.

Two additional metametric quality measures, one of intelligibility (I) and one for pleasantness (P), are computed as follows:

$$I = C_1^I(I_R) + C_2^I(I_R)^2 + C_3^I \tag{3.10}$$

$$P = C_1^P(P_R) + C_2^P(P_R)^2 + C_3^P \tag{3.11}$$

where I_R and P_R are fully adjusted listener ratings of intelligibility and pleasantness, and the C^I's and C^P's are regression coefficients computed in a manner similar to those in Eq. (3.9).

The final, and most influential, score is composite acceptability (CA) which is a rating of overall system quality. It is computed as

$$CA = \frac{(\sum_{j=1}^{10} b_j^{CA} \bar{S}_j + C_1^{CA} A + C_2^{CA} I + C_3^{CA} P)}{(\sum_{j=1}^{10} b_j^{CA} + C_1^{CA} + C_2^{CA} + C_3^{CA})} \tag{3.12}$$

where $\bar{S}_j$ is the fully adjusted listener rating of composite acceptability and b_j^{CA}, C_1^{CA}, C_2^{CA}, C_3^{CA} are weights which are proportional to the statistical reliability of the associated quality scores.

References

[3.1] W. D. Voiers, "Diagnostic Acceptability Measure for Speech Communication Systems," *Proc. 1977 IEEE ICASSP*, pp. 204–7, May 1977.

[3.2] T. P. Barnwell III and W. D. Voiers, "An Analysis of Objective Measures for User Acceptance of Voice Communication Systems," Final Report, DCA100-78-C-0003, Sep. 1979.

[3.3] W. D. Voiers, personal communication, Feb. 1985.

[3.4] W. H. Kruskal and J. M. Tanur (Eds.), *International Encyclopedia of Statistics.* New York: The Free Press, 1978.

[3.5] W. D. Voiers and C. P. Smith, "Diagnostic Evaluation of Intelligibility in Present-day Digital Vocoders," AFCRL-72-0120, 1972.

[3.6] W. D. Voiers, "Diagnostic Evaluation of Intelligibility in Present-Day Digital Vocoders," Record, *1972 Conf. on Speech Communication and Processing,* Newton, Mass., pp. 170-4, Apr. 1972.

[3.7] C. E. Williams, H. L. Hecker, K. N. Stevens and B. Woods, "Intelligibility Test Methods for the Evaluation of Communication Systems," Final Report, Contract No. AF 19(628)-5659, ESD-TR-66-677, Dec. 1966.

[3.8] B. J. McDermott, "Multidimensional Analysis of Circuit Quality Judgments," *J. Acoust. Soc. Amer., vol. 45,* no. 3, pp. 774–81, 1969.

[3.9] W. D. Voiers, "Methods of Predicting User Acceptance of Voice Communication Systems," Final Report, DCA100-74-C-0056, Jul. 1976.

[3.10] S. R. Quackenbush, "Objective Measures of Speech Quality," Ph.D. Thesis, Georgia Institute of Technology, 1985.

Chapter 4

Testing Objective Measures

The purpose of this chapter is to present the basic techniques used to design and test objective measures in this research. Because they form the foundation of the design process, the speech quality data bases are described first. The procedure for using these data bases to design and test quality measures is presented next. Finally, a number of analysis tools used throughout the speech quality research are discussed.

Three data bases are required for the testing of objective speech quality measures. The first contains undistorted speech, the second contains distorted speech, and the third contains subjective evaluations of the qualities of the distorted speech. The contents of the *undistorted*, the *distorted*, and the *subjective quality* data bases will be described in detail in the next three sections.

4.1 The Undistorted Data Base

The undistorted speech data base used in this research was originally generated by Voiers [4.34]. It consists of forty-eight different sentences, recorded in a quiet environment, with each of four sets of twelve spoken by a different person. Three speakers were male (denoted by CH, LL, and RS) and one was female (QQ). Each sentence is six syllables in length and the ensemble is phonetically balanced. In particular, each contains at least one vowel from each of the categories front, mid,

Table 4.1: Sentence set used in the undistorted speech data base.

Speaker CH (Male)

1. I read the news today.
2. Wrap the bones in tin foil.
3. He stole all her hats.
4. I hated the poems.
5. You are the biggest man.
6. We all ate fresh mushrooms.
7. The cold pond made me sneeze.
8. Don't permit that cheap talk.
9. The boat sailed off the edge.
10. You kids get off the fence.
11. Dirt was blown in my face.
12. Doubtless he was too thin.

Speaker LL (Male)

1. This road forks up ahead.
2. Jack was lost in the woods.
3. Do not present the prize.
4. Don't feed that vicious hawk.
5. The trees grew ten feet tall.
6. Snow fell all last winter.
7. Let's sit in the cool bar.
8. This box is much too flat.
9. You should be there on time.
10. The cook can boil beef stew.
11. Look at those peacocks strut.
12. The wind blows hard at sea.

Speaker RS (Male)

1. That hose can wash her feet.
2. The goose laid an odd egg.
3. That quiz was much too hard.
4. Those are pudgy old men.
5. We cracked all those pecans.
6. He has the bluest eyes.
7. Don met me first today.
8. These shoes were black and brown.
9. They are too loud in church.
10. We saw a bad movie.
11. I suggest you leave now.
12. The rabbits and dogs drowned.

Speaker QQ (Female)

1. Go and sit on the bed.
2. The book is about trash.
3. Moths still turned yellow.
4. Fire consumed the paper.
5. The bowl dropped from his hands.
6. Don't thrash around that way.
7. Those children are dirty.
8. Dress sleeves are much too warm.
9. Tractor plowed the fields.
10. We made some fine brownies.
11. They broke out of prison.
12. He drank our lusty brew.

and back listed in Table 3.3 and at least one consonant from each of the categories sibilant, stop, and fricative listed in Table 3.4. The undistorted speech was bandlimited to 3.2 kHz and uniformly quantized to twelve bits resolution at a sampling rate of 8 kHz, maintaining "toll quality." In addition, each set of twelve sentences was adjusted so that one sentence was spoken every four seconds, with the total duration for each set being 49 seconds or roughly 400,000 time samples. The four sets of sentences in the undistorted speech data base are listed in Table 4.1.

4.2 The Distorted Data Base

For historical reasons, the distorted speech data base is divided into two groups: the original, and an extended data base. This division is indicated by the rightmost two columns in Table 4.2. The original portion is composed of forty-four types of distortions, while the extension is composed of nine new or revised types. (Each of the narrowband distortions actually represents six distortions—this will be explained shortly.) The distortions in the extension were intended either to correct shortcomings in the original data base or to introduce into the data base distortions representative of new coder techniques.

Six *levels*, or degrees of distortions were implemented for each of the fifty-three distortion types, for a total of 318 *distortion systems.* The distorted data base was produced by passing the four sets of sentences in the undistorted data base through each of the 318 distortion systems. Five of the distortion groups, however, represent narrowband distortions, with each group consisting of six distortion levels within each of six frequency bands, giving a total of thirty-six distortions per group. The fifty-three distortions can be divided into two categories: *coding* distortions, which are simulations of digital speech coding systems, and *controlled* distortions, in which specific, perceptually relevant distortions are applied to the speech. In this latter category the distortions can be considered "idealized" speech coder distortions in that they are controlled realizations of the types of distortion which might result from digital speech coders. A key feature of these controlled distortions is that each is designed to simulate only one well defined aspect of the complex distorting effect of digital speech coders. The motivation for this procedure was to permit one aspect of a coder's distortion to be studied in isolation from other aspects, and thus simplify the problem of understanding the perceptual impact of complex coding distortions. The total duration of the sentences in the original distorted data base was over 17 hours, excluding anchors and probes. The extensions to the data base increased this figure by approximately twenty percent. The original data base contained 1056 systems, while the extended data base contained 1272 systems.

In all cases the distortions were simulated on a digital computer in the Georgia Tech Digital Signal Processing Laboratory [4.1] so that they could be precisely controlled. In addition, the distorted speech

Table 4.2: Distortions in the distorted speech data base.

CODING DISTORTIONS:	Original	Extended
CVSD	X	
Adaptive Delta Modulation	X	
Adaptive PCM	X	
Adaptive Differential PCM (ADPCM)	X	
ADPCM Noise Feedback		X
Linear Predictive Coding (LPC)	X	
Adaptive Predictive Coding	X	
Voice Excited Vocoder, low bit rate	X	
Voice Excited Vocoder, high bit rate	X	
Multipulse LPC		X
Adaptive Transform Coder - I	X	
Adaptive Transform Coder - II		X
Subband Coder		X
Channel Vocoder		X
CONTROLLED DISTORTIONS:		
Time Domain		
Additive Noise	X	
Interruption, short period	X	
Interruption, long period	X	
Clipping	X	
Center Clipping	X	
Quantization	X	
Echo	X	
	X	
Frequency Domain		
Lowpass Filter	X	
Highpass Filter	X	
Bandpass Filter	X	
Narrowband		
Narrowband Additive Noise	X	
Narrowband Angular Pole Distortion-I	X	
Narrowband Radial Pole Distortion-I	X	
Narrowband Pole Distortion-II		X
Narrowband Frequency Distortion	X	

was adjusted, when necessary, so that there was no time delay between it and the undistorted speech signal. This is an extremely important issue, since many objective measures compare the two waveforms on a sample-by-sample or frame-by-frame basis, and a time lag between the two would introduce artificial errors. Distortions which, to a large degree, retain the structure of the original waveform were aligned to within a sample, while coding distortions which reconstruct only a perceptually equivalent waveform, such as LPC, were aligned to within a single analysis frame.

Because this research focused on speech quality as opposed to speech intelligibility, distortions ranging from "no distortion" to "moderate" distortion were applied to the original speech. Intelligible speech was produced in all cases, with scores ranging from 40 (moderately distorted), to 80 (nearly undistorted), on most DAM quality scales. The maximum possible score on the Diagnostic Acceptability Measure is approximately 85. A "soft" upper limit which is less than 100 is used so that subjective evaluators can always rate a new speech signal as "better" than previous speech and still not exceed the numerical limit of 100 in their ratings.

The coding distortions can be divided into four groups. The first group contains simple waveform coders, which are represented by continuously variable slope delta modulation, adaptive delta modulation, adaptive pulse code modulation, adaptive differential pulse code modulation, and adaptive differential pulse code modulation with noise feedback. The second group is composed of coders which utilize an all-pole model of the vocal tract. This group includes an adaptive predictive coder, a linear predictive coder, two variations of a voice-excited (or residual-excited) vocoder, and a multi-pulse linear predictive coder. The third group, which contains two adaptive transform coders and one channel vocoder, uses the frequency domain representation of the speech for coding. Finally, in a group by itself, is the subband coder, which codes the speech in both the time and frequency domains.

Although they overlap to some degree, the controlled distortions can be divided into three groups: time domain, frequency domain, and narrowband distortions. The last group contains distortions which are characterized by their frequency domain transfer function, but are set apart from the rest of the controlled distortions because they affect the speech signal only in specified frequency bands. These frequency bands

Table 4.3: Frequency bands for the narrowband distortions.

Band	Range
1	0 – 400 Hz
2	400 – 800 Hz
3	800 – 1300 Hz
4	1300 – 1900 Hz
5	1900 – 2600 Hz
6	2600 – 3400 Hz

are listed in Table 4.3. Each of the narrowband distortions is actually composed of six similar applications of the same distortion. These six applications differ in that, for each, the distortion is limited to one of the six frequency bands. For each of these applications, or within each frequency band, there are six distortion levels, producing thirty-six distorted speech signals for each narrowband distortion, as opposed to only six for the other distortions. These distortions, which comprise a large part of the controlled data base, are included for two reasons. First, they facilitate the measurement of the subjective importance of specific distortions when applied in different bands. Second, they offer an environment which permits the objective importance of each band to be studied separately.

There are two features of the speech quality data base which are particularly important, and should be stressed. The first is the size of the data base: it contains a very large number of distorted speech signals. Statistically, this is desirable since larger amounts of data lead to parameter estimates which are more stable and accurate in the objective measures models. Also, the data base is large enough to be divided into two sets to permit cross-validation of objective measure regression models, while still retaining accurate parameter estimates.

The second important feature is that the data base represents an extremely diverse mix of distorting speech systems. This means that objective measures derived from this data base will have widespread applicability, and not be limited to a single class of speech distortion. This data base is therefore a tool useful for exploring the fundamental objective features of speech quality.

The remainder of this section will be devoted to describing the individual distortions that are included in the data base.

4.2.1 Coding Distortions

In this study, there were four systems which were classed as *simple waveform coders*: Continuously Variable Slope Delta Modulation (CVSD), Jayant's [4.1] Adaptive Delta Modulation (ADM), Adaptive Pulse Code Modulation (APCM), and Adaptive Differential Pulse Code Modulation (ADPCM). All of these systems can be considered special cases of the general adaptive waveform coding system illustrated in Fig. 4.1. In all cases, the decimation and interpolation, where used, was implemented using zero phase FIR interpolation filters. The *channel simulation* in the system models those kinds of transmission errors introduced by actual communication systems. However, since each of the coding distortions could have had several levels of bit error rate, it was infeasible to compile and present a complete set of channel error distortions. Therefore, no results from this form of distortion will be reported.

4.2.1.1 Continuously Variable Slope Delta Modulation

Since Continuously Variable Slope Delta Modulation (CVSD) is a delta modulator, the quantizer always has exactly two levels and the coder is always one bit. The main feature of CVSD as implemented for this data base is in the way $\delta(n)$ is computed. (see Fig. 4.1). Since $e(n)$ is the output of a one bit quantizer, it may be thought of as a series of ± 1's. $\delta(n)$ is computed as

$$\delta(n) = \beta\delta(n-1) + \Delta(n) \tag{4.1}$$

where $\Delta(n)$ is equal to one of two constants, depending on whether or not all of the last three values of $e(n)$ were equal to one another.

$$\Delta(n) = \begin{cases} A \text{ if } e(n-1) = e(n-2) = e(n-3) \\ B \text{ otherwise} \end{cases} \tag{4.2}$$

$B/(1-\beta)$ is known as the *minimum step-size* for CVSD. The corresponding maximum step-size is given by $A/(1-\beta)$.

CVSD is hence characterized by five features: the input speech sampling rate, the value of the predictor parameter, α, the value of the

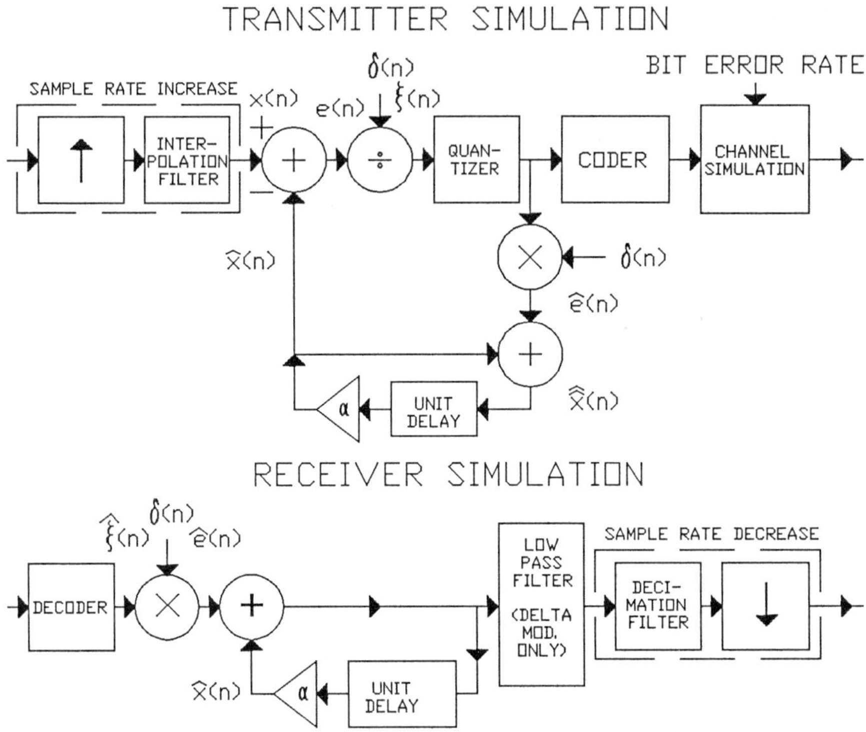

Figure 4.1: General system for describing waveform coders.

step integrator parameter, β, the value of the minimum step-size, B, and the value of A. A is usually not given, but is rather represented as an *expansion ratio*, which is the maximum step-size divided by the minimum step-size: A/B.

In terms of its basic parameters, the CVSD systems used in this study are summarized in Table 4.4.

4.2.1.2 Adaptive Delta Modulation (ADM)

The adaptive delta modulator used in this study was essentially that suggested by Jayant [4.2]. Like CVSD, the ADM is a delta modulator, implying that the bit rate is the sampling rate. The quantizer has two levels, and the coding is one bit per sample. For this delta modulator,

Table 4.4: Parameters for CVSD.

	Predictor Constant (α)	Step Size Integration (β)	Minimum Step size (B)	Expansion Ratio	Bit Rate
1	.8600	.9922	10	166	8 kbps
2	.9696	.9922	10	166	12 kbps
3	.9800	.9922	10	166	16 kbps
4	.9900	.9922	10	166	24 kbps
5	.9950	.9922	10	166	32 kbps
6	uncoded speech				

Table 4.5: Parameters for adaptive delta modulator.

	Predictor Constant (α)	Single Bit Multiplier (A)	Rate
1	.86	1.10	8 kbps
2	.90	1.06	12 kbps
3	.96	1.03	16 kbps
4	.98	1.03	24 kbps
5	.99	1.03	32 kbps
6	uncoded speech		

$$\delta(n) = \Delta(n)\delta(n-1) \tag{4.3}$$

where $\Delta(n)$ takes on one of two values: A when $\hat{e}(n) = \hat{e}(n-1)$ and B otherwise. In general, A is greater than one and B is less than one. For this study,

$$A = 1/B \tag{4.4}$$

The ADM is hence characterized by only three parameters: the input speech sampling rate, the value of the predictor parameter α, and the value of the quantizer control parameter A. In terms of these parameters, the ADM distortions used in this study are summarized in Table 4.5.

Table 4.6: Parameter for adaptive pulse code modulation.

	Quantizer Integration (β)	Quantizer Multiplier (Q)	No. of Levels (N)	Bit Rate (BPS)
1	.92	1	3	12679
2	.92	1	5	18575
3	.92	1	7	22453
4	.92	1	9	25359
5	.92	1	11	27675
6	.92	1	13	29603

4.2.1.3 Adaptive Pulse Code Modulation (APCM)

In terms of the system of Fig. 4.1, APCM has three main characteristics: first, it uses a multilevel quantizer; second, it operates at the Nyquist rate, and hence the interpolation and decimation filters are not used; and third, it has no prediction loop, i.e., $\alpha = 0$. The quantizer control sequence is controlled exponentially from

$$\Delta(n) = \beta\delta(n-1) + (1-\beta)|\hat{e}(n)| \tag{4.5}$$

This can be thought of as an exponentially integrated estimate of the energy in the quantized error signal, $\hat{e}(n)$. From this,

$$\delta(n) = \frac{4Q}{N}\Delta(n) \tag{4.6}$$

where Q is a control parameter and N is the number of levels in the quantizer. This realization is, therefore, completely controlled by three parameters: the quantizer integration factor β, the quantizer multiplier Q, and the number of levels N. In terms of these parameters, the APCM distortions used in this study are given in Table 4.6.

4.2.1.4 Adaptive Differential Pulse Code Modulation

The Adaptive Differential Pulse Code Modulation (ADPCM) used in this study was exactly the same as the APCM previously described except the value of α was not zero. The operation of this system

Table 4.7: Parameters for ADPCM.

	Predictor Constant (α)	Quantizer Integration (β)	Quantizer Multiplier (Q)	No. of Levels	Bit Rate (BPS)
1	.9	.92	1	3	12679
2	.9	.92	1	5	18575
3	.9	.92	1	7	22458
4	.9	.92	1	9	25359
5	.9	.92	1	11	27675
6	.9	.92	1	13	29603

is hence characterized by four parameters: the quantizer integration factor β, the quantizer multiplier Q, the number of quantizer levels N, and the feedback parameter α. In terms of these parameters, Table 4.7 describes the ADPCM systems used in this study.

4.2.1.5 Linear Predictive Coder (LPC)

The operation of the LPC vocoder used in this study is illustrated in Fig. 4.2. This procedure is a framed analysis and is characterized by a frame spacing, I. At each frame interval the input signal, $x(n)$, is windowed, to give

$$x_n(m) = x(m)w(m - nI) \tag{4.7}$$

where $w(m)$ is a window function, n is the frame number and m is the time index. For this study, only Hamming windows and a 10th order analysis were used. A set of autocorrelation functions is estimated by

$$R(k) = \sum_{m=-\infty}^{+\infty} x_n(m)x_n(m-k) \tag{4.8}$$

Using the well known Durbin's recursion, a set of predictor coefficients, $a(1), \ldots, a(10)$, a set of PARCOR coefficients, $K(1), \ldots, K(10)$, and a gain, given by [4.3]

$$G = \left[R(0) - \sum_{k=1}^{10} a(k)R(k)\right]^{1/2} \tag{4.9}$$

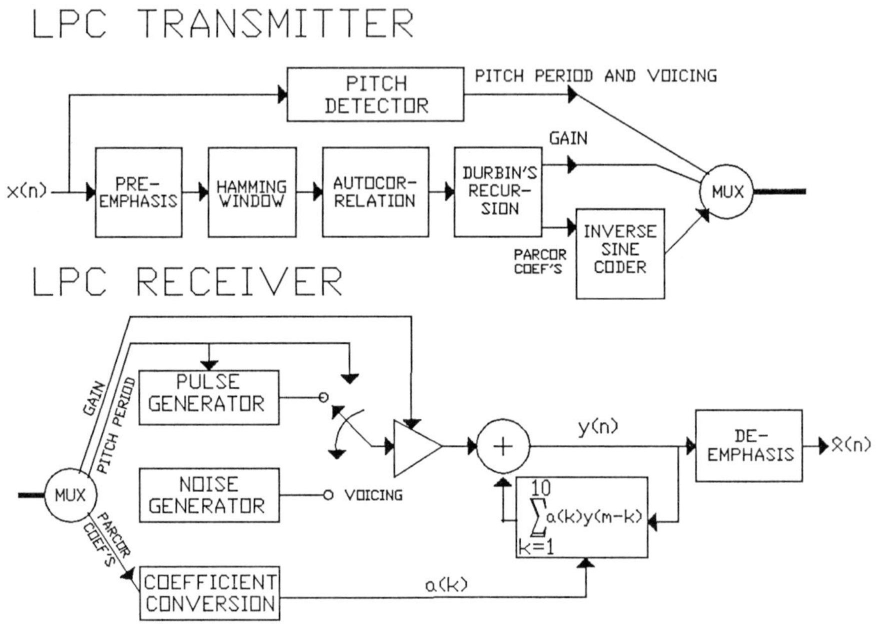

Figure 4.2: Block diagram of LPC distortion.

are computed. Since the spectral sensitivity to quantization errors increases when the PARCOR coefficients have values close to ± 1, an inverse sine transformation of the parameters is used.

The pitch detector used is a form of *Homomorphic* or *Cepstrum* pitch detector [4.4], [4.5]. The pitch and voicing output from the pitch detector is multiplexed in with the vocal tract information for transmission.

There are four parameters which characterize the LPC vocoder distortion. They are the window length W, the number of bits per frame for the PARCOR coefficients, the frame increment I, and the number of bits for pitch and gain coding. The LPC distortions used in this study are described in terms of these parameters in Table 4.8.

Table 4.8: Parameters for the LPC vocoder.

	Window Length (ms)	Bits/ Frame (vocal tract)	Pitch and Gain	Frame Interval (ms)	Bit Rate (BPS)
1	30	unquantized	7	15	—
2	30	58	7	15	4333
3	30	48	7	15	3666
4	30	38	7	15	3000
5	30	29	7	15	2400
6	30	20	7	15	1800

4.2.1.6 Adaptive Predictive Coder (APC)

The operation of APC used in this study is illustrated in Fig. 4.3. In this system, a framed LPC analyzer is applied to the input speech waveform. The analyzer is the same as that described in Sec. 4.2.1.5, and produces a vector of predictor coefficients, $a(k)$ for $k = 1, \ldots, 10$. This information, as before, is coded to some fixed bit rate using inverse sine PARCOR quantization [4.3] and is then used to control a time varying prediction filter with the z-transform

$$P(z) = 1 \quad \sum_{k=1}^{10} a(k) z^{-k} \tag{4.10}$$

The $a(k)$ coefficients are also transmitted to the receiver, and the adaptive predictor, inside the prediction loop, is used to estimate the input sequence $x(n)$. The error signal, $e(n)$, the difference between the input sequence and the output of the predictor, is then adaptively quantized by following an AGC with a fixed quantizer. In this simulation, $\delta(n)$ was taken to be the *look ahead* frame energy average, given by

$$\delta(n) = \left[\frac{1}{W} \sum_{m=-\infty}^{+\infty} x_n^2(m) \right]^{1/2} \frac{4Q}{N} \tag{4.11}$$

where W is the window length, Q is a quantizer control parameter, N is the number of levels in the uniform quantizer, and $x_n(m)$ is the windowed speech signal for the nth frame (Eq. (4.7)).

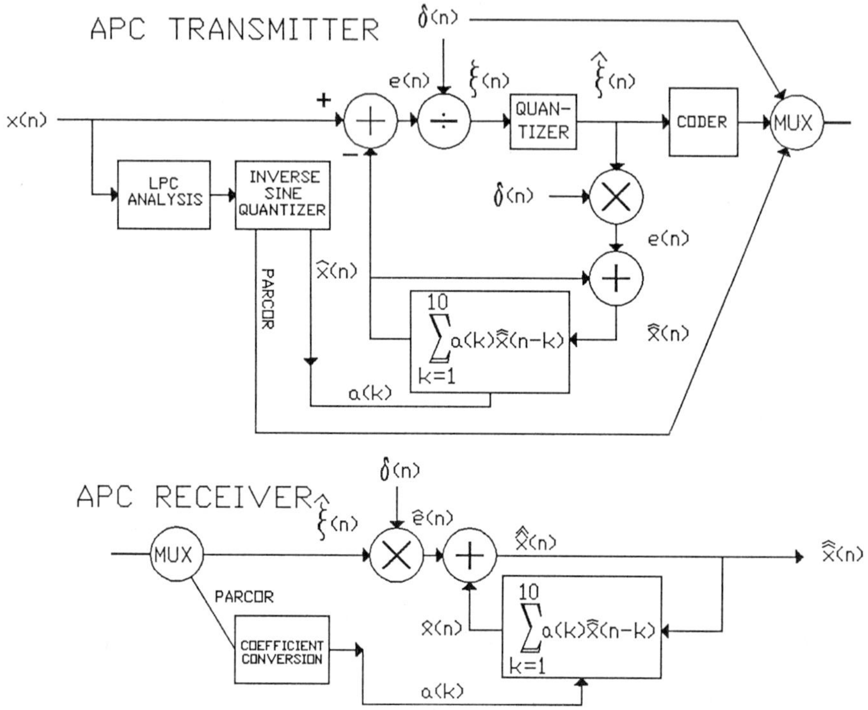

Figure 4.3: Block diagram of APC distortion.

The total operation of this APC is then characterized by five parameters: the number of levels in the quantizer N, the frame rate, the window length, the number of bits per frame in the predictor coding, and the quantizer control factor Q. In terms of these parameters, the APC distortion set used for this study is given in Table 4.9.

4.2.1.7 ADPCM with Noise Feedback

In this context, *noise feedback* refers to a class of analysis procedures, introduced by Atal and Schroeder [4.31], which can be applied at the transmitter of either an APC and ADPCM speech coding system in order to control the spectral shape of the coding noise generated at the receiver systematically. The reason for doing this is to take advantage of *auditory noise masking*, in which a strong signal will mask a weaker

Table 4.9: Parameters for the adaptive predictive coder.

	Window Length	Bits/ Frame	Level (N)	Frame Interval	Bit Rate (BPS)
1	30	unquantized	3	15	—
2	30	58	3	15	15867
3	30	48	3	15	15200
4	30	38	3	15	14533
5	30	29	3	15	13933
6	30	20	3	15	13333

signal in the same frequency region. Hence, it is desirable to shape the coding noise in such a way that the noise energy is concentrated into regions of the spectrum where speech signal energy is high.

The fundamentals of the noise feedback technique are illustrated in Fig. 4.4. One key feature is that this technique only alters the transmitter of APC or ADPCM systems, and unmodified APC or ADPCM receivers can be used. Both APC and ADPCM encode a residual signal, $e(n)$, which is obtained by passing the original signal through either a variable (APC) or fixed (ADPCM) whitening filter. In the traditional system, $P(z) = F(z)$, so that after quantization $E(z)$, the z-transform of the residual signal is given by

$$E(z) = [1 - P(z)]S(z) + [1 - P(z)]\Delta(z) \tag{4.12}$$

where $P(z)$ is the transfer function of the prediction filter, $S(z)$ is the z-transform of the original speech signal, and $\Delta(z)$ is the z-transform of the quantization noise signal, d(n). At the receiver, an estimate of the original signal, $\hat{S}(z)$, is created by passing the transmitted residual signal through the inverse whitening filter, giving

$$\hat{S}(z) = \frac{E(z)}{[1 - P(z)]} = S(z) + \Delta(z) \tag{4.13}$$

Hence, in an ordinary ADPCM or APC, the output signal is the sum of the input and the quantization noise signals. The quantization noise, which is nearly white, is distributed evenly across the entire frequency band, independent of the short-time frequency spectrum of the speech.

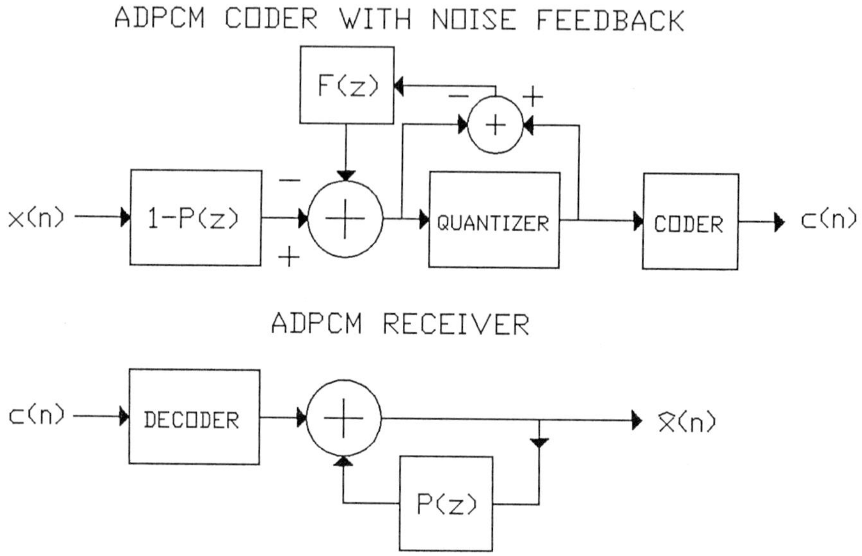

Figure 4.4: Block diagram of ADPCM with noise feedback.

In a noise feedback approach (Fig. 4.4), the quantization noise is explicitly filtered separately from the speech signal, and the residual signal can be written as

$$E(z) = [1 - P(z)]S(z) + [1 - F(z)]\Delta(z) \tag{4.14}$$

giving an estimated speech signal at the receiver of

$$\hat{S}(z) = \frac{E(z)}{[1 - P(z)]} = S(z) + \left[\frac{1 - F(z)}{1 - P(z)}\right] \Delta(z) \tag{4.15}$$

Hence the approximately white noise signal, $\Delta(z)$, is passed through a filter whose transfer function is given by $[1 - F(z)]/[1 - P(z)]$. By varying the characteristics of $F(z)$ on a frame-by-frame basis (since $P(z)$ is always known whether it is fixed or time-varying), it is possible to shape the noise to any desired form. An important point here is that the minimum noise energy always occurs for no noise shaping, i.e., $F(z) = P(z)$. Hence, noise feedback shapes the noise, but also increases the overall noise energy.

Table 4.10: Parameters for ADPCM with noise feedback.

Coder	Quantizer Levels	γ	Number of LPC Taps	Predictor Coefficient
NF-1	4	.8	10	.9
NF-2	6	.8	10	.9
NF-3	8	.8	10	.9
NF-4	12	.8	10	.9
NF-5	16	.8	10	.9
NF-6	32	.8	10	.9

The coding system used an ADPCM coder with a single tap fixed predictor, and the noise feedback filter was designed so that

$$\frac{1-F(z)}{1-P(z)} = \frac{1}{1-\sum_{n=1}^{10} a(n)\gamma^n z^{-n}} \tag{4.16}$$

where γ is a control parameter, and $P(z) = .9z^{-1}$. The control parameters used for this distortion are shown in Table 4.10.

4.2.1.8 Voice Excited Vocoder (VEV)

The voice excited vocoder used in this study is illustrated in Fig. 4.5. Its operation is essentially similar to the APC described in Sec. 4.2.1.6 except for a few important features. Instead of sending the entire residual signal, $\hat{e}(n)$, a lowpass version is sent. In this step, data rate compression is achieved by coding and decimating this lowpass signal to the Nyquist rate appropriate to its bandwidth. At the receiver, the residual is simulated by passing the lowpass residual through a full-wave rectification and LPC flattening to regenerate the high frequency components.

The VEV vocoder implemented in this study is characterized by five parameters: the frame increment I, the window length W, the ADPCM transmission rate, the voice band bandwidth, and the vocal tract parameter bit rate. Table 4.11 describes the VEV distortions used in this study as a function of these parameters.

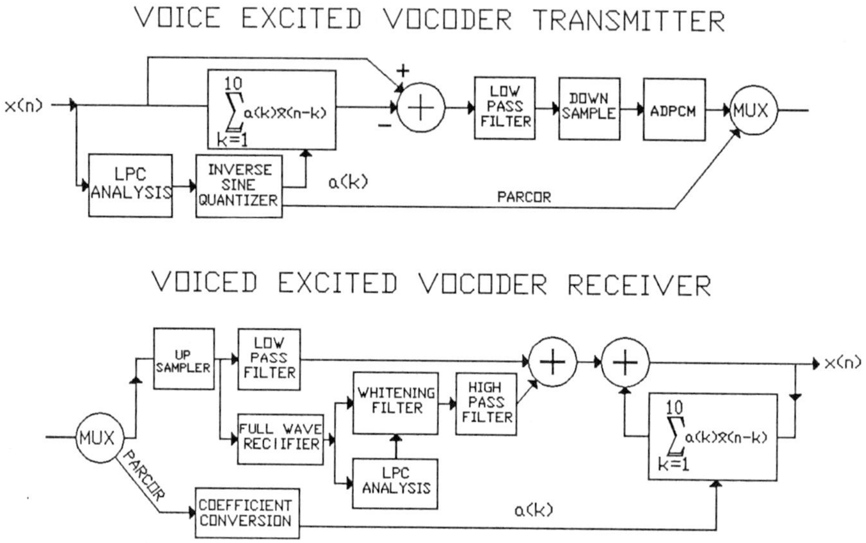

Figure 4.5: Block diagram of VEV distortion.

4.2.1.9 Multi-Pulse Linear Predictive Coder (MPLPC)

Since its introduction in 1981 [4.21], the Multi-Pulse Linear Predictive Coder (MPLPC) has been one of the most extensively reported and studied techniques for medium-to-low bit rate speech coding [4.22]–[4.24]. Although researchers had been searching for ways to improve the quality of speech at the bit rates between the medium-bit-rate waveform coders (down to about 16 kbps) and the low-bit-rate pitch-excited vocoders (down to about 2.4 kbps) for nearly a decade, little progress had been made. MPLPC was the first technique to show real promise in this area.

MPLPC can be considered a form of residual excited vocoder where the excitation information is generated and encoded in a special way. MPLPC derives its advantage from extensive utilization of the speech model and the LPC-estimated vocal tract transfer function. A block diagram of the MPLPC vocoder used in this study is shown in Fig. 4.6. In this system, the speech signal is first divided into two

Table 4.11: Parameters for the voice excited vocoder.

	Frame Interval (I) (ms)	Window Length (W) (ms)	ADPCM Rate (BPS)	Voice Band (Hz)	Vocal Tract Data	Total Data Rate
1	15	30	5615	1000	3867	9482
2	15	30	5615	1000	3200	8815
3	15	30	5615	1000	2533	8148
4	15	30	5615	1000	1933	7548
5	15	30	5615	1000	1333	6948
6	15	30	5615	1000	1000	6615
7	15	30	7400	1000	3867	11267
8	15	30	7400	1000	3200	10600
9	15	30	7400	1000	2533	9933
10	15	30	7400	1000	1933	9333
11	15	30	7400	1000	1333	8733
12	15	30	7400	1000	1000	8400

channels: the analysis channel, in which the LPC analysis and coding is performed, and the residual channel, in which the residual coding is performed. In the analysis channel, the first step is to apply a pre-emphasis filter of the form

$$H(z) = \frac{1}{1 - b(1)z^{-1} - b(2)z^{-2}} \tag{4.17}$$

where the coefficients of the filter, $b(1)$ and $b(2)$, have been set so as to estimate the spectral shape effect of the glottal pulse [4.9]. The output from this filter is then used as input to an autocorrelation LPC analysis routine which performs a tenth order LPC analysis and gives an estimated vocal tract filter of the form

$$V(z) = \frac{1}{1 - \sum_{n=1}^{10} a(n)z^{-n}} \tag{4.18}$$

This tenth order transfer function is then both coded for transmission and, in a separate operation, corrected to include the spectral shaping effects of the pre-emphasis filter, giving the twelfth order transfer

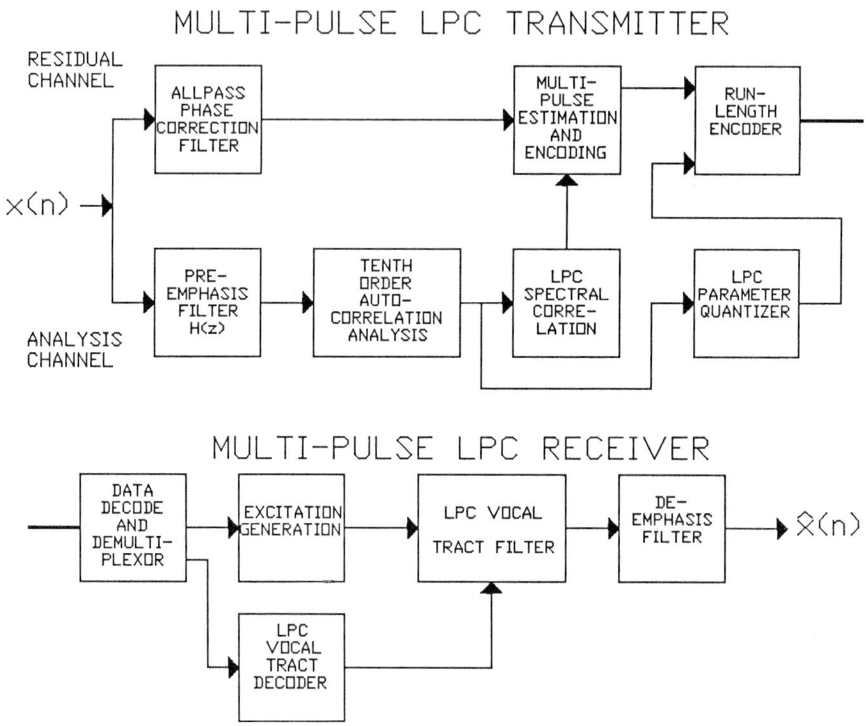

Figure 4.6: Block diagram of MPLPC distortion.

function

$$V'(z) = \frac{1}{(1 - \sum_{n=1}^{10} a(n)z^{-n})(1 - b(1)z^{-1} - b(2)z^{-2})} \tag{4.19}$$

In the residual channel, the original sampled speech signal is first passed through an all-pass filter whose transfer function is given by

$$A(z) = \frac{b(2) + b(1)z^{-1} + z^{-2}}{1 + b(1)z^{-1} + b(2)z^{-2}} \tag{4.20}$$

This filter has the effect of approximately correcting for the non-minimum phase components of the original speech signal [4.9], which in turn, has the effect of both making the speech signal more peaky in appearance and also making the vocal tract model, $V'(z)$, more nearly correct with regard to spectral phase and magnitude.

The heart of the MPLPC is the *Multi-Pulse Estimation and Encoding* functions shown in the analysis channel in Fig. 4.6. This function uses the phase corrected speech signal, $x'(n)$, and the spectrally corrected vocal tract parameters, $a'(1), \ldots, a'(12)$, in an iterative procedure to choose a set of residual pulses to be coded and transmitted. In this study, the entire procedure is performed in frames (one frame every 60 samples) and only a small number of pulses (2 to 10) are kept for transmission. Because of the sparse nature of the multi-pulse signal, run-length coding can be used to reduce the bit rate in the MPLPC residual signal.

Finding the optimal pulse positions and amplitudes globally is not mathematically tractable. Hence a suboptimal iterative procedure which locates the pulses one at a time is used, and can be summarized as follows. First, the ordinary LPC residual signal $e_0(n)$ is formed over the range $[0, M-1]$, giving

$$e_0(n) = x'(n) - \sum_{k=1}^{12} a'(k)x'(n-k) \tag{4.21}$$

Next, the *modified vocal tract impulse response*, $h_w(n)$, is computed as

$$h_w(n) = \begin{cases} 1 & \text{for } n = 0 \\ \sum_{k=1}^{12} a'(k)\gamma^k h_w(n-k) & \text{for } 1 < n \leq M-1 \end{cases} \tag{4.22}$$

where γ and M are control parameters of the coder. In this study M was set to 120. Then the *modified vocal tract autocorrelation filter*, $r_w(n)$, is computed as

$$r_w(n) = h_w(n) * h_w(-n) \tag{4.23}$$

Using $r_w(n)$ and $h_w(n)$, the pulse locations and pulse amplitudes are computed in the following iterative procedure. First, the pulse index, p, is set to zero and $f_p(n)$ is computed as

$$f_p(n) = e_0(n) * r_w(n) \tag{4.24}$$

where "$*$" denotes convolution. Then the time index which maximizes $|f_p(n)|$ is found giving N_p, the location of the pth pulse (for $p = 0$ first). The approximate amplitude of the pth pulse is then computed as

$$A_p = \frac{f_p(N)}{\sum_{m=0}^{M-1} h_w^2(m)} \tag{4.25}$$

Once A_p is computed, the pulse index p is incremented, and then $f_p(n)$ is computed as

$$f_p(n) = f_{p-1}(n) - A_p r_w(n - N_p) \tag{4.26}$$

The above steps are repeated until the desired number of pulse locations, $N_0, \ldots, N_{p-1}$, are found in the range [0,59]. For the given ensemble of pulse positions, new pulse amplitudes which optimize the error criterion for those positions are computed in a single step [4.21].

In this study, the intent was to generate a class of distortions which were typical of MPLPC, and not specifically to implement any particular algorithm. Hence, no actual run-length coding was performed and no precise bit rates were computed. In addition, the unquantized LPC vocal tract parameters were used to generate the synthetic speech.

Another feature of the MPLPC is that once an estimate of the multi-pulse residual signal is known, it is possible to use that signal to obtain an improved estimate of the LPC vocal tract parameters. In this study, three different pulse rates (2/60, 6/60, and 10/60) were combined with original and improved LPC vocal tract parameters in order to form the six members of the MPLPC distortion set.

4.2.1.10 Adaptive Transform Coder (ATC)

One of the more successful methods for frequency domain speech coding is the adaptive transform coder (ATC). The basic concept on which the ATC is based involves encoding a spectral representation of the speech rather than the time domain waveform. The steps involved in the coding in this study are:

1. Windowing and transforming a segment of speech.
2. Producing a model of the spectrum from LPC analysis and pitch detection.
3. Dynamically allocating a predetermined number of bits among the transform coefficients using the model spectrum.
4. Adaptively quantizing the coefficients to the number of bits allocated.

The decoder requires both the quantized transform coefficients and the quantized LPC parameters of the model spectrum in order to synthesize a speech waveform. From these parameters, the bit allocations and adaptation parameters which are used in the quantizers can be computed. Synthesis results from decoding of the transform, inverse transformation, and overlap-add combination of adjacent segments.

Adaptive Transform Coder - I When the distorted data base was first created, adaptive transform coding was a relatively new technique as applied to speech [4.6], [4.7]. Therefore, it was not a goal to produce high quality ATC speech, because that was still a subject of active research. Rather the goal was to include in the data base a distortion which was qualitatively *like* that produced by ATC.

The original ATC coding system used is illustrated in Fig. 4.7. In this coder, the speech is first windowed to 256 samples using a rectangular window and a frame interval of 256 points. The windowed speech is then transformed using the discrete cosine transform (DCT) and a linear prediction analysis is done. An approximate spectrum is computed from the predictor coefficients as

$$V(\theta_l) = \left| \frac{1}{1 - \sum_{k=1}^{10} a(k) e^{-jk\theta_l}} \right| \tag{4.27}$$

and it is these spectral values which are coded. The number of levels at spectral sample $\theta_l, 0 < l < 255$, is allocated by

$$\text{level}(\theta_l) = \text{TOTAL LEVELS} \cdot V(\theta_l) \tag{4.28}$$

(since $\sum_{l=0}^{255} V(\theta_l) = 1$), and

$$\text{TOTAL LEVELS} = 2^B \tag{4.29}$$

where B is the total number of bits allocated. The individual quantizers are uniform with a range, $r(l)$ given by

$$-GV(\theta_l) < r(l) < GV(\theta_l) \tag{4.30}$$

where G, the gain, is given by Eq. (4.9).

The operation of this transform coder is characterized by four parameters: the frame interval and window length, which must be the same, the order of the LPC analysis, the number of bits per frame used to code LPC vocal tract parameters, and B, the number of bits per frame for the transform coding. The distortions used in this ATC system are summarized in terms of these parameters in Table 4.12.

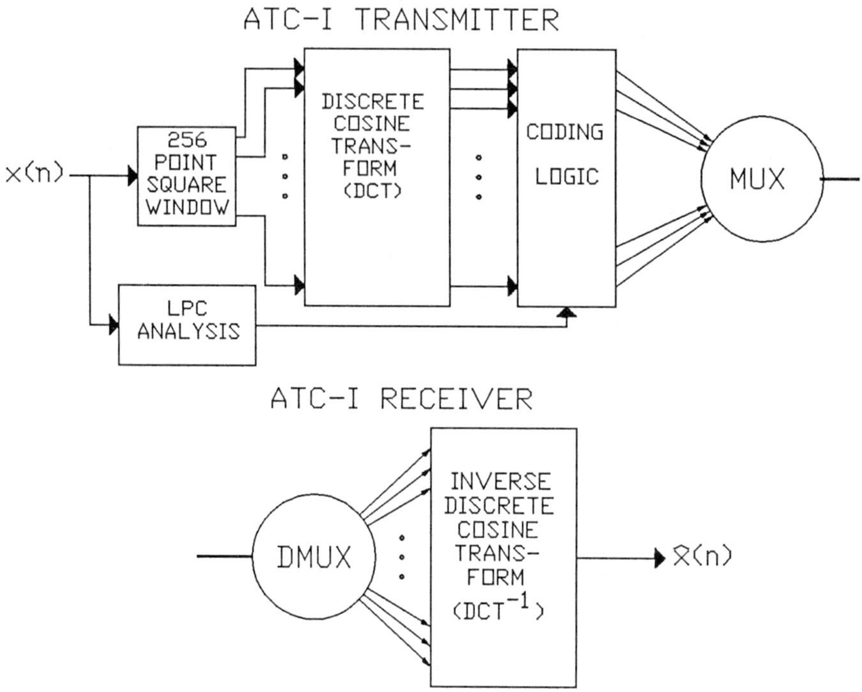

Figure 4.7: Block diagram of the ATC - I distortion.

Table 4.12: Parameters for the adaptive transform coder.

	Window Length	LPC Order	LPC Bits/ Frame	Trans Bits/ Frame	Bit Rate (BPS)
1	256	10	4333	15667	20000
2	256	10	3666	12334	16000
3	256	10	3000	9000	12000
4	256	10	2400	8600	11000
5	256	10	1800	7800	9600
6	256	10	1500	6500	8000

Adaptive Transform Coder - II The revised ATC procedure closely follows that of Tribolet and Crochiere [4.25] with some modifications. The transform used in analysis is the discrete cosine transform (DCT) which is defined by

$$V_c(k) = \sum_{n=0}^{M-1} v(n)c(k)\cos[(2n+1)\pi k/2M] \tag{4.31}$$

The inverse DCT is defined as

$$v(n) = \frac{1}{M}\sum_{k=0}^{M-1} V_c(k)c(k)\cos[(2n+1)\pi k/2M] \tag{4.32}$$

where in both formulas:

$$k = 0, 1, \ldots, M-1$$

and

$$c(k) = \begin{cases} 1 \text{ if } k = 0 \\ 2 \text{ if } k = 1, 2, \ldots, M-1 \end{cases}$$

Note that this transform is real, and involves computation of M equally spaced frequency components from zero to the sampling frequency. The reasons for this particular transform's use include the fact that its coefficients are always real, it is relatively simple to compute (efficient algorithms involving FFT's exist), and it minimizes windowing effects when quantized.

For the balance of the discussion, an 8 kHz sampling rate will be assumed for the digitized speech, since this was the case for all of the speech materials used in this study. The windows used in the analysis are 256 point trapezoids with a value of one for the center 240 points, and linear tapering to zero on both sides. Adjacent segments are overlapped by 16 points, yielding an overall rate of one frame every 30 ms. A 256-point DCT is applied to each segment.

In addition to the DCT, another analysis is performed independently on the speech data for the purpose of spectrum modeling. A twelfth order LPC analysis using 256 point windows is performed every 30 ms, and pitch detection is performed by an interactive, semi-automatic procedure so as to minimize the probability of pitch and voicing errors. These two components give rise to a smooth spectrum,

$\sigma_f(k)$, and a pitch spectrum, $\sigma_p(k)$, which are combined to form a model spectrum $\sigma_s(k) = \sigma_f(k)\sigma_p(k)$. The estimate $\sigma_f(k)$ is computed using a discrete Fourier transform (DFT) of the quantized linear prediction model over the first half of the unit circle. The pitch spectrum, $\sigma_p(k)$, is computed by windowing and then taking the DFT of

$$p(n) = \sum_{m=0}^{\infty} (G^{m/2})\delta(n - mL) \tag{4.33}$$

where L is the pitch period and G is the ratio of the Lth autocorrelation lag to the zeroth lag in the speech segment and

$$\delta(n) = \begin{cases} 1 \text{ if } n = 0 \\ 0 \text{ otherwise} \end{cases}$$

The bit assignment is a function of a weighted version of the log of $\sigma_s(k)$ and is formulated so as to hide some of the quantization noise under the high energy spectral peaks. The algorithm is iterative and results in allocating B bits over M samples, according to the formula:

$$b(k) = \max\{0, \min[\text{int}[\log_2(\sigma_s(k)\ \sigma_f^{-.125}(k)) + \delta], N_{max}]\} \tag{4.34}$$

where $b(k)$ is the number of bits assigned to coefficient $V_c(k)$, int$[a]$ truncates a to an integer, $\max[a, b]$ and $\min[a, b]$ take the maximum and minimum respectively of the two arguments, N_{max} is the maximum number of bits allowed for any one coefficient, and δ is the parameter which is iteratively adjusted to make

$$B = \sum_{k=0}^{M-1} b(k) \tag{4.35}$$

The parameters N_{max} and B depend on the desired bit rate for the coding.

Since it can be argued that $V_c(k)$ is a zero mean Gaussian random variable with variance equal to $\sigma_s(k)$, the quantization procedure consists of normalizing $V_c(k)$ by $\sigma_s(k)$ and then applying a non-uniform $b(k)$-bit quantizer optimized for a Gaussian process of unit variance. Parameters for the quantizer were taken from Max [4.26].

In all, N bits per segment are allowed, giving a rate of $(N \times 8000)/240 = N \times 33.3$ bits per second. Of these, B bits are "main information" and $N - B$ bits are "side information," which include inverse

Table 4.13: Parameters for the revised adaptive transform coder.

Bit Rate	Number Bits In Side Information Per Frame	Maximum Number of Bits Per Coefficient	Number of Bits For Transform Quantization
16 kb/s	51	5	445
12 kb/s	44	4	316
9.6 kb/s	44	4	244
8 kb/s	44	4	204
6 kb/s	44	4	136
4.8 kb/s	44	4	100

sine quantized LPC reflection coefficients, LPC gain, pitch gain (G from Eq.(4.9)), and pitch.

In resynthesis, $b(k)$, $\sigma_s(k)$, $\sigma_f(k)$, and $\sigma_p(k)$, are all computed from the side information in a manner identical to that used in the analysis portion. With the quantizer characteristics now known, the quantized version of $V_c(k)$ can be computed from the main information. An inverse DCT is then computed, and an overlap add performed with the previous reconstructed segment. The parameters used to control the adaptive transform coder are summarized in Table 4.13.

4.2.1.11 Subband Coder

In recent years, subband coders for digital speech coding at medium bit rates have been widely studied in the literature [4.27], [4.28]. In the basic subband coding procedure (Fig. 4.8), the speech is first split into frequency bands using a bank of bandpass filters. The individual bandpass signals are then decimated and encoded for transmission. At the receiver, the channel signals are decoded, interpolated, and added together to form the received signal. The subband coder derives its quality advantage by limiting the quantization noise from the encoding/decoding operation largely to the band in which it is generated, thereby taking advantage of known properties of aural perception [4.29].

The basic component of octave-band tree-structured subband coders is the two-band analysis/reconstruction system shown in Fig. 4.9. In this system, the analysis is performed by the two frequency

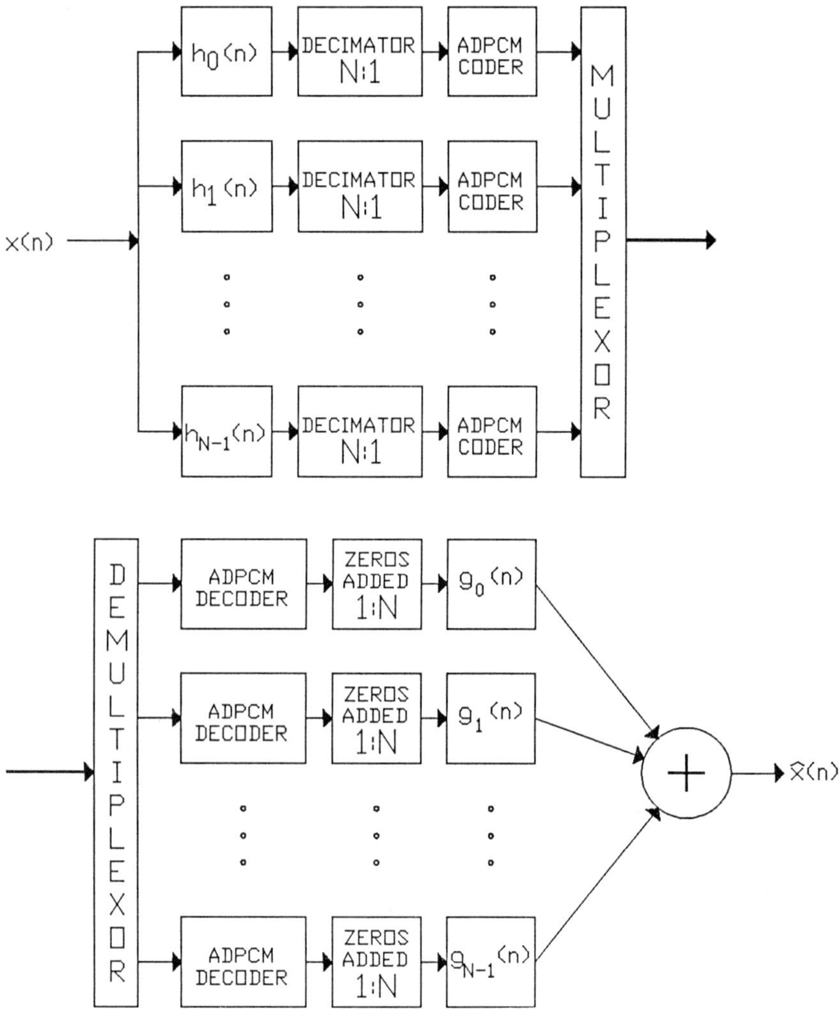

Figure 4.8: Block diagram of subband coder distortion.

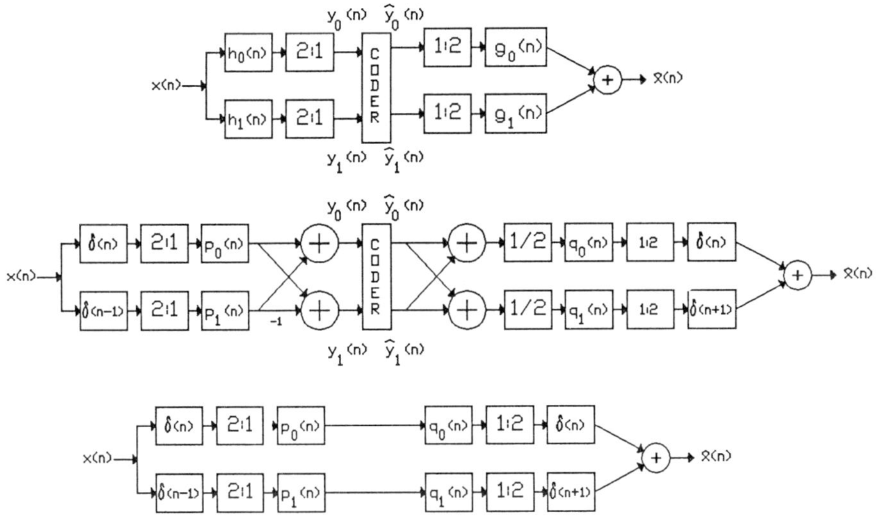

Figure 4.9: Subband analysis/reconstruction system.

selective filters, $H_0(e^{j\omega})$ and $H_1(e^{j\omega})$, which are nominally half-band lowpass and half-band highpass filters respectively. In order that the system sampling rate might be preserved, both channels are maximally decimated at a rate of two-to-one, resulting in the two sub-sampled signals, $Y_0(e^{j\omega})$ and $Y_1(e^{j\omega})$, given by

$$Y_0(e^{j\omega}) = (1/2)[H_0(e^{j\omega/2})X(e^{j\omega/2}) + H_0(-e^{j\omega/2})X(-e^{j\omega/2})] \quad (4.36)$$

$$Y_1(e^{j\omega}) = (1/2)[H_1(e^{j\omega/2})X(e^{j\omega/2}) + H_1(-e^{j\omega/2})X(-e^{j\omega/2})] \quad (4.37)$$

In the reconstruction section, the bands are recombined, giving

$$\begin{aligned}\hat{X}(e^{j\omega}) = (1/2)[H_0(e^{j\omega})G_0(e^{j\omega}) + H_1(e^{j\omega})G_1(e^{j\omega})]X(e^{j\omega}) \\ +(1/2)[H_0(-e^{j\omega})G_0(e^{j\omega}) + H_1(-e^{j\omega})G_1(e^{\omega})]X(-e^{j\omega})\end{aligned} \quad (4.38)$$

The frequency response of the two-band linear system component is contained in the first term of Eq. (4.38), while the second term contains the aliasing. In the quadrature mirror filter (QMF) solution, the aliasing is removed by defining the reconstruction filters as

$$G_0(e^{j\omega}) = H_1(-e^{j\omega}) \quad (4.39)$$

Table 4.14: Control function F[ ] for APCM coders.

	Magnitude of Code Word [$\|c(n)\|$]							
Bits/Sample	0	1	2	3	4	5	6	7
4	.9	.9	.9	.9	1.2	1.6	2.0	2.4
3	.85	.9	1.4	2.0				
2	.85	1.9						

$$G_1(e^{j\omega}) = -H_0(-e^{j\omega}) \tag{4.40}$$

This assignment forces the aliasing to zero, and results in a total system frequency response, $C(e^{j\omega})$, of

$$C(e^{j\omega}) = (1/2)H_0(e^{j\omega})H_1(-e^{j\omega}) - (1/2)H_1(e^{j\omega})H_0(-e^{j\omega}) \tag{4.41}$$

In the conventional solution, the highpass filter $[H_1(e^{j\omega})]$ and lowpass filter $[H_0(e^{j\omega})]$ are chosen to be versions of each other, shifted by π in frequency so that

$$H_1(e^{j\omega}) = H_0(-e^{j\omega}) \tag{4.42}$$

For this class of analysis/reconstruction system, exact reconstruction requires that

$$H_0(e^{j\omega}) - H_1(e^{j\omega}) = 2 \tag{4.43}$$

A number of authors using various methods have designed FIR filters which approximate this condition. The analysis/reconstruction systems used in this study all were based on quadrature mirror filters as described by Johnston [4.30], and the systems were simulated as described by Barnwell [4.28]. The coding of subband signals used APCM (see Sec. 4.2.1.3) and were based on work by Jayant [4.2]. The adaptive quantizers in these systems are controlled by the dynamic step-size $\Delta(n)$, given by

$$\Delta(n) = \Delta(n-1)F[c(n-1)] \tag{4.44}$$

where $c(n)$ is the nth code word and F[ ] is a preset control function. The control functions for the APCM coders used in this study are given in Table 4.14, while the control parameters for the individual systems are shown in Table 4.15.

Table 4.15: Parameters for the subband coder.

Coder	No. of Bands	1	2	3	4	5	Harmonic Scaling	Bit Rate
SUB-1	5	4	4	2	2	2	No	16000
SUB-2	5	3	3	2	2	2	No	14000
SUB-3	4	4	3	2	2		No	12000
SUB-4	5	4	4	2	2	2	Yes	8000
SUB-5	5	3	3	2	2	2	Yes	7000
SUB-6	4	4	3	2	2		Yes	6000

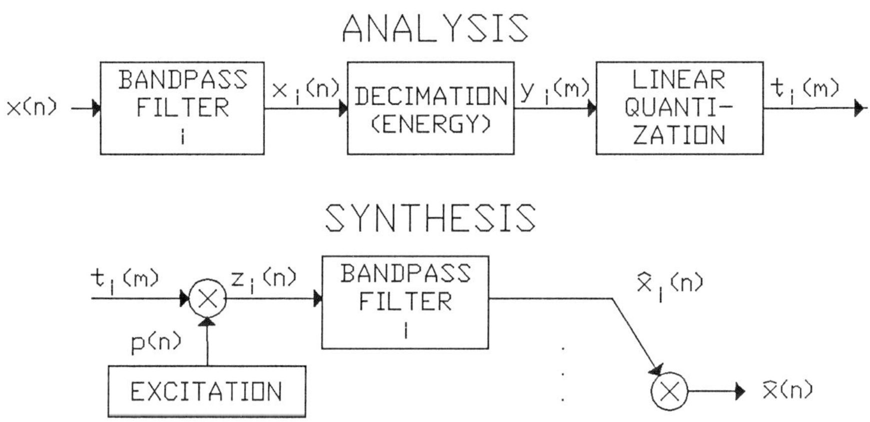

Figure 4.10: A single channel of channel vocoder.

4.2.1.12 Channel Vocoder

The channel vocoder was realized as a thirty band system which spans the frequency range of 0–3.6 kHz. A block diagram for one of the channels (analysis and synthesis ports) is given in Fig. 4.10.

The filters in both the analysis and synthesis filter banks are realized using recursive elliptic filters implemented as second order cascaded sections. The filters, whose characteristics are given in Table 4.16 have bandwidths of 120 Hz. Exactly the same filters are used for analysis and synthesis.

The short-time RMS value of $y_i(m)$, the output of filter channel

Table 4.16: Characteristics of the channel vocoder filter bank.

Filter No.	Low Cutoff Frequency (kHz)	High Cutoff Frequency (kHz)	Order
1	0	0.12	8
2	0.12	0.24	12
3	0.24	0.36	12
4	0.36	0.48	12
5	0.48	0.60	12
6	0.60	0.72	12
7	0.72	0.84	12
8	0.84	0.96	12
9	0.96	1.08	12
10	1.08	1.20	12
11	1.20	1.32	12
12	1.32	1.44	12
13	1.44	1.56	12
14	1.56	1.68	12
15	1.68	1.80	12
16	1.80	1.92	12
17	1.92	2.04	12
18	2.06	2.16	12
19	2.16	2.28	12
20	2.28	2.60	12
21	2.60	2.52	12
22	2.52	2.64	12
23	2.64	2.76	12
24	2.76	2.88	12
25	2.88	3.00	12
26	3.00	3.12	12
27	3.12	3.24	12
28	3.24	3.36	12
29	3.36	3.48	12
30	3.48	3.60	12

i, is computed using the formula

$$y_i(m) = \left| \frac{\sum_{n=1}^{N} [w(n)x(n)]^2}{\sum_{n=1}^{N} w(n)^2} \right| \tag{4.45}$$

where $w(n)$ is an N point Hamming window. In this study N is set to 215, and $y_i(m)$ is sampled every 120 points, or 15 ms.

A uniform quantizer is used for coding the positive signal $y_i(m)$.

Table 4.17: Parameters for the channel vocoder.

System Number	Bits Per Channel	Bit Rate per Channel (Bits/Second)
1	9	600
2	10	667
3	11	733
4	12	800
5	14	933
6	16	1067

The numbers of quantizer bits used to generate the distortion set are 9, 10, 11, 12, 14 and 16.

The pitch period estimations used for the channel vocoder are the same as those used for the revised adaptive transform coder (see Sec. 4.2.1.10), which used a semi-automatic technique to minimize pitch and voicing errors. The pitch periods are estimated every 120 samples (15 ms). The excitation signal, $p(n)$, is generated as follows: for unvoiced sounds, a uniformly distributed white random process with standard deviation G_N is used; for voiced sounds, a periodic pulse train with the correct period and amplitude G_p is used. The choice of the gains G_N and G_p is critical, and a ratio $G_p/G_N = 10$ was found to be appropriate.

In the receiver, the excitation $p(n)$ is multiplied by the transmitted signal $t_i(m)$ to create $z_i(n)$. This signal, in turn, is filtered to generate the channel signals, $s_i(n)$, which are summed together to create the output speech signal. The control parameters for the channel vocoder are summarized in Table 4.17.

4.2.2 Controlled Distortions

4.2.2.1 Additive Noise

In the additive noise distortions, white Gaussian noise was added to each sample of the undistorted signal, as

$$x_d(m) = x_\phi(m) + A \cdot v(m) \tag{4.46}$$

where $v(m)$ is a zero mean unit variance white noise sequence, A is a multiplicative constant, $x_d(m)$ are the undistorted speech samples

Table 4.18: Paramters for additive noise distortion.

	Signal-to-Noise Ratio (dB)
1	30
2	24
3	18
4	12
5	6
6	0

and $x_\phi(m)$ are the distorted speech samples. This distortion is well characterized by its signal-to-noise ratio (SNR) as shown in Table 4.18, where SNR is defined by

$$10\log_{10}\frac{\sum_m x^2(m)}{\sum_m A^2 v^2(m)} \quad . \tag{4.47}$$

4.2.2.2 Interruption

The interruption distortion was characterized by two numbers: a *keep* number, KP, and a *discard* number, DR. The interrupt distortion operated on frames of length $KP + DR$. Within the frame, the first KP samples were undisturbed, while the last DR were set to zero. Table 4.19 summarizes the interrupt distortions in this study.

4.2.2.3 Clipping

Clipping is a non-linear distortion given by

$$x_d(m) = \begin{cases} CL & \text{if } |x_\phi(m)| \geq CL \\ x_\phi(m) & \text{if } |x_\phi(m)| < CL \end{cases} \tag{4.48}$$

where the constant CL is called the *clipping constant*. The constant is compared to the *maximum average magnitude*, MAM, for an utterance, given by

$$MAM = \max[M(m)] \tag{4.49}$$

Table 4.19: Parameters for interrupt distortion.

	KP Keep Constant	DR Discard Constant
1	300	10
2	300	25
3	300	50
4	300	75
5	300	110
6	300	150
7	1024	16
8	1024	32
9	1024	64
10	1024	128
11	1024	256
12	1024	512

Table 4.20: Parameters for clipping distortion.

	Clipping Constant
1	.152
2	.076
3	.038
4	.0305
5	.0153
6	.0076

where $M(m)$ is given by

$$M(m) = (1 - \alpha)M(m - 1) + \alpha|x_\phi(m)| \tag{4.50}$$

and α is an exponential integration constant set to have an effective length of approximately 30 ms. All input speech was scaled such that $-1 \leq x_\phi(m) \leq 1$ and the resulting MAM was equal to 0.122. In these terms, the clipping constants for the distortions are shown in Table 4.20.

Table 4.21: Parameters for center clipping distortion.

	Center Clipping Constant
1	.0019
2	.0038
3	.0076
4	.0190
5	.0380
6	.0760

4.2.2.4 Center Clipping

Center clipping is a non-linear distortion given by

$$x_\phi(m) = \begin{cases} x_\phi(m) & \text{if } |x_\phi(m)| \geq CN \\ 0 & \text{if } |x_\phi(m)| < CN \end{cases} \tag{4.51}$$

where CN is the *center clipping constant.* In computing this distortion, the original speech is scaled in the same manner as for the clipping distortion, prior to comparison to the center clipping constant. Table 4.21 gives the parameters for the center clipping distortion.

4.2.2.5 Quantization

The quantization distortion is just a PCM system which is non-adaptive and which uses relatively coarse quantization. Linear quantizers were used, with a range of twice the MAM of the input signal (defined in Eq. (4.49)). The quantization distortion is described in terms of the number of levels in the quantizer and the associated bit rate, as listed in Table 4.22.

4.2.2.6 Echo

The Echo distortion was implemented by

$$x_d(m) = 1/2\{x_\phi(m) + x_\phi(m - EC)\} \tag{4.52}$$

Although this is certainly not the only way in which an echo can be implemented, the result is very clearly a subjective echo. The distortion

Table 4.22: Parameters for quantization distortion.

	Number of Levels in Quantizer	Bit Rate
1	64	48,000.0
2	48	44,697.7
3	32	40,000.0
4	24	36,679.7
5	16	32,000.0
6	12	28,679.7

Table 4.23: Parameters for echo distortion.

	Echo Constant
1	10
2	50
3	100
4	200
5	500
6	1000

is entirely characterized by the *echo delay*, EC, and is described in Table 4.23.

4.2.2.7 Filtering

There were three filtering distortions included: lowpass filtering, highpass filtering, and bandpass filtering. The filters were implemented digitally using kth order recursive elliptic filters. Table 4.24 gives the orders of the filters used along with the band limits for each distortion.

4.2.2.8 Narrowband Additive Noise

The narrowband additive noise distortion was meant to approximate waveform coder distortions in a frequency variant way. This distortion system is illustrated in Fig. 4.11. White Gaussian noise is first filtered into six bands giving an output signal $N_b(m)$, where b is the band

Table 4.24: Characteristics of filters used in filter distortions.

	Lowpass Filters	
	Order	Band Limit
1	6	400
2	6	800
3	7	1300
4	7	1900
5	7	2600
6	5	3400

	Highpass Filters	
	Order	Band Limit
1	4	0
2	6	400
3	7	800
4	7	1300
5	7	1900
6	7	2600

	Bandpass Filters		
	Order	Lower Band Limit	Upper Band Limit
1	6	0	400
2	9	400	800
3	9	800	1300
4	9	1300	1900
5	9	1900	2600
6	11	2600	3400

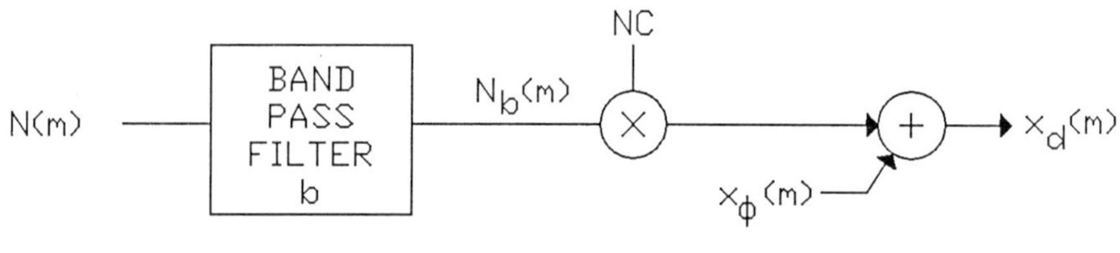

Figure 4.11: Block diagram of narrowband noise distortion.

number and m is the time index. Then the banded noise is added to the input speech using a noise constant, NC, giving

$$x_d(m) = x_\phi(m) + NC \cdot N_b(m) \tag{4.53}$$

The bandpass filters were all elliptic with a unity gain in the passband (see Sec. 4.2.2.7). Table 4.11 gives a summary of the narrowband additive noise distortions.

Table 4.25: Parameters for narrowband noise distortion.

		Noise Constants					
	Band	1	2	3	4	5	6
1	0-400 Hz	.305	.152	.076	.038	.019	.009
2	400-800 Hz	.305	.152	.076	.038	.019	.009
3	800-1300 Hz	.305	.152	.076	.038	.019	.009
4	1300-1900 Hz	.305	.152	.076	.038	.019	.009
5	1900-2600 Hz	.305	.142	.076	.038	.019	.009
6	2600-3400 Hz	.305	.152	.076	.038	.019	.009

4.2.2.9 Narrowband Pole Distortion

Over the past decade, linear predictive analysis has become one of the most widely used techniques in speech coding, and has been used in many coding systems operating at many different bit rates. These systems include the pitch-excited LPC vocoders, the vector-quantized pitch-excited LPC vocoders, the residual-excited LPC vocoders, adaptive predictive coders, the multi-pulse excited LPC vocoders, adaptive transform coders, and many more. All of these systems have the common feature that, as part of the speech coding procedure, they quantize and transmit frames of LPC coefficients in some form. In the systems where this is done, this quantization generally causes some perceivable distortion. Because the quantization of LPC coefficients is such a common feature in modern speech coding systems, it is clear that understanding how to predict subjective responses to this class of distortion should be an important goal. A serious problem, however, is that the relation between linear prediction quantization distortion and human perception is not a simple one. Linear predictive coding generally involves a transformation of the linear prediction coefficients, to inverse sine transformed PARCOR coefficients, log area ratios, or line spectral pairs, and the quantization of these transformed coefficients. The resulting distortions are rarely frequency localized and are usually spread over the entire frequency range of the signal. When viewed in the frequency domain, LPC quantization has the effect of moving the roots of the LPC polynomial, and hence the poles of the LPC vocal tract transfer function in both radius and angle affecting bandwidth and frequency). Small variations in pole angles, though easily perceivable, produce relatively mild distortions. Small pole radii variations, however, can actually have dramatic perceptual effects, for example, those too close to the unit circle can cause clearly perceivable "chirps," while those which are too far cause the speech to sound "muffled." Pole positions which vary too much cause fluttering sounds.

In actual coding systems, the LPC coefficient quantization distortions always encompass the entire frequency range and perturb all the poles. If the perceptual effects of this distortion are to be well understood, then controlled distortions need to be generated which present components of LPC quantization distortion *in isolation* and *in a frequency variant way.* In the original data base, the distorting system shown in Fig. 4.12 was used to generate pole distortions. In this system, the speech is first pre-emphasized using a second order filter, and then a framed LPC analysis is performed. The results of the LPC analysis are then used to inverse filter the original speech, giving an approximation of the glottal wave excitation [4.10]. Following the inverse filtering operation, the poles of the vocal tract function are found by factoring the LPC polynomial. Then the narrowband pole distortion is applied by first identifying all poles within a fixed frequency range, and then moving the poles slightly in either frequency (angle) or bandwidth (radius), or both. This "jittering" of the poles is controlled by two uniform random number generators. The "frequency range" factor, F, gives the range of frequency, in Hz, within which the poles are allowed to move. The "bandwidth factor," B, is a multiplicative factor controlling the bandwidth motion by

$$\text{distorted angle} = (\text{undistorted angle})(F)(r/2) \tag{4.54}$$

$$\text{distorted radius} = (\text{undistorted radius})[1 + (B)r] \tag{4.55}$$

where r is a uniform random number which ranges between plus one and minus one. Once the pole locations are distorted, they are recombined to form a new set of predictor coefficients, $a'(k)$. These coefficients are then used to implement a new vocal tract filter to create the distorted speech. The pole distortions included in the original distortion ensemble are summarized in Table 4.26 and the results of the DAM analysis of these distortions are shown in Figs. 4.42–4.47.

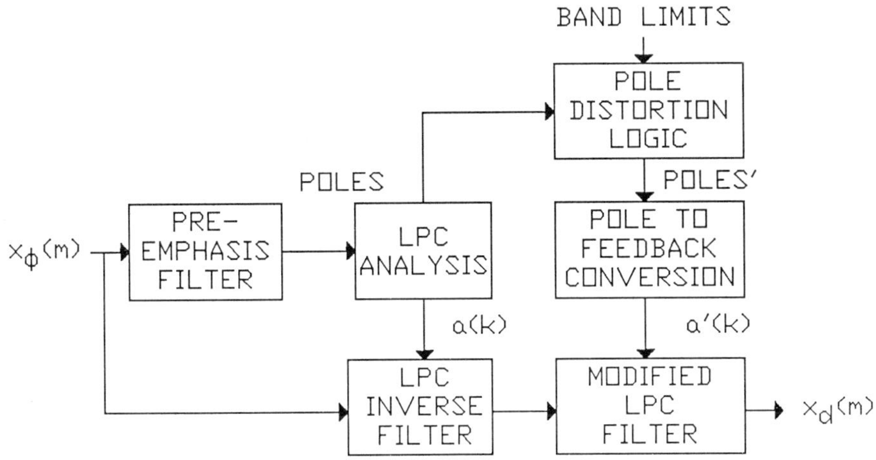

Figure 4.12: System of generating original pole distortions.

Table 4.26: Parameters for original pole distortions (I).

	Frequency Distortion (F)					
Band (Hz)	1	2	3	4	5	6
200–400	20	40	60	80	100	120
400–800	20	40	60	80	100	120
800–1300	50	90	130	170	210	250
1300–1900	50	90	130	170	210	250
1900–2600	100	150	200	250	300	350
2600–3400	150	200	250	300	350	400

	Bandwidth Distortion (B)					
Band (Hz)	1	2	3	4	5	6
0–400	.025	.05	.075	.1	.2	.3
400–800	.025	.05	.075	.7	.2	.3
800–1300	.025	.05	.075	.1	.2	.3
1300–1900	.025	.05	.075	.1	.2	.3
1900–2600	.025	.05	.075	.1	.2	.3
2600–3400	.025	.05	.075	.1	.2	.3

Table 4.27: Parameters for revised pole distortion (II).

			(F)			
Band (Hz)	1	2	3	4	5	6
50–420	10	20	30	40	50	55
420–900	20	40	60	80	100	120
900–1600	25	50	75	100	125	150
1600–3200	80	160	240	320	400	500

			(B)			
Band (Hz)	1	2	3	4	5	6
50–420	.01	.02	.04	.08	.16	.32
420–900	.01	.02	.04	.08	.16	.32
900–1600	.01	.02	.04	.08	.16	.32
1600–3200	.01	.02	.04	.08	.16	.32

A study of the DAM results shown in Figs. 4.42 and 4.47 reveals some basic problems with the distortions used in the original study. One problem is that certain of the distortion classes did not exhibit an adequate variation in perceived distortion. This is particularly true for the case of frequency distortion in the ranges 200–400 Hz, 1900–2600 Hz, and 2600–3400 Hz, but is also true for radial distortion in the range of 2600–3400 Hz. An examination of the control parameters for the pole distortion shown in Table 4.26 indicates that this is a fundamental problem, since the frequency variations used were already very large when compared to the dimensions of the frequency bands. In short, the bands used were too narrow for clearly perceivable distortions to be generated.

Based on these observations, a new set of narrowband pole distortions was generated (pole distortion - II) using only four bands which were chosen to have approximately equal frequency content on a mel scale. The control parameters for this study are shown in Table 4.27. Notice that the pole distortions were chosen so as to exhibit both pole-frequency and pole-bandwidth variations.

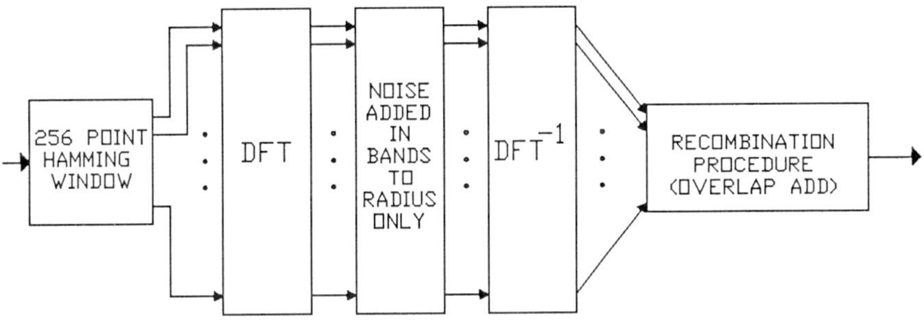

Figure 4.13: Block diagram for narrowband frequency distortion.

4.2.2.10 Narrowband Frequency Distortion

The *Narrowband Frequency Distortion* was designed to approximate the distortions found in ATC and adaptive subband coders in a frequency variant way. The operation of this distortion is illustrated in Fig. 4.13. The speech is first windowed using overlapping Hamming windows, with a window length of twice the frame interval, and a frame interval of 128 points. The speech is then transformed using a 256 point FFT. In the frequency domain, noise is added to the samples in bands, with the noise having random magnitude but with a phase equal to the phase of the original speech. Finally, the samples are inverse transformed back into the time domain and recombined using overlapped adds. The parameters controlling the banded frequency distortion are the band limits and the standard deviation of the added noise, which is white and Gaussian. Table 4.28 summarizes the banded distortions used in this study.

4.3 The Subjective Data Base

As previously noted, the subjective data base was formed by applying the Diagnostic Acceptability Measure (DAM) to the distortions in the distorted data base. The primary purpose of the subjective data base was to provide a basis for evaluating the predictive potential of the various physical measures to be considered. For this purpose it was essential that representatives of widely diverse forms of degra-

Table 4.28: Parameters for narrowband frequency distortion.

		Standard Deviation of Noise					
	Band (Hz)	1	2	3	4	5	6
1	0-400	.1	.2	.4	.6	.8	1.0
2	400-800	.5	.8	1.1	1.4	1.7	2.0
3	800-1300	2.0	2.2	2.4	2.6	2.8	3.0
4	1300-1900	2.0	2.2	2.4	2.6	2.8	3.0
5	1900-2600	3.5	4.0	4.5	5.0	5.5	6.0
6	2600-3400	10.0	13.0	16.0	19.0	22.0	25.0

dation be included among the conditions evaluated. Among the distortions considered are those inherent in the simplest types of analog speech transmission as well as those associated with some of the more complex digital voice coding and transmission techniques in use today. Only with such diversity could any assurance be had that correlations between specific physical voice measurements and various subjective criteria would be valid for more than a narrow class of distortions.

Four speakers, three males and one female, were used for all evaluations of the DAM. The ordering of experimental system conditions varied from one speaker to the next in a systematic manner designed to minimize time-order effects on the data for any system condition. Twenty-four system conditions, two anchors, and four probes were evaluated in each testing session. The anchors and probes were always presented at the beginning of each series of system conditions involving a given speaker, and the ordering of anchors and probes was randomly determined in each instance. Whenever possible, several distinct types of degradation were represented in a given session. System conditions were randomly ordered within a session for one speaker, and then systematically reordered for the remaining speakers to provide some amount of counterbalancing and, thus, to soften the effects of any inter-condition influences.

Presented in the following sections are DAM score patterns for the various forms of degradation. For each class of degradation the diagnostic patterns are presented in separate panels for male (average of three) and female speakers. Except where pronounced sex differences are evident, the discussion will be addressed primarily to the results for the male speakers. Primary interest in these figures attaches to the Composite Acceptability (CA) and the Acceptability (A) scores, Intelligibility (I), and Pleasantness (P).

Of considerable, but secondary, interest are the "diagnostic patterns" of perceptual quality scores. Depending on the form of degradation involved, diagnostic score patterns for coding or communication systems may provide important insights leading to appropriate remedial action. Here, they serve primarily to enhance the basic knowledge of the perceptual consequences of a given speech degradation.

Two administrations of the DAM, separated by intervals of four to six weeks, were performed for all the system conditions except those involving pole distortions, band distortions, and the system conditions in the extension to the data base. Thus, most results presented in the following sections are averages based on response data from two independent evaluations.

4.3.1 Coding Degradation

Treated in this section are cases of distortion which are intrinsic to various speech coding techniques and, in a sense, reflect the inadequacies of such techniques. In one category are various broadband waveform preserving techniques in which a major source of degradation is quantization of the speech signal. In the second category are various, more complex coders. The DAM subjective quality scales referred to in the descriptions of the distortions are shown in Table 3.1.

4.3.1.1 Continuously Variable Slope Delta Modulation

Five realizations of Continuously Variable Slope Delta Modulation (CVSD) techniques, which differed only with respect to data rate, were treated in this investigation. A control condition involving essentially unprocessed speech was included within the same DAM testing session. The DAM results are presented in Fig. 4.14. In all major respects they are typical of previous DAM results for CVSD [4.32]. Except in the case of the lowest data rate, background quality is negligibly affected by CVSD. Listeners evidently do not confuse quantization "noise" with true noise; rather, they correctly perceive it as distortion. The SD scale of the DAM is the most sensitive of the perceptual quality measures.

The present results differ somewhat from those of previous studies in that they show consistent, though not pronounced, reductions in scores on the SH, SL and SN scales as data rate is reduced. Such results are most typical of conditions involving audio passband restriction and may, therefore, have a rational basis in the present case.

For reasons that are not immediately obvious, scores for the control case and for the case of 32 kbps CVSD are generally somewhat lower than those previously obtained for nominally comparable conditions, though scores for the 16 kbps case are very close to historical norms for this condition.

4.3.1.2 Adaptive Delta Modulation (ADM)

Figure 4.15 presents DAM results for the case of adaptive delta modulation. Predictably, perhaps, they are quite similar to those for CVSD at most corresponding data rates. An exception is the case of ADM at 8 kbps where a severely depressed score on the SI (signal interrupted) scale can be observed. It is quite possible that this result can be attributed to an experimental artifact, but further investigation will be needed to resolve this issue. This is not of great interest since no one has seriously suggested using such a coding procedure at this rate.

4.3.1.3 Adaptive Pulse Code Modulation (APCM)

Figure 4.16 presents DAM results for the case of adaptive pulse code modulation techniques. As in the two previous cases the subjective consequences of this type of coding are confined almost exclusively to signal quality. Here their general form is quite similar to those for the cases of CVSD and ADM, except for a small, though consistent, depression of scores on the SI scale. However, the general level of scores for APCM is substantially lower than that for CVSD and ADM, but this is attributable to the lower associated bit rate.

4.3.1.4 Adaptive Differential Pulse Code Modulation

Figure 4.17 shows DAM scores for adaptive differential pulse code modulation (ADPCM). The results for this condition are quite similar to those for APCM. It would appear that adaptive differential pulse code modulation does not significantly improve acceptability over APCM at

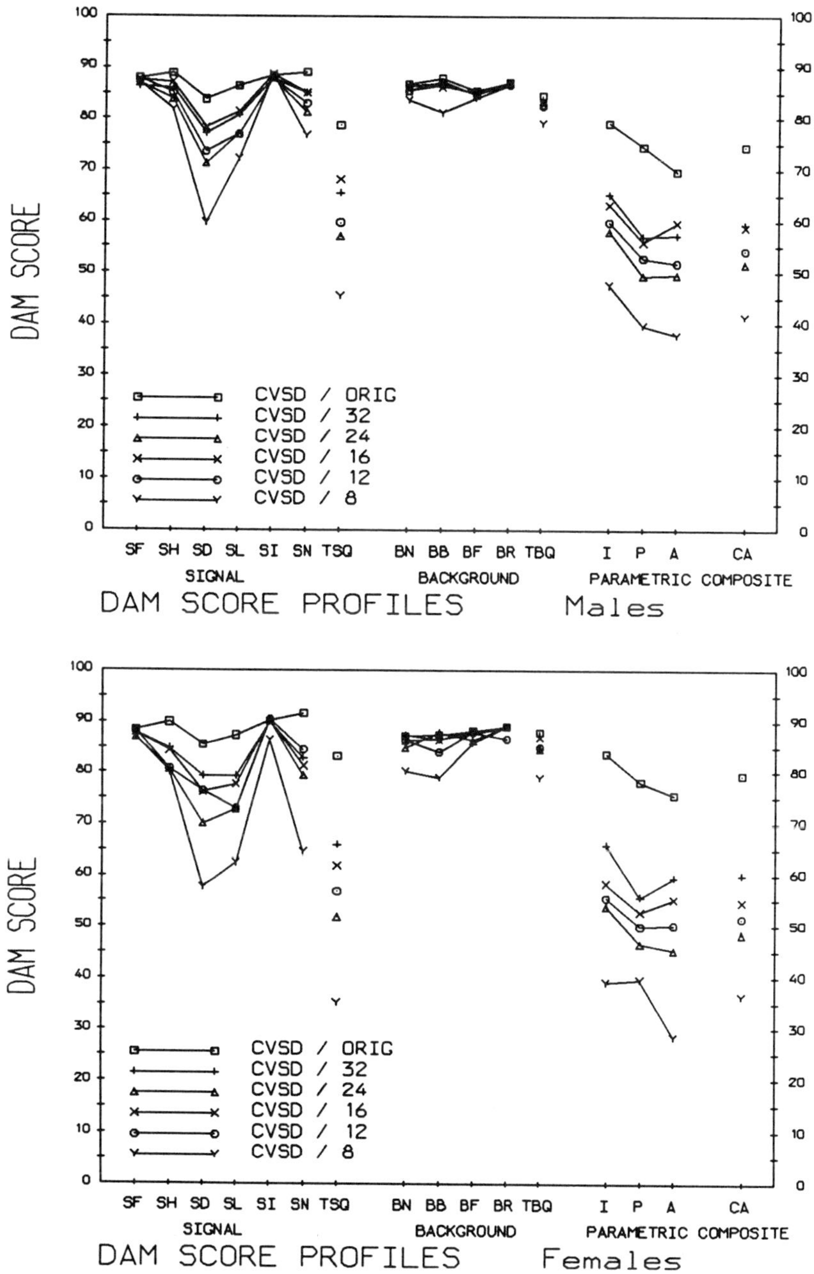

Figure 4.14: Effects of CVSD distortion.

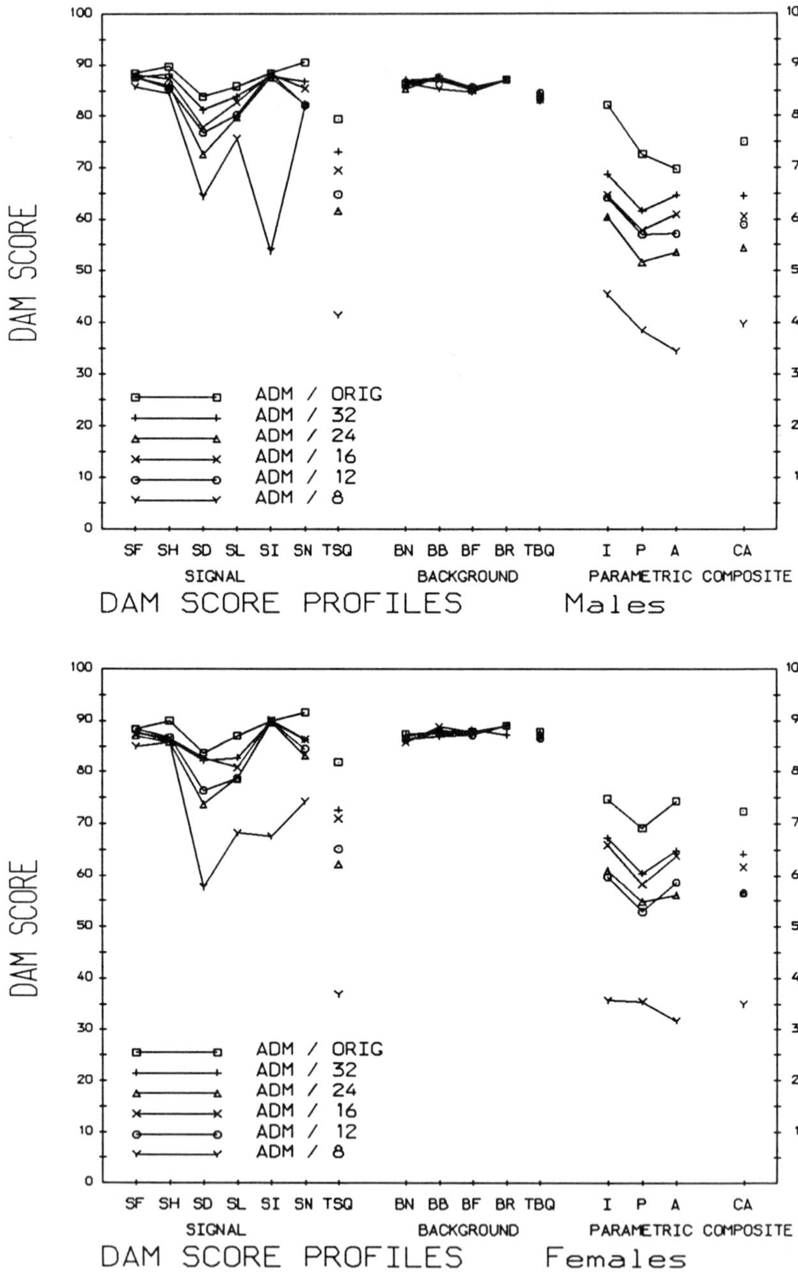

Figure 4.15: Effects of ADM distortion.

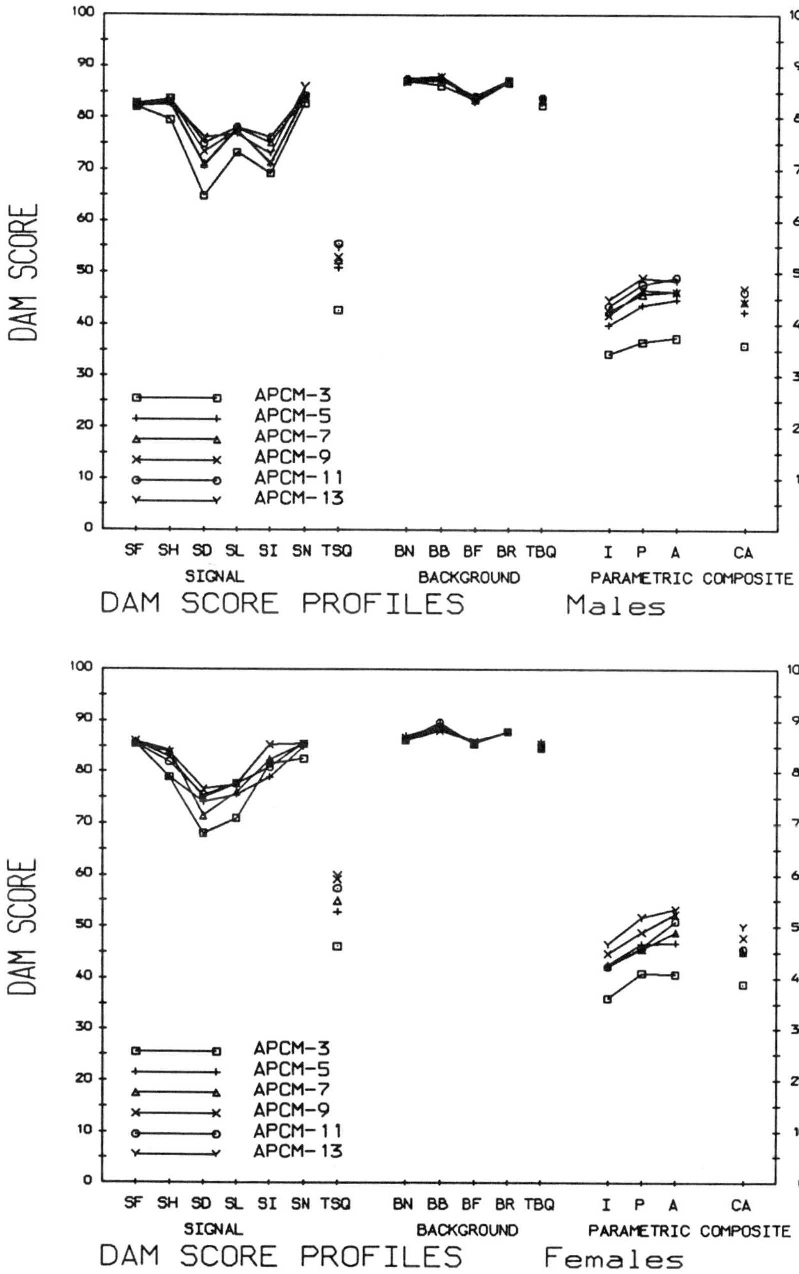

Figure 4.16: Effects of APCM distortion.

comparable degrees of quantization, but may, however, at comparable transmission data rates with optimal channel coding. Qualitatively, ADPCM appears to sound somewhat less distorted but more interrupted than APCM.

4.3.1.5 Linear Predictive Coder (LPC)

Figure 4.18 shows that the present realization of LPC in the range of 2–2.9 kbps yields DAM score patterns and overall levels very similar to those of the normative 2.4 kbps LPC as reported by Voiers [4.32]. Normative DAM results for the higher data rates are not available for systems without error correction, so that comparisons are not possible for these. However, the present results indicate that increasing the data rate to 3.9 kbps significantly improves the quality of LPC-processed speech, though further increases do not appear to be beneficial.

4.3.1.6 Adaptive Predictive Coder (APC)

Figure 4.19 shows the effects of APC on DAM scores. The parameter in these graphs is bits/frame which is associated with data rates of from 13333 to 15867 bps.

Perceptual quality score patterns for APC are quite similar to those for LPC. Though score levels are generally somewhat higher for APC, this superiority is achieved only at a large cost in transmission data rate. Listeners perceive significant amounts of signal and background "flutter" (SI scale) and "raspiness" (SD scale).

4.3.1.7 ADPCM with Noise Feedback

Figure 4.20 shows the results of the subjective quality tests applied to the ADPCM-NF distortion. As can be seen from Fig. 4.20, this distortion exhibits a good range of subjective responses. Almost all of the distortion shows up on the SD (signal distorted) parametric scale, as is typical of many waveform coder systems. One of the claims made for the noise feedback approach is that for equivalent bit rate systems, noise feedback generally results in improved quality over systems without noise feedback. Figure 4.21 shows the results of subjective tests applied to equivalent ADPCM systems without noise feedback for the

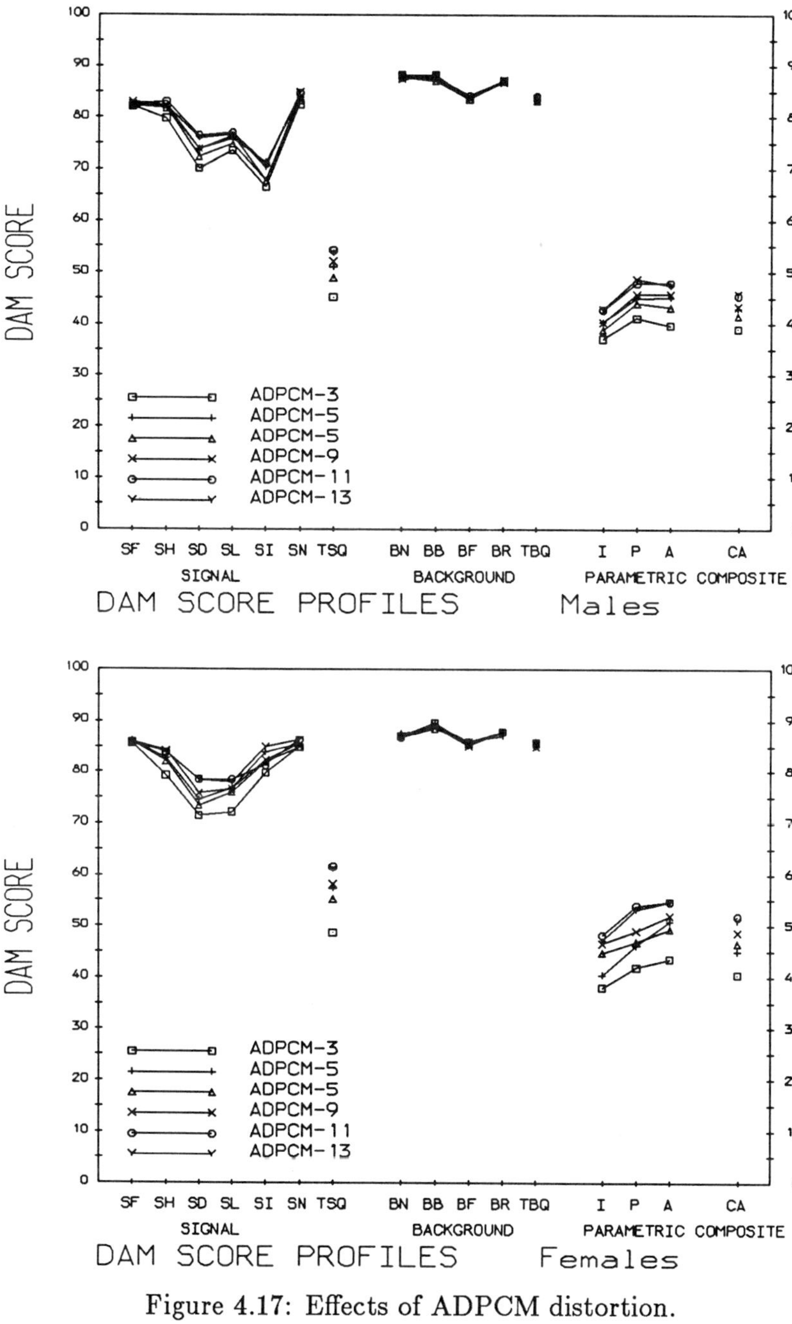

Figure 4.17: Effects of ADPCM distortion.

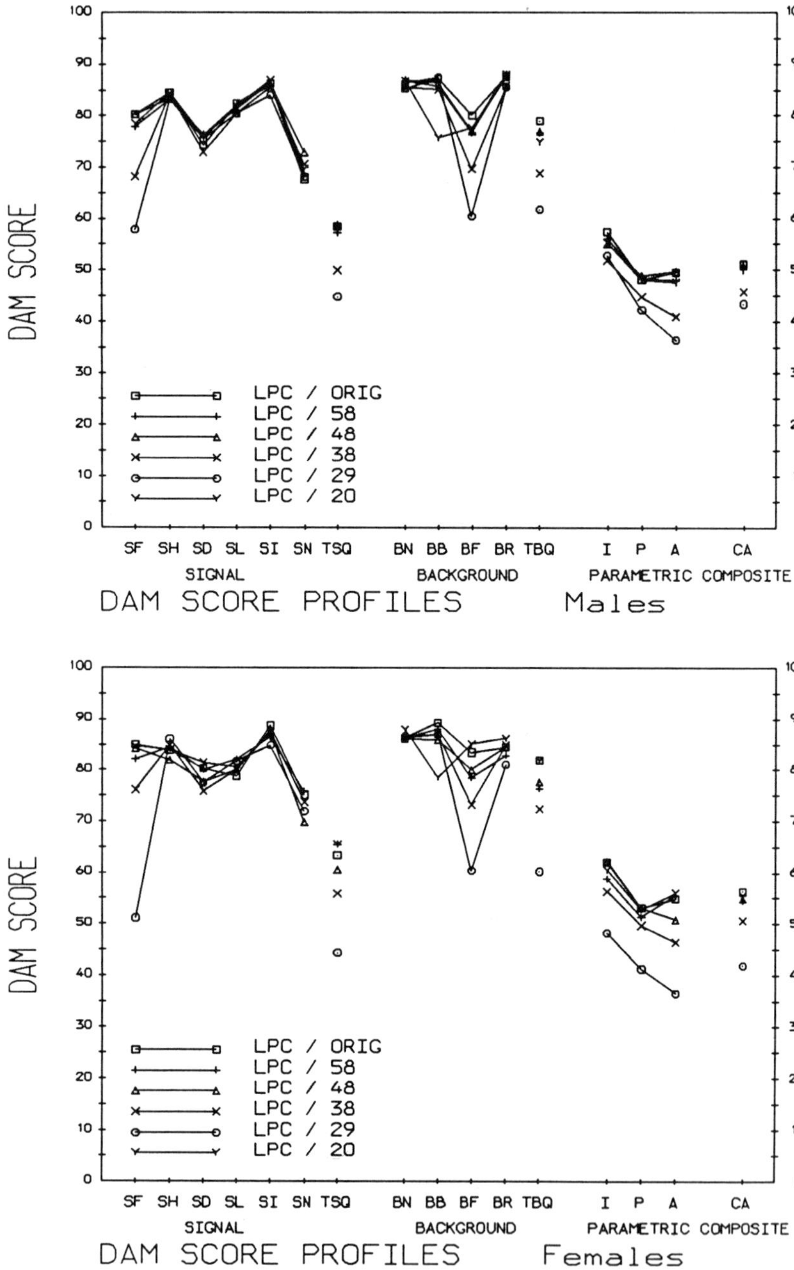

Figure 4.18: Effects of LPC distortion.

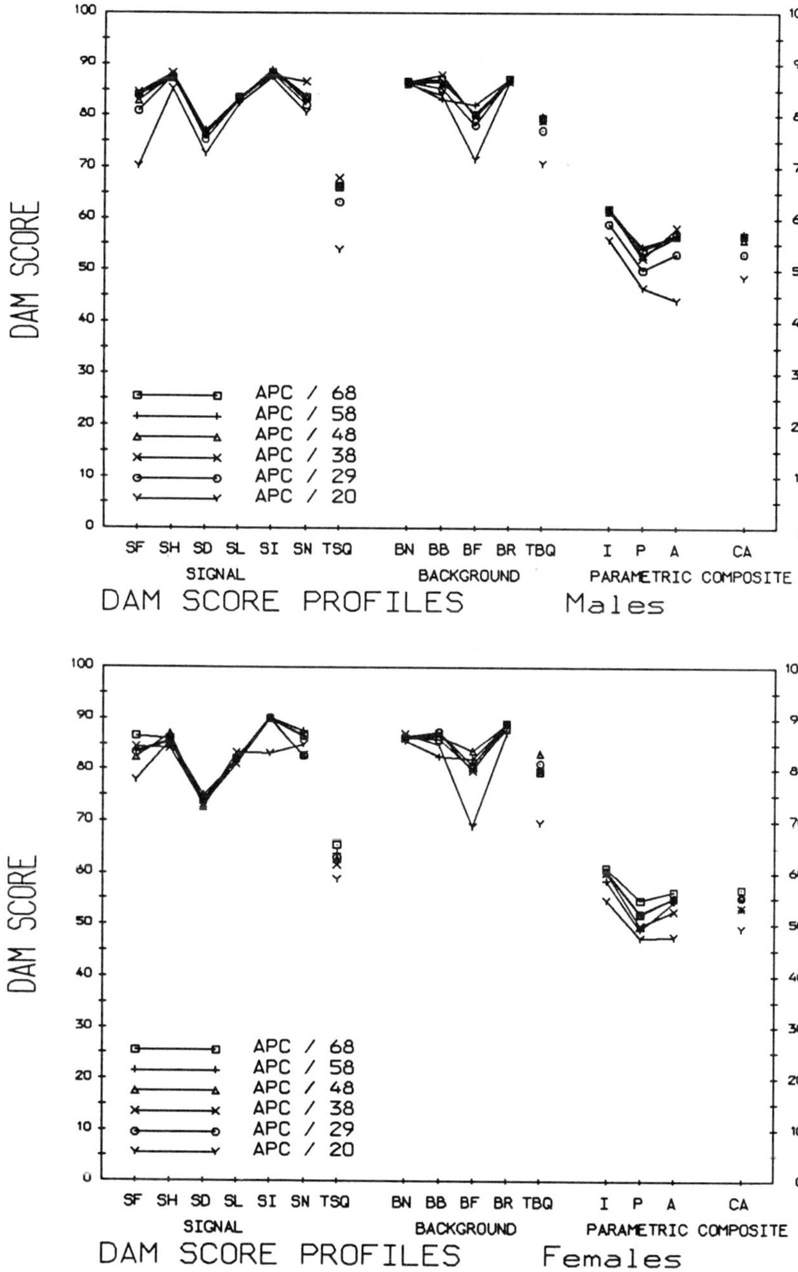

Figure 4.19: Effects of APC distortion.

four lowest bit rate systems. Clearly, from these tests it appears that there is no measurable advantage to using noise feedback.

4.3.1.8 Voice-Excited Vocoder (VEV)

Two realizations of VEV were examined here. In the first (Fig. 4.22) the voice band had seven-level quantization, in the second (Fig. 4.23), thirteen-level quantization. The parameter in each case is PARCOR frame rate. Differences between the subjective effects of these two techniques are very small. Listeners possibly perceive a slightly greater degree of signal and background flutter with coarser quantization in the case of male speakers, but this trend is absent in the case of the female speaker. Generally, more background flutter is perceived in the case of VEV than in the case of APC.

4.3.1.9 Multi-Pulse LPC (MPLPC)

The effects of Multi-Pulse LPC on subjective responses are illustrated in Fig. 4.24. There are several points which should be noted here. First, the Multi-Pulse LPC is capable of generating quite high quality systems at relatively low bit rates. In fact, the only coding system in this study which resulted in better quality was an ATC which operated at about twice the equivalent bit rate of the Multi-Pulse LPC. Second, the technique of using the estimated excitation function to improve the LPC analysis (systems 2, 4, and 6 of the MPLPC distortion) gives a consistent improvement for the lowest bit rates (2 pulses every 60 samples) but has little impact at the higher rates. Third, the MPLPC tends to excite a broad class of parametric distortion scales, including SF (signal fluttering), SH (signal highpass), SL (signal lowpass), and SD (signal distorted) as well as BF (background fluttering). However, on many of these scales the responses are bimodal depending on whether or not there are enough pulses in the residual representation to support the true pitch. If this effect is corrected, most of the perceived distortion occurs on the SF and BF scales.

4.3.1.10 Adaptive Transform Coder (ATC - I)

The original adaptive transform coder distortion excited primarily the SF, BF and BR parametric scales. The coded speech most likely has the

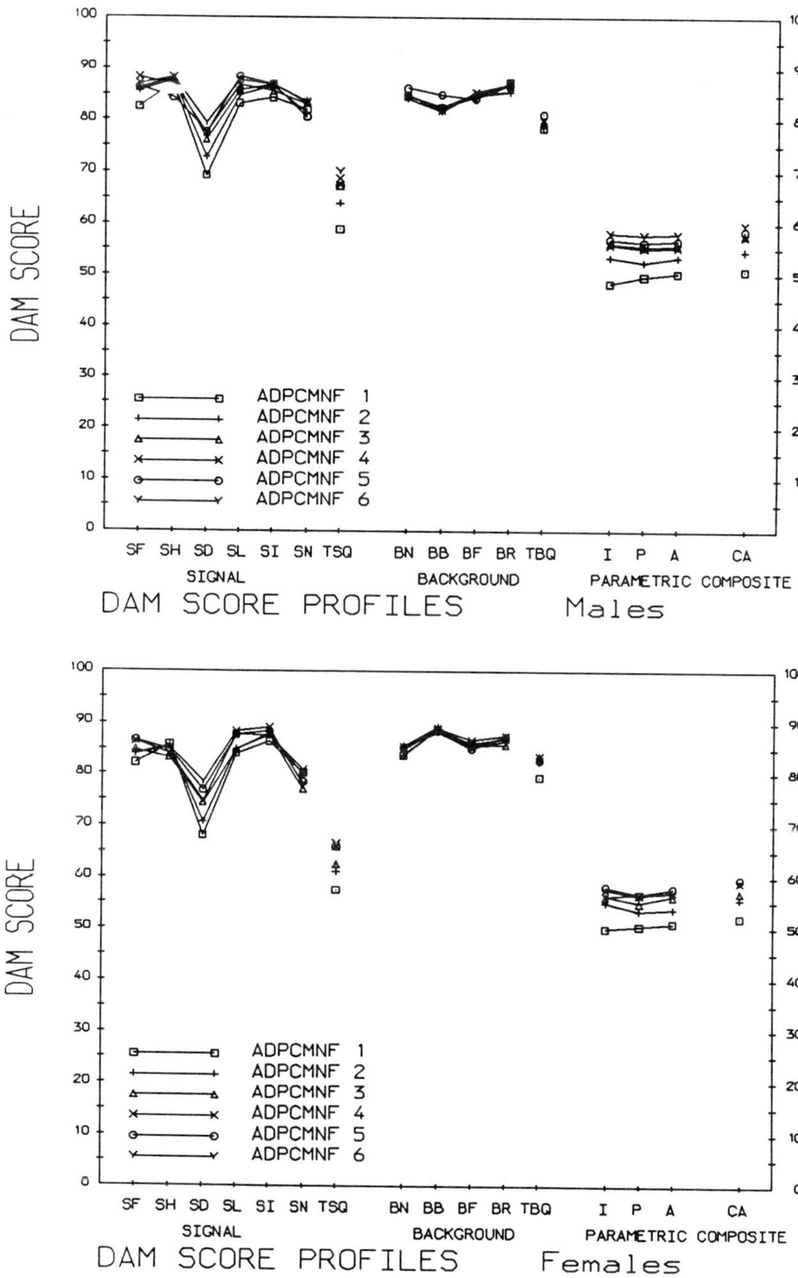

Figure 4.20: Effects of ADPCM-NF distortion.

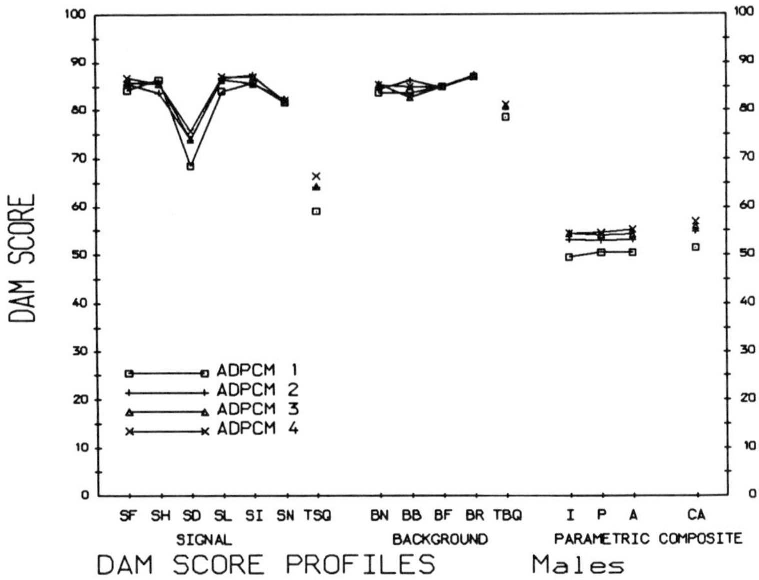

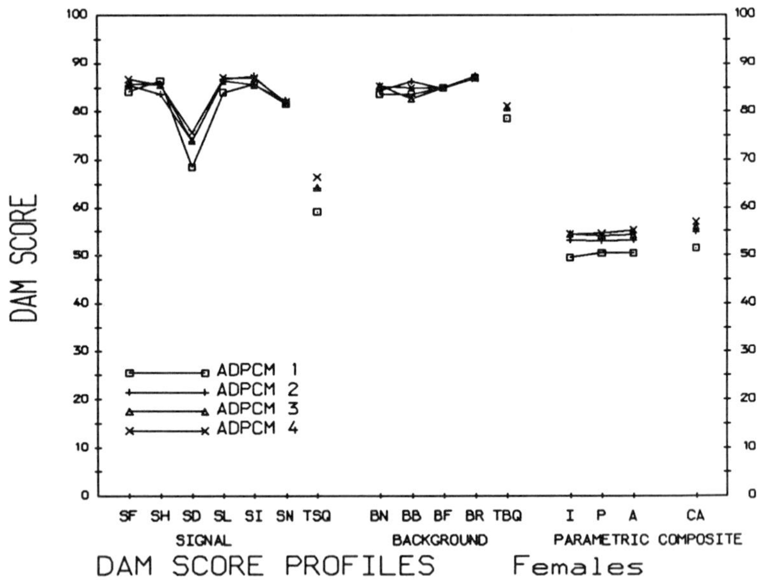

Figure 4.21: Effects of equivalent ADPCM system.

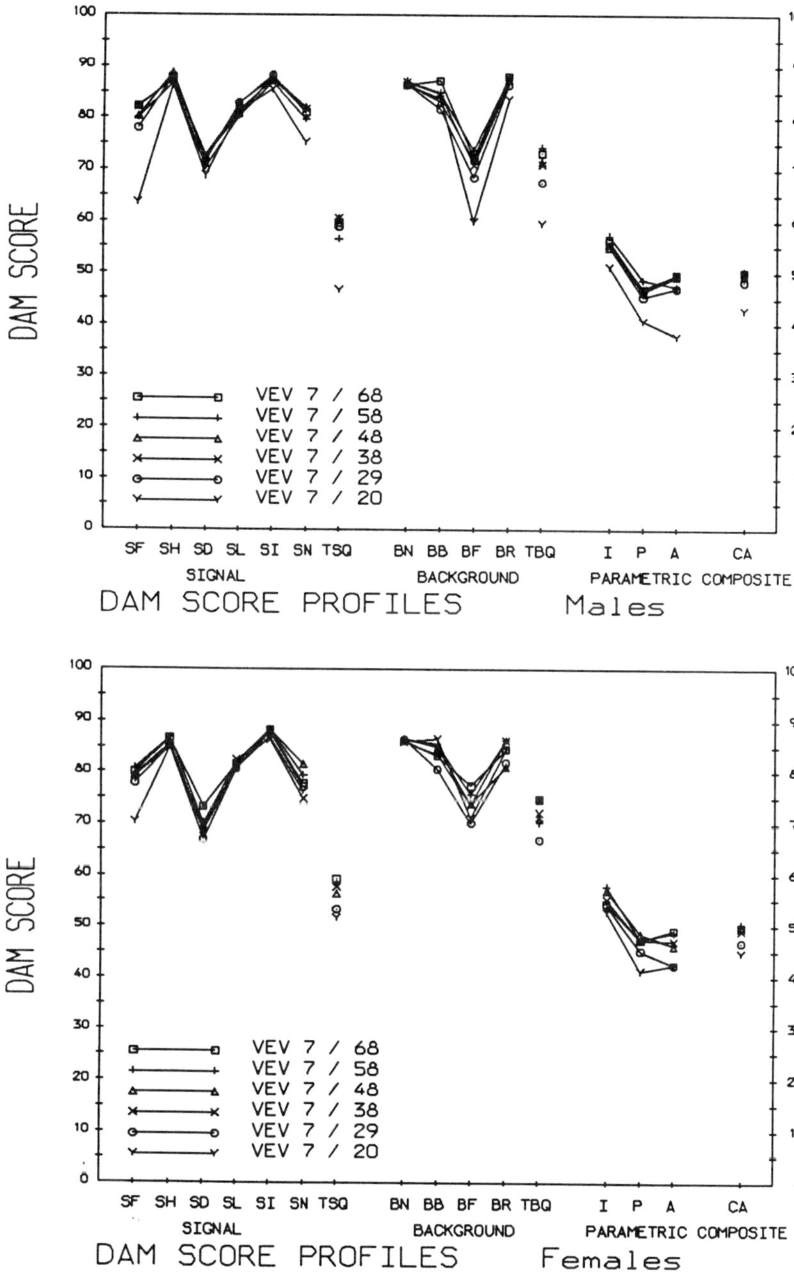

Figure 4.22: Effects of seven-level VEV distortion.

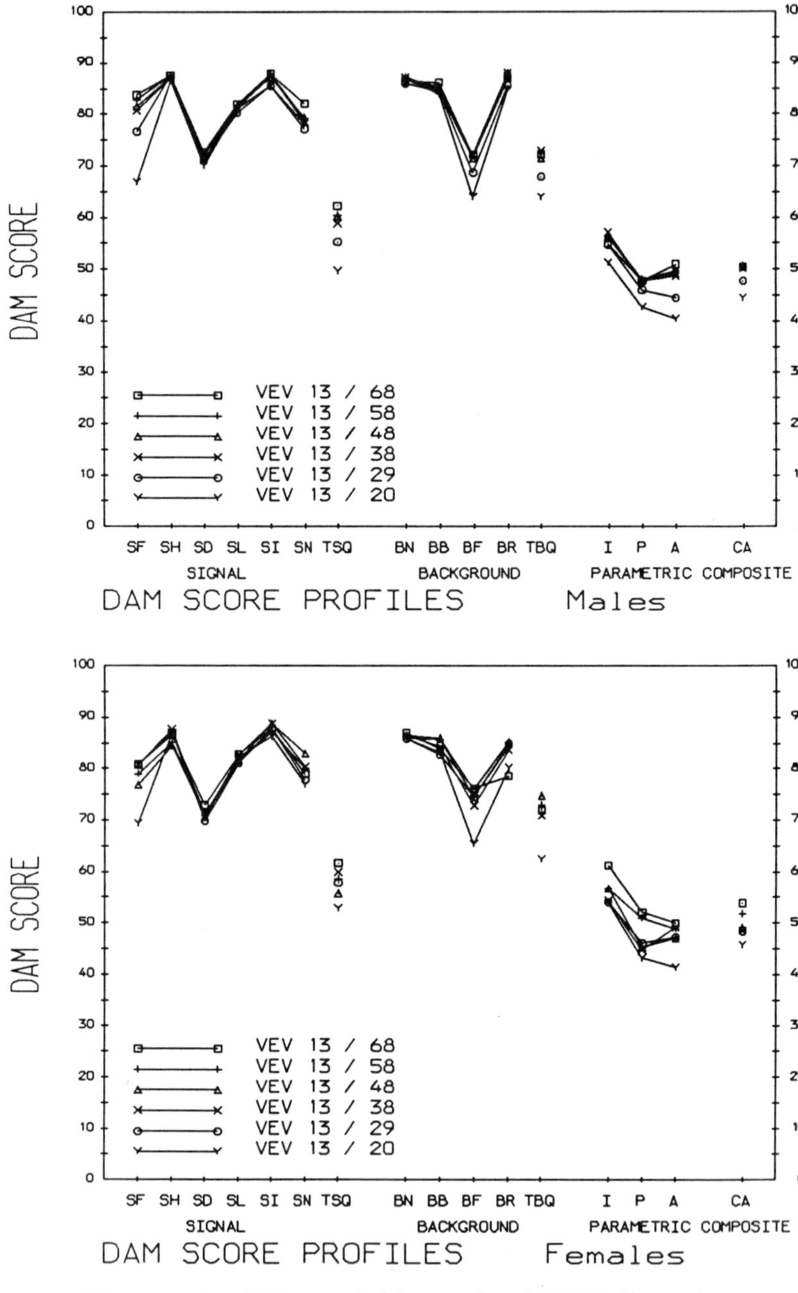

Figure 4.23: Effects of thirteen-level VEV distortion.

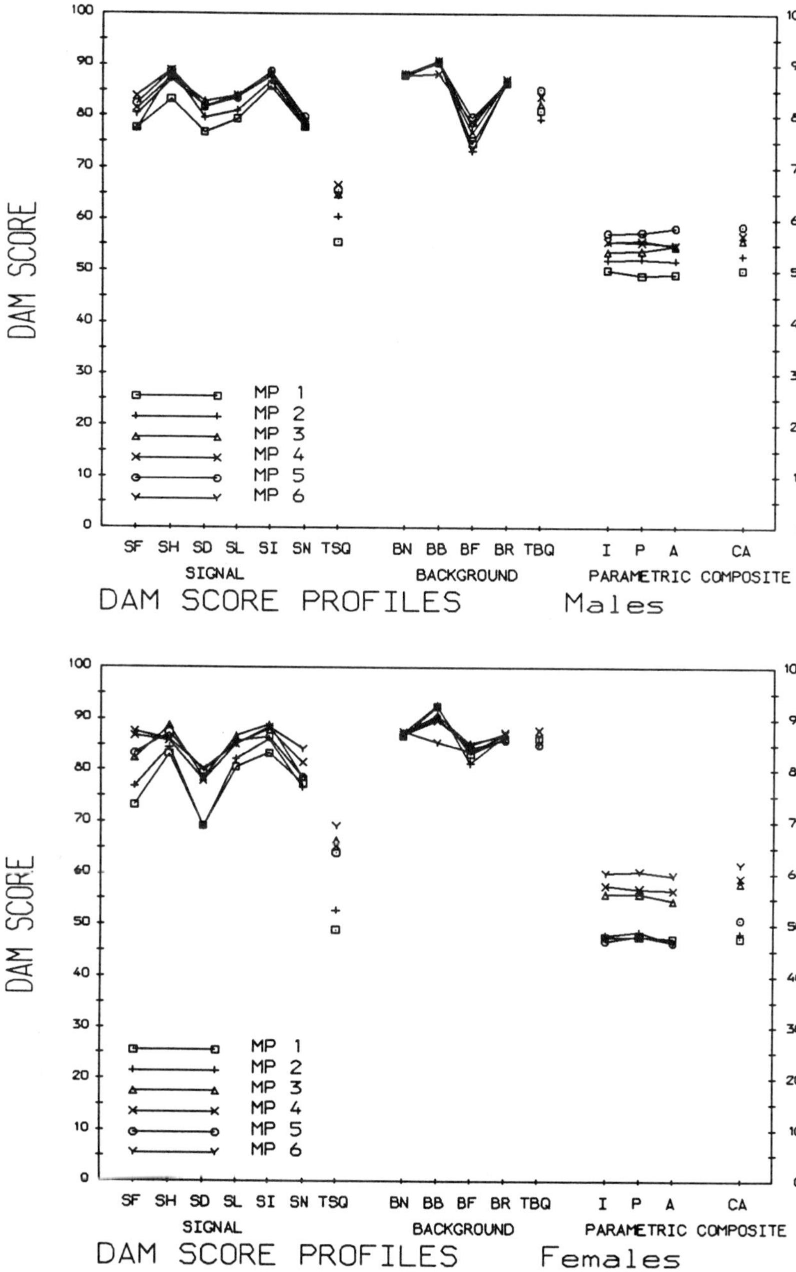

Figure 4.24: Effects of MPLPC distortion.

signal fluttering quality because the coder operates on framed speech segments, as do the LPC and VEV coders, which also excite the SF scale. Similarly, such coders also excite the BF scale. However, it is unlike the various linear prediction coders in that its speech has a "rumbling, thumping" quality, as measured by the BR scale.

4.3.1.11 Revised Adaptive Transform Coder (ATC - II)

The results of the subjective quality evaluation of the ATC is shown in Fig. 4.25. The ATC is a a high quality coder for medium bit rates, with near toll quality performance at 16 kbps. Like the MPLPC, the ATC excites a number of parametric quality scales. Clearly, the ATC distortion is mostly perceived as SF (signal fluttering) and BF (background fluttering). However there are also non-trivial deviations shown on the SN (signal nasal), SD (signal distorted), and SL (signal lowpass) scales. One of the primary reasons for including a revised adaptive transform coder in the extended data base was that the spread of subjective quality results for the original transform coder was rather small. As can be seen from Fig. 4.26, the revised ATC has corrected this shortcoming.

4.3.1.12 Subband Coder

Figure 4.27 shows the results of the subband coder distortions on subjective quality. Like all of the previous distortions, the subband coder distortion exhibits a good range of subjective responses. The subband coder also exhibits a distinct bimodal behavior for a number of parametric scales, specifically SF (signal fluttering), SN (signal nasal), and BF (background fluttering). This is a direct reflection of the inclusion or exclusion of time domain harmonic scaling in the subband coding system. The basic subband coder distortion shows up mostly on the SD (signal distorted) scale, while the TDHS excites mostly the SF (signal fluttering), SN (signal nasal), and BF (background fluttering) parametric scales.

4.3.1.13 Channel Vocoder

The subjective results for the Channel Vocoder distortion are shown in Fig. 4.28. Of all the coding distortions in this study, the channel

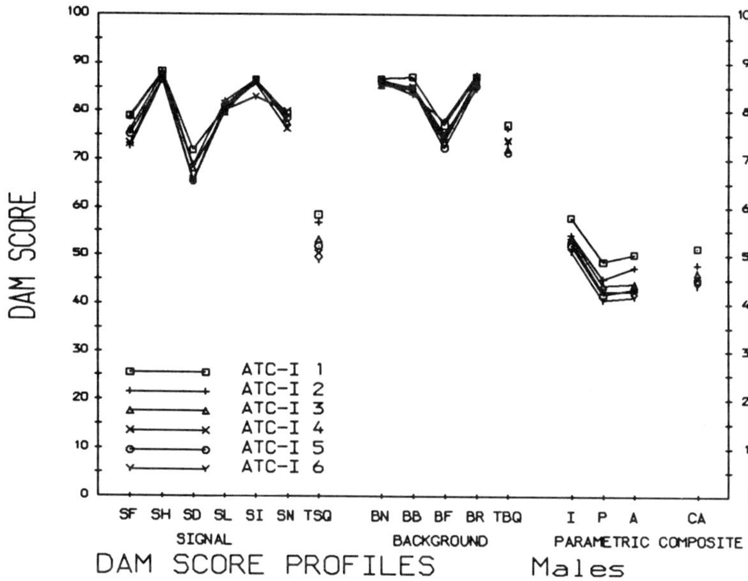

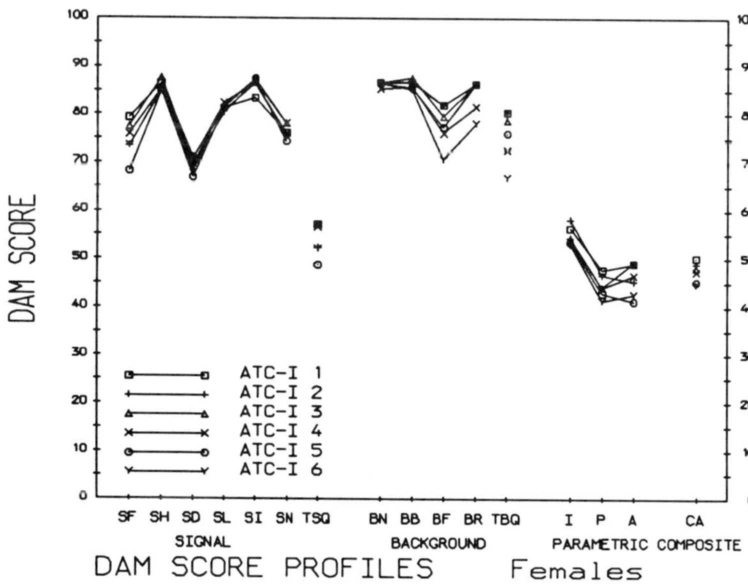

Figure 4.25: Effects of ATC - I distortion.

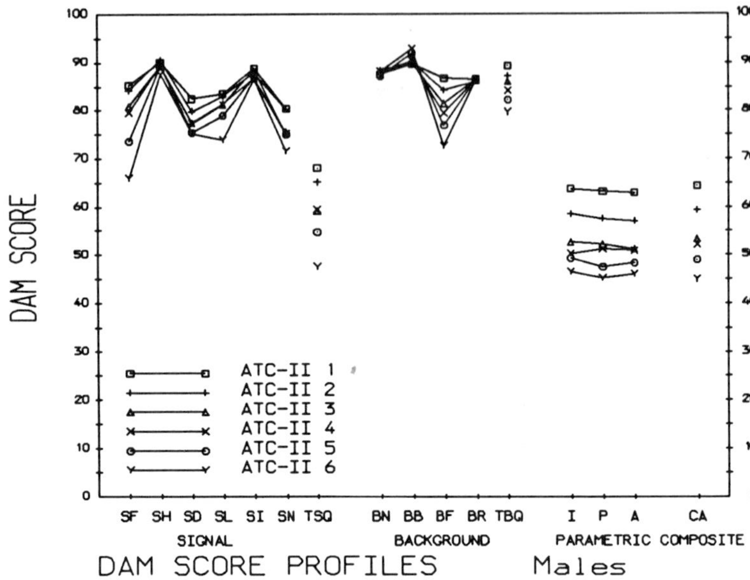

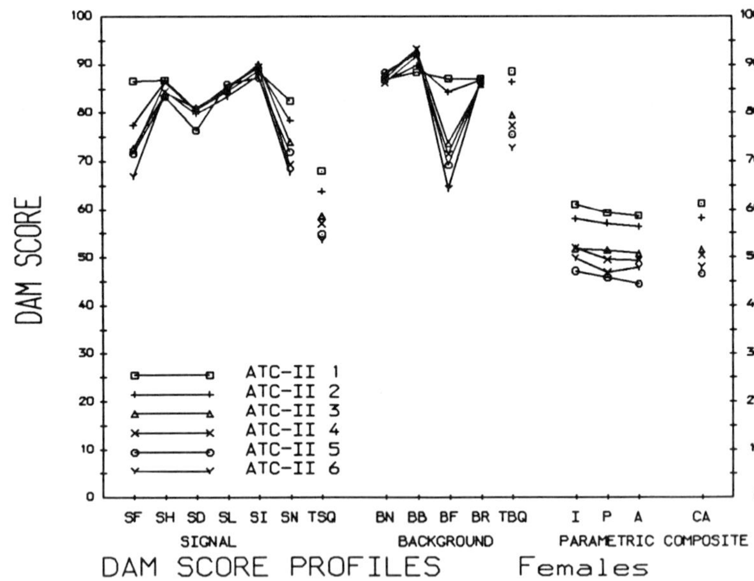

Figure 4.26: Effects of ATC - II distortion.

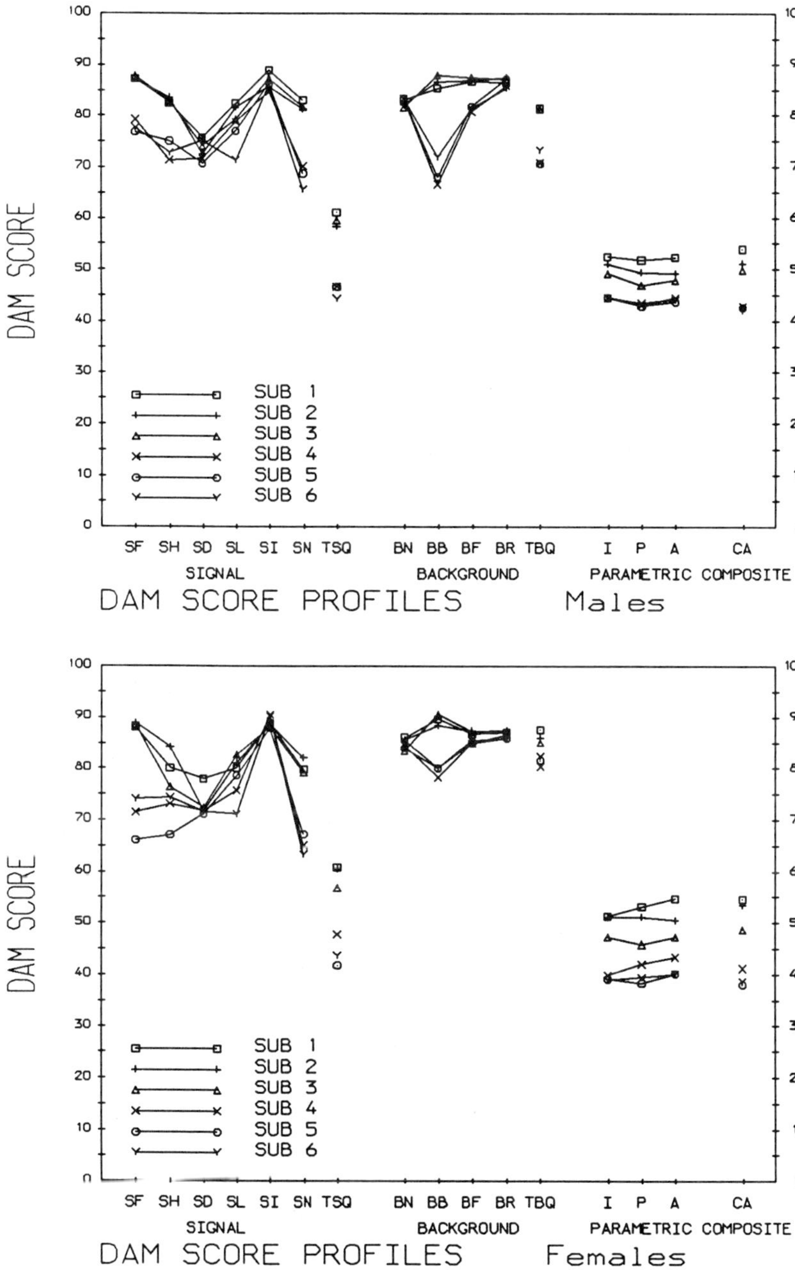

Figure 4.27: Effects of subband coder distortion.

vocoder was the least successful in generating a good range of subjective responses. However, the results are still adequate for use in the subjective data base. It is clear from Fig. 4.28 that most of the channel vocoder distortion shows up on the SN (signal nasal) and BN (background noisy) scales.

4.3.2 Controlled Degradation

Treated in this section are basic types of speech degradation, one or several of which may be present in most speech communication systems. These distortions generally do not occur in isolation in speech coding systems, but rather occur in varying combinations and intensities as a result of various coding techniques or channel characteristics.

4.3.2.1 Additive Noise

Noise is the most ubiquitous of all factors which hamper efficient communication. Accordingly, its effects on DAM scores merit special consideration. Figure 4.29 shows DAM diagnostic profiles for six conditions of signal-to-noise ratio (4 kHz passband for the speech and noise). From the figure it is clear that the SN scale is the most sensitive to additive Gaussian noise, but the results again illustrate an important principle of the perception of speech. In few instances are the consequences of degradation with respect to a single stimulus parameter confined to a single elementary psychological description. It has long been known, for example, that although the perception of the elementary psychological parameter, *pitch*, depends primarily on stimulus periodicity, it can also vary with stimulus *intensity*, *duration*, and *complexity*. In the present case, the mechanism that causes values on the SL scale (normally most sensitive to high frequency attenuation) to decrease with signal-to-noise ratio is easily explained. The long-term spectrum of typical speech is highest in the region of 500 Hz but decreases at approximately 9 dB per octave both above and below that region. One effect of the addition of spectrally flat noise, therefore, is that of passband restriction through masking (particularly at the upper end of the speech spectrum). Less readily predicted, but by no means counterintuitive, is a slight reduction of the SD scale (the scale most sensitive to amplitude distortion). With extremely unfavorable signal-to-noise ra-

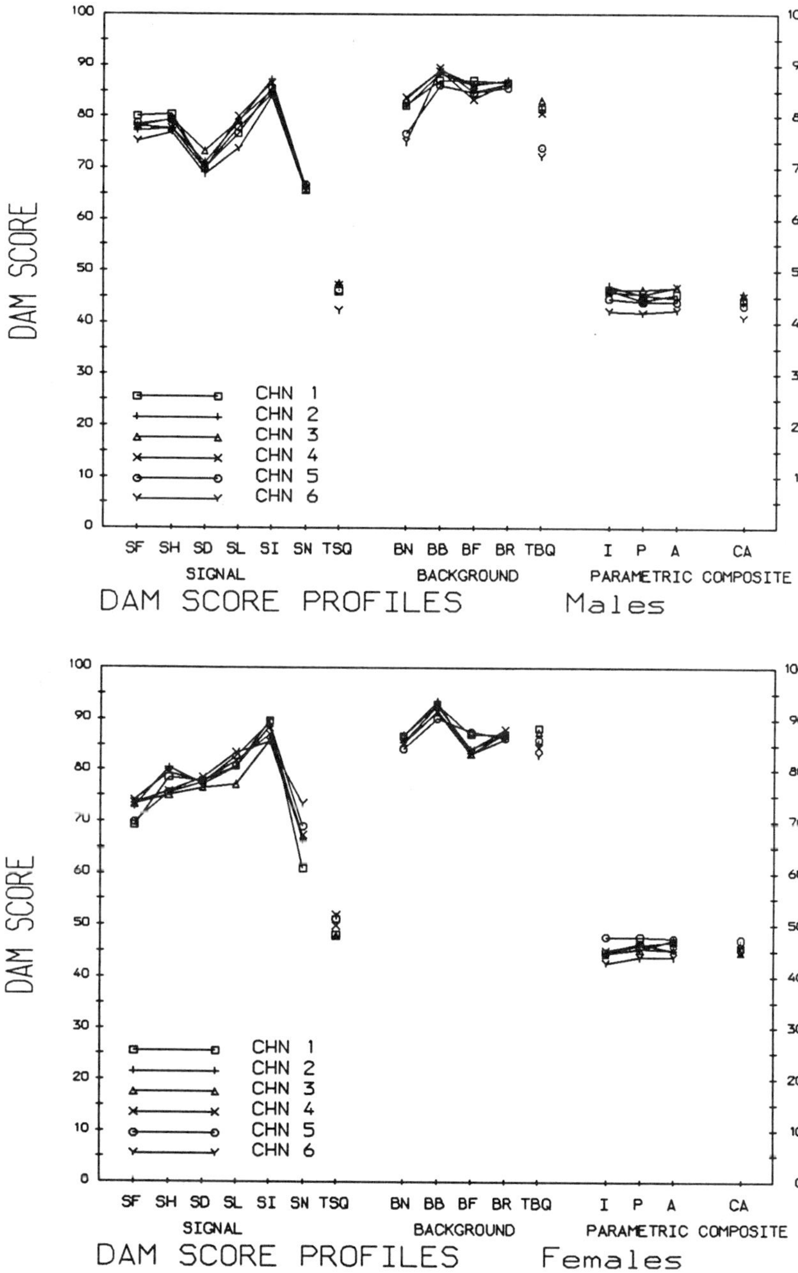

Figure 4.28: Effects of channel vocoder distortion.

tios, listeners are evidently not able to make the noise versus distortion distinction with the same ease exhibited under less severe conditions of degradation.

4.3.2.2 Periodic Interruption

For each of two rates of interruption, six levels of severity, or decreasing signal duty factors, were imposed. In the first case, the signal was interrupted once every 300 samples or 26.6 times per second. The duration of each interruption was then varied from 10 to 150 samples, that is, from 1.25 ms to 18.75 ms In the second case, the signal was interrupted once every 1024 samples or 7.8125 times per second. Duration of these interruptions was varied from 2 ms to 64 ms.

Figure 4.30 shows that the predominant perceptual quality associated with the more rapid interruption rate is "signal flutter," which becomes increasingly pronounced as signal duty factor decreases. Although listeners perceive the signal to be increasingly interrupted with increasing duration of interruption, the interrupted quality is less salient than the fluttering quality. Figure 4.31 shows the effects of less frequent interruption. The fluttering quality is still pronounced but the interruption is now more apparent and in fact dominates in the case of the lowest signal duty factor (0.50). Moreover, listeners appear less inclined in this case to perceive the background as fluttering than they do in the case of more rapidly interrupted speech.

4.3.2.3 Peak Clipping

Two forms of amplitude distortion are potentially present in many voice communication channels. They can be simply described as *peak clipping* (clamping) and *center clipping*, (as might result from intermodulation or crossover distortion). It was out of concern for the first of these that the SD scale of the DAM was developed. However, no special provision for the latter was made in the design of the DAM.

Figure 4.32 shows the effect of six levels of peak clipping on the DAM score patterns. The values associated with each condition indicate levels at which peak clipping occurred on a scale where the RMS amplitude of the unclipped speech signal was approximately 1350. Two perceptual quality scales SD and BB appear sensitive to this form of

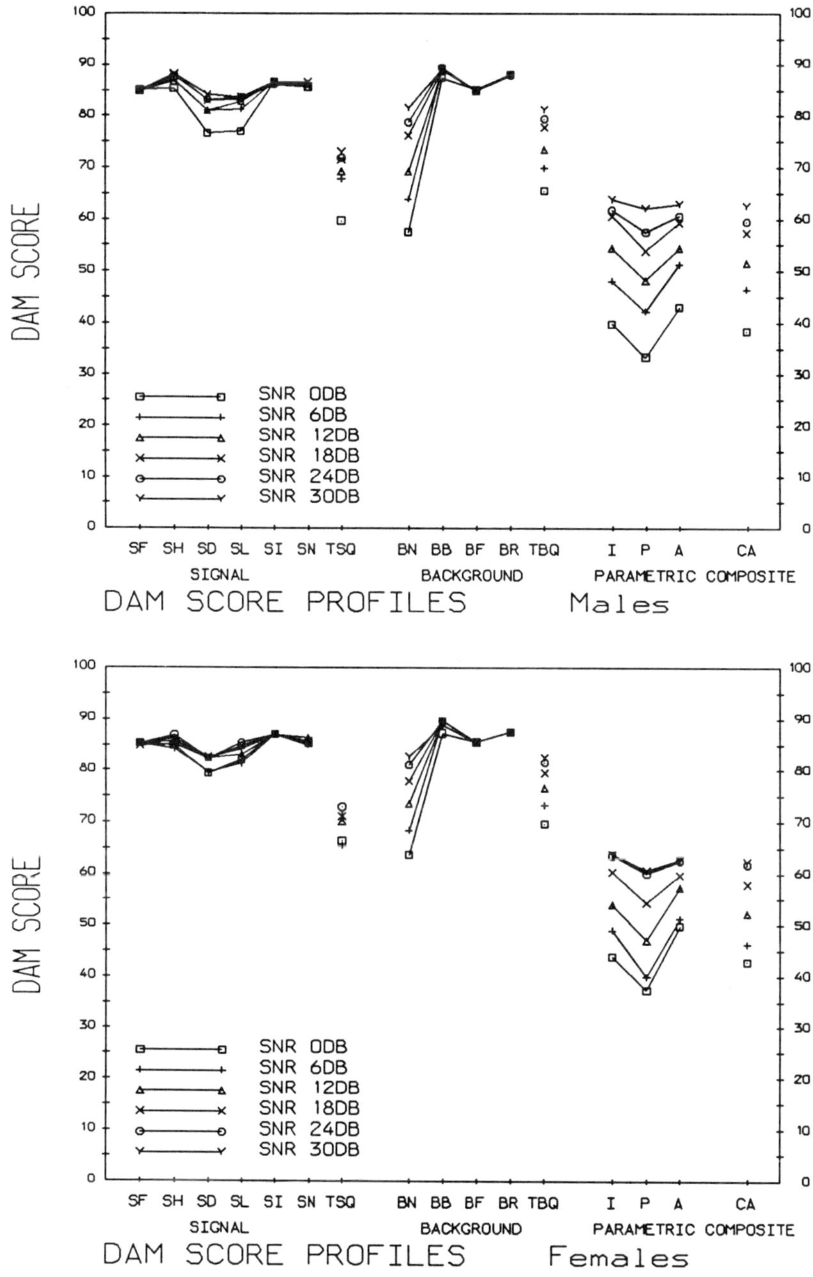

Figure 4.29: Effects of additive noise distortion.

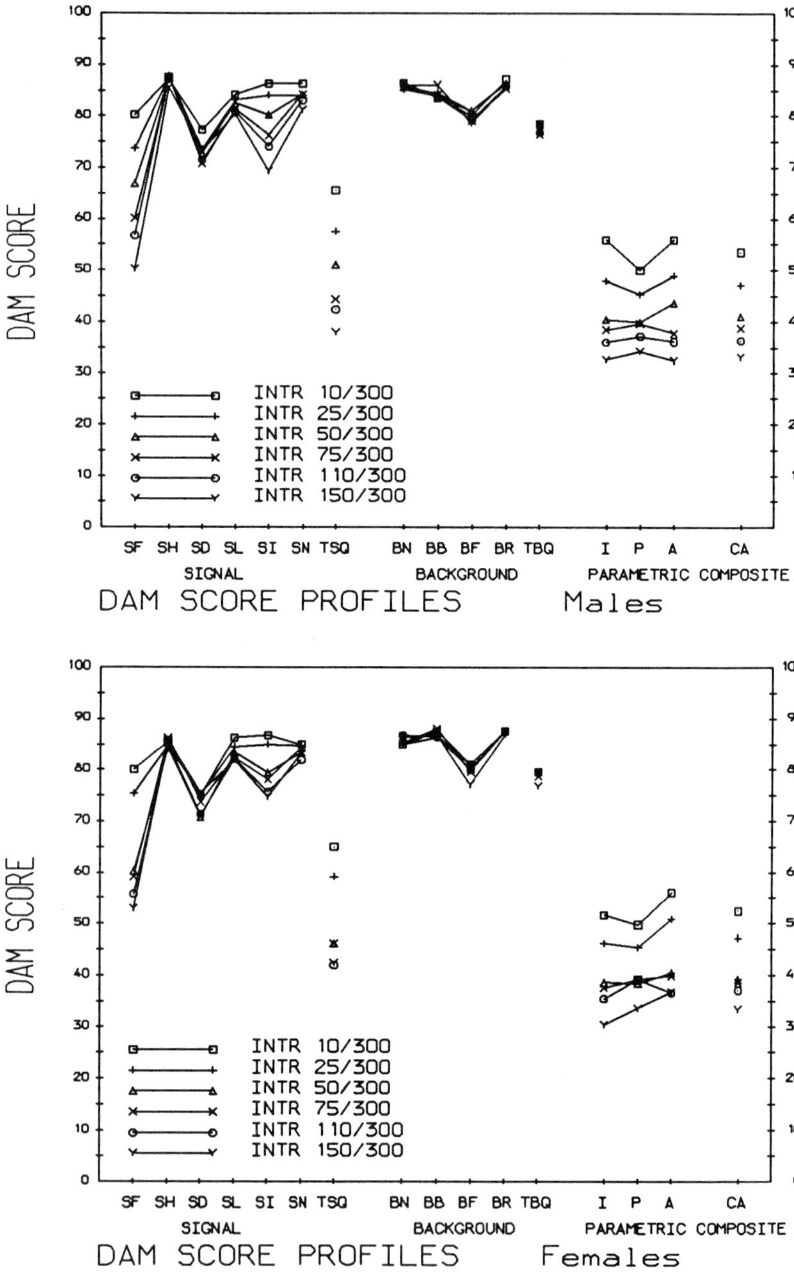

Figure 4.30: Effects of rapid periodic interruption.

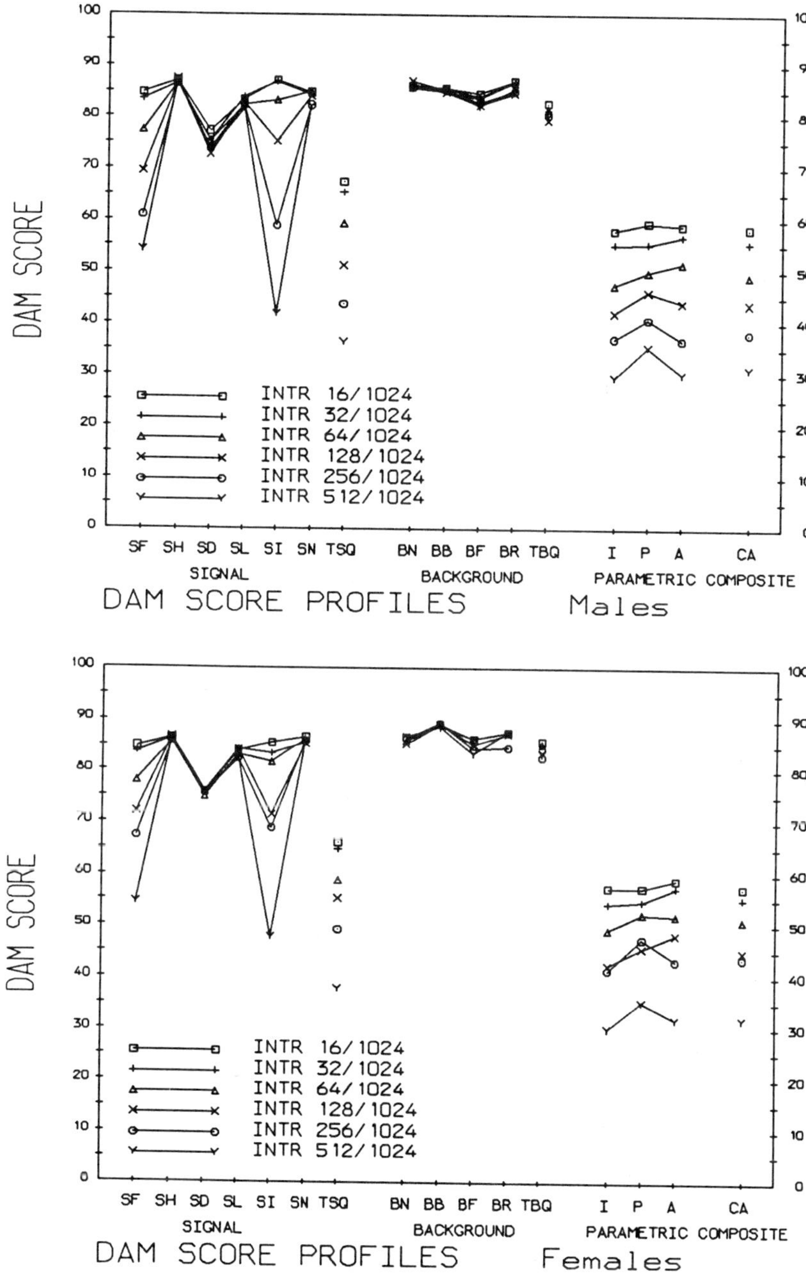

Figure 4.31: Effects of slow periodic interruption.

degradation. However, an experimental artifact is possibly involved in the case of the latter scale. As noted earlier, the BB scale is quite sensitive to 60 Hz hum, which might have been expected to increase as audio gain was increased to compensate for the effects of clipping on signal level.

4.3.2.4 Center Clipping

The effects of peak clipping and center clipping on speech intelligibility have long been known, but no attempt has thus far been made to quantify their effects on acceptability. Licklider [4.33] observed that peak clipping could actually enhance intelligibility under certain circumstances but that center clipping is detrimental to intelligibility under all circumstances. The reasons for this are readily found in the fact that the low energy segments of the speech signal that are removed by center clipping are generally those involving consonant sounds, which are, in turn, the major information carriers in speech.

Since many of the low energy components of speech have energy concentrations in high frequency range of the speech spectrum, one would predict the perceptual effects of center clipping to be considerably more complex than those of peak clipping. Fig. 4.33 shows this to be the case. From the figure it appears first that the perceptual consequences of center clipping are confined completely to perceived signal qualities, i.e., listeners perceive virtually no background effects. Among the six signal qualities, however, the effects of center clipping are quite diverse, with all but one (SH) highly sensitive to this form of degradation. The reason for this diversity of effects is easily understood, once it is recalled that removing the low energy components of speech serves to interrupt and to lowpass the speech.

4.3.2.5 Quantization

Amplitude quantization is an essential step in a number of modern speech coding techniques, although in most cases the quantization is extremely fine. Because the SD scale was originally designed for sensitivity to these techniques it is of some interest to know how the DAM generally and the SD scale in particular respond to relatively coarse quantization.

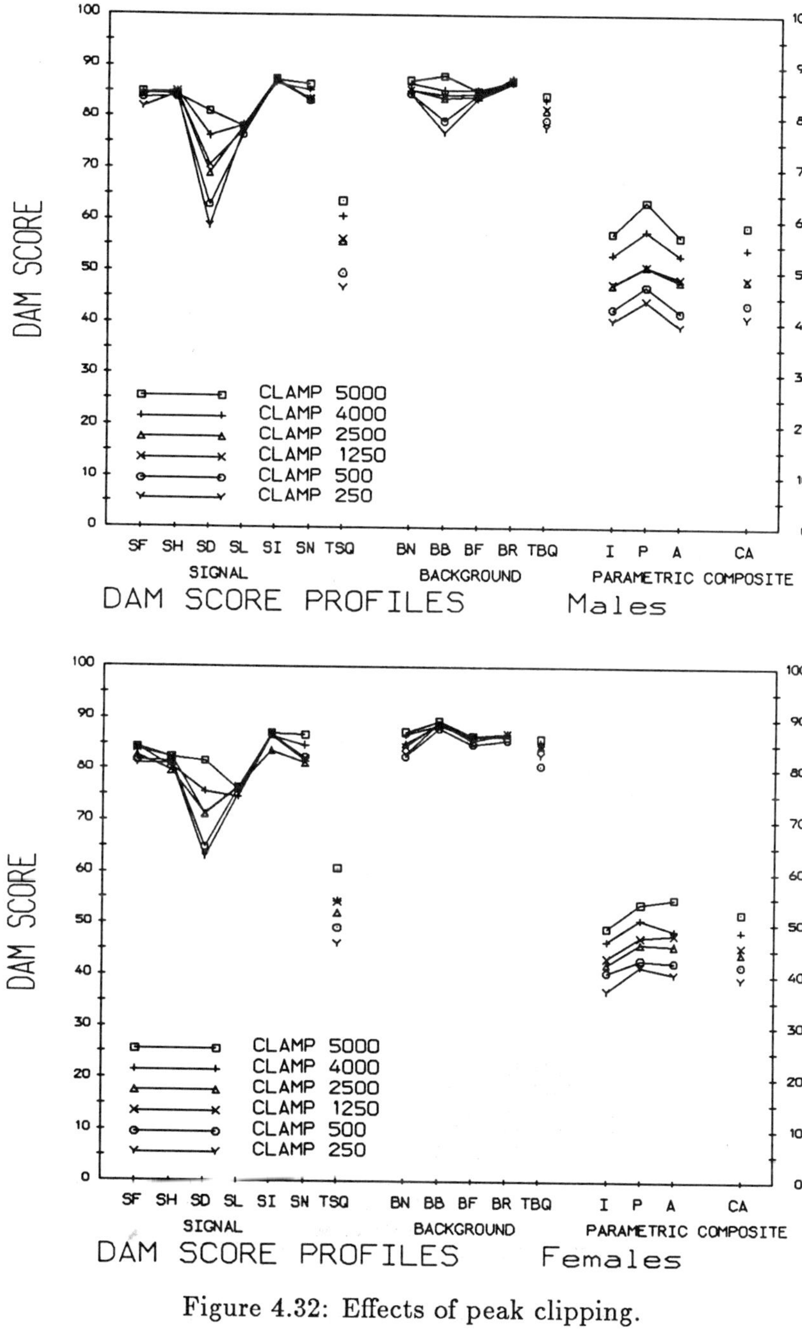

Figure 4.32: Effects of peak clipping.

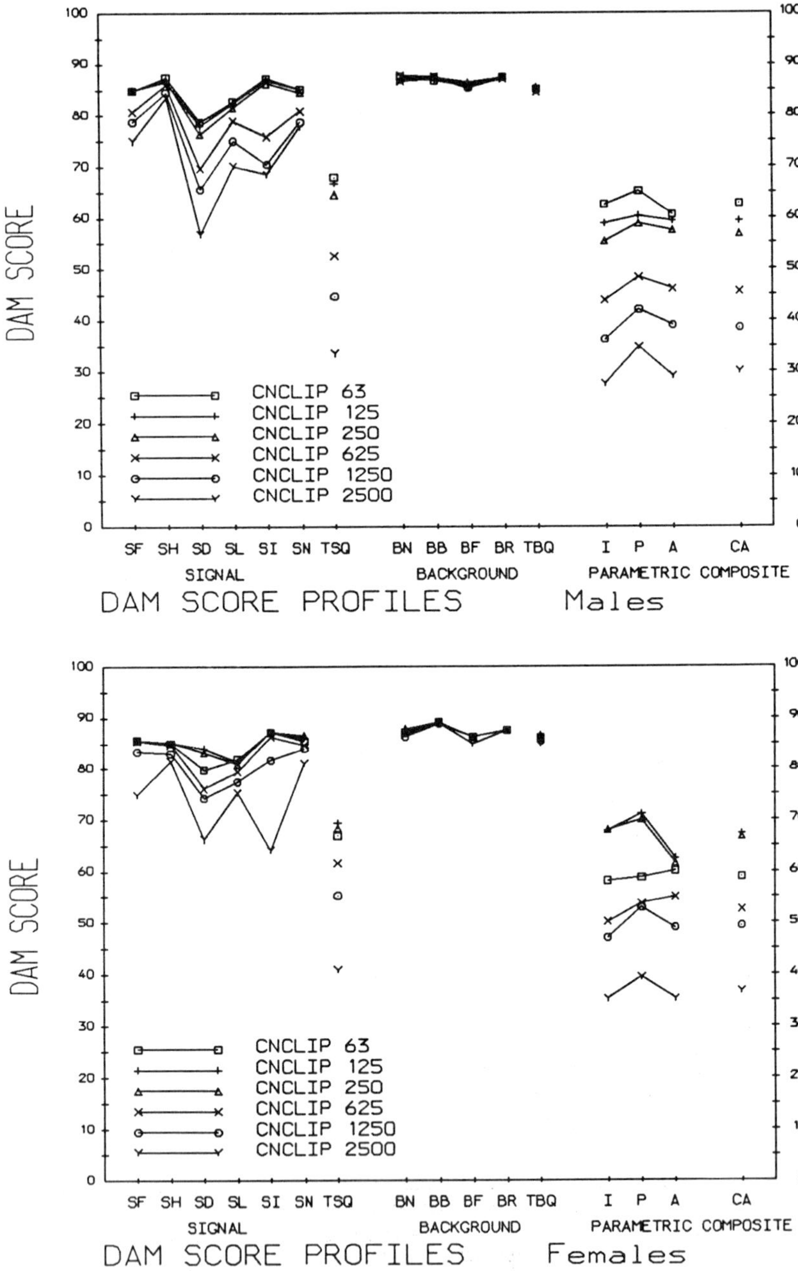

Figure 4.33: Effects of center clipping.

Figure 4.34 shows that the SD scale is extremely sensitive to this form of distortion, and also shows that several background scales are somewhat sensitive. Buzzing (BB) is the background scale most affected, indicating that the low frequency components of quantization noise are perceived as a background process. Of additional interest is the difference between scores for the higher order qualities, intelligibility and pleasantness. Listeners perceive quantized speech to possess relatively high intelligibility, but apparently find it unacceptable on aesthetic grounds, as evidenced by the low ratings they give it on pleasantness.

4.3.2.6 Echo

As noted in Sec. 4.2.2.6 the echoic conditions treated here were somewhat unrealistic, but were selected to ensure a clear subjective effect. As observed elsewhere, the DAM in its present version does not make explicit provision for echo: no single rating scale pertains unequivocally to this phenomenon. From Fig. 4.35, however, listeners were evidently able to find the means of distinguishing between the various delays through a combination of perceptual quality scales. It appears, moreover, that listeners experienced no uncertainty as to whether echo should be characterized as a signal distortion or a background disturbance. They agreed that it should be the former and indicated their perceptions primarily via the SL and SI scales, with the result that all of the higher order perceptual quality scales "tracked" in a consistent manner.

4.3.2.7 Filtering Degradation

The results of research leading to the development of the DAM served in many instances to confirm the principle that even the simplest forms of signal degradations have relatively complex subjective consequences. Although the effects of a given form of distortion may be most pronounced in one subjective dimension, they are usually evident in two or more dimensions. On the other hand, many forms of degradation may have a common perceptual effect, as well as unique perceptual consequences. In fact, perceptual quality scales SH, Sl, and SN were

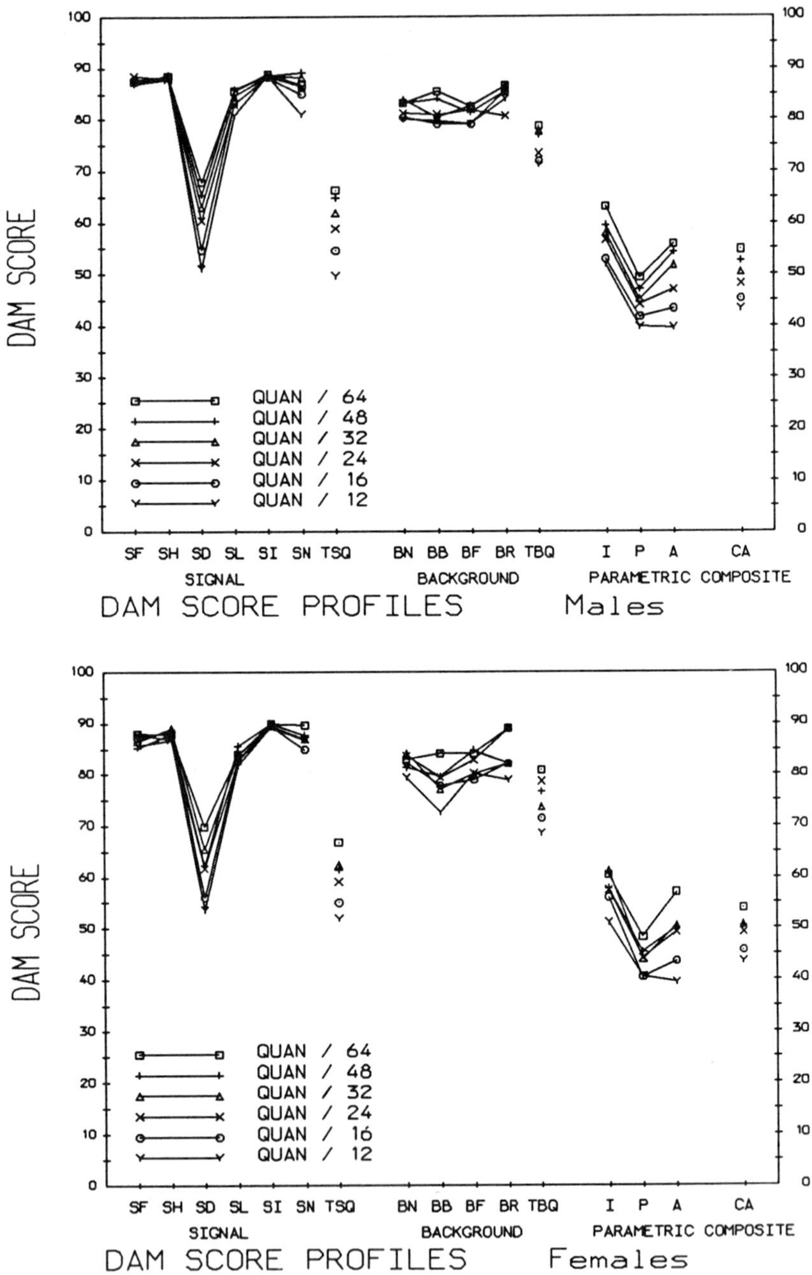

Figure 4.34: Effects of quantization distortion.

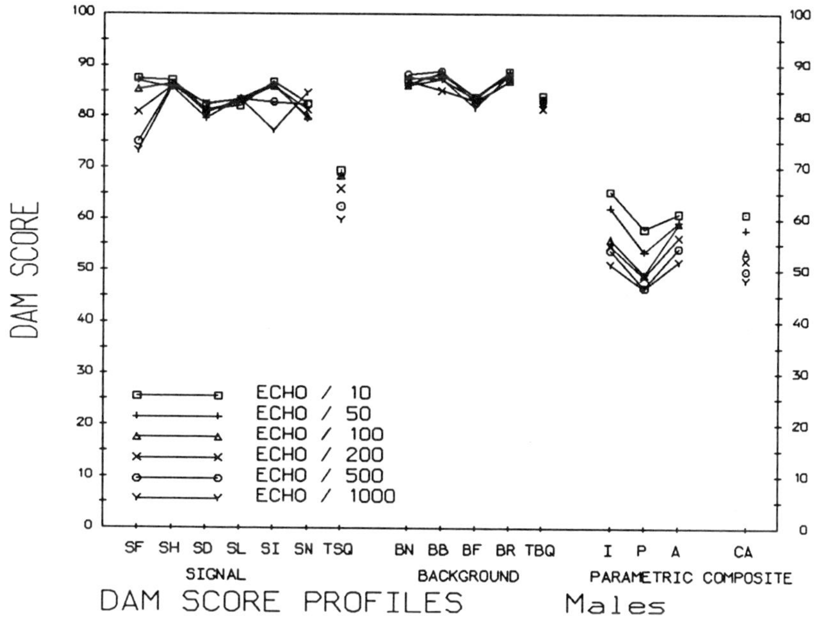

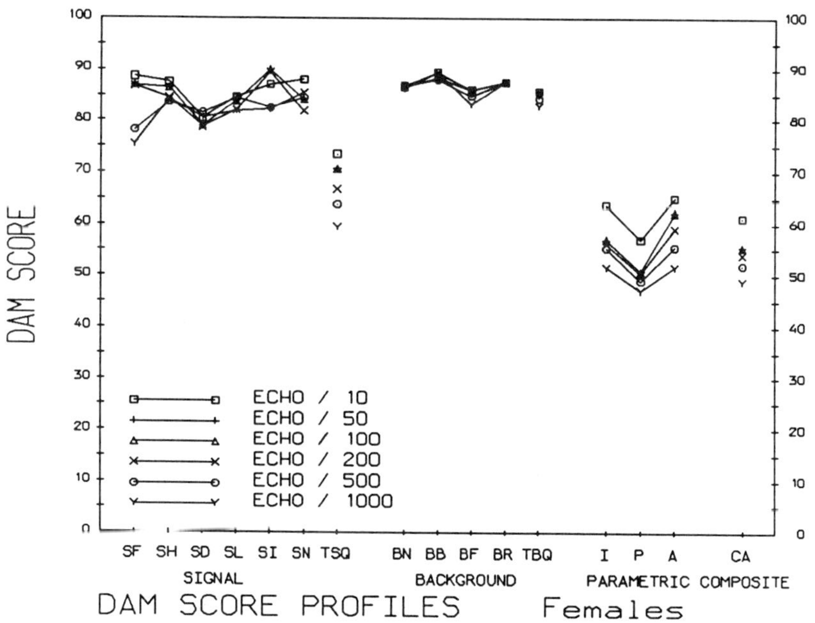

Figure 4.35: Effects of echo distortion.

found to be sensitive in varying degrees to the effects of three major forms of frequency distortion.

All forms of passband restriction have previously been observed to affect the SL scale, which is associated with the perceptual qualities of muffledness, dullness, etc. The effects of high frequency attenuation were found to be confined primarily to this DAM parameter, though some depression of scores on the SH scale was observed with extremely severe high frequency attenuation.

The effects of low frequency attenuation, such as highpass filtering, were observed to be most pronounced in the case of the SH scale, which was in fact designed primarily to sense such effects. However, the SL scale was also observed to be sensitive to highpass filtering in lesser degree. A third scale, (SN) which is concerned with the perceptual quality of nasality, was found to be sensitive to passband width restriction more or less without regard to the location of the passband. Present results will be seen to provide strong confirmation of findings of the earlier validation studies of the DAM, though sharper filtering was achieved here than previously. In the present investigation, frequency filtering was achieved by means of elliptic digital filters with 40 dB or less ripple and transition bandwidths equal to 5 percent of the nominal cutoff frequencies.

Bandpass Filtering Figure 4.36 shows the effects of bandpass filtering on DAM score patterns. Consistent with previous findings, three of the primary perceptual quality scales (SH, SL, and SN) are sensitive, but in different ways, to this form of signal distortion. Differences between the trends of the SH scores and SL scores are best rationalized in terms of the character of the rejected band(s) associated with each condition. To the extent that high frequency rejection predominates, SL scores suffer greatest reduction, whereas SH scores reflect the predominance of low-frequency rejection. Scores on the SN scale vary in a more complex manner with the location of the passband, being highest for the high and low extremes, lowest for those passbands near the middle of the frequency scale, in particular those covering the frequency range of the second formant. Generally, the scales pertaining to background qualities are little affected by passband restriction. The one case in which a background exhibits depression (BB scale in the case of the 2600-3400 Hz condition) is quite possibly due to an increase in hum associated with the higher gains needed for the high frequency bands.

Lowpass Filtering From Fig. 4.37 it is evident that the effects of lowpass filtering are confined primarily to the SL scale, a result consistent with the purpose for which this scale was designed. Although some variation in other signal quality scales is evident, no consistent trends emerge. None of the background quality scales are sensitive to this form of degradation. It is interesting to note that neither SL nor overall acceptability are substantially affected until attenuation of frequencies below 2 kHz occurs. The fact DAM evaluations of intelligibility are more uniformly affected by high frequency attenuation is consistent with results on the effects of high frequency attenuation on objectively measured intelligibility. It is perhaps consistent with common intuition that parametric pleasantness is the least affected of the higher order qualities.

Highpass Filtering Four perceptual quality scales appear sensitive to highpass filtering as shown in Fig. 4.38. As expected, the SH scale ultimately exhibits the greatest depression, but two other scales, SL and SN, are more sensitive to moderate degrees of highpass filtering. Only after frequencies as high as 800 Hz are attenuated does the signal appear to acquire the characteristic "highpass" quality.

Again, a decrease in BB scores with increased highpass filtering is possibly an experimental artifact. The fact that no comparable trend in BB scores is evident in the case of the female speaker adds credibility to such an explanation.

4.3.2.8 Narrowband Additive Noise

Figures 4.39–4.41 permit a comparison of the effects of noise in six different frequency regions on DAM score patterns. (Due to the large number of narrowband distortions, and because the scores for the males and for the female are similar, only the average scores for the males is shown for this and all other narrowband distortions.) In the lowest band, depressed scores on the BN (background noise) and BR (background rumble) scales conform with intuitive expectations. However, the reason for somewhat depressed scores on several signal quality scales and on the scale for total signal quality is not clear. As the

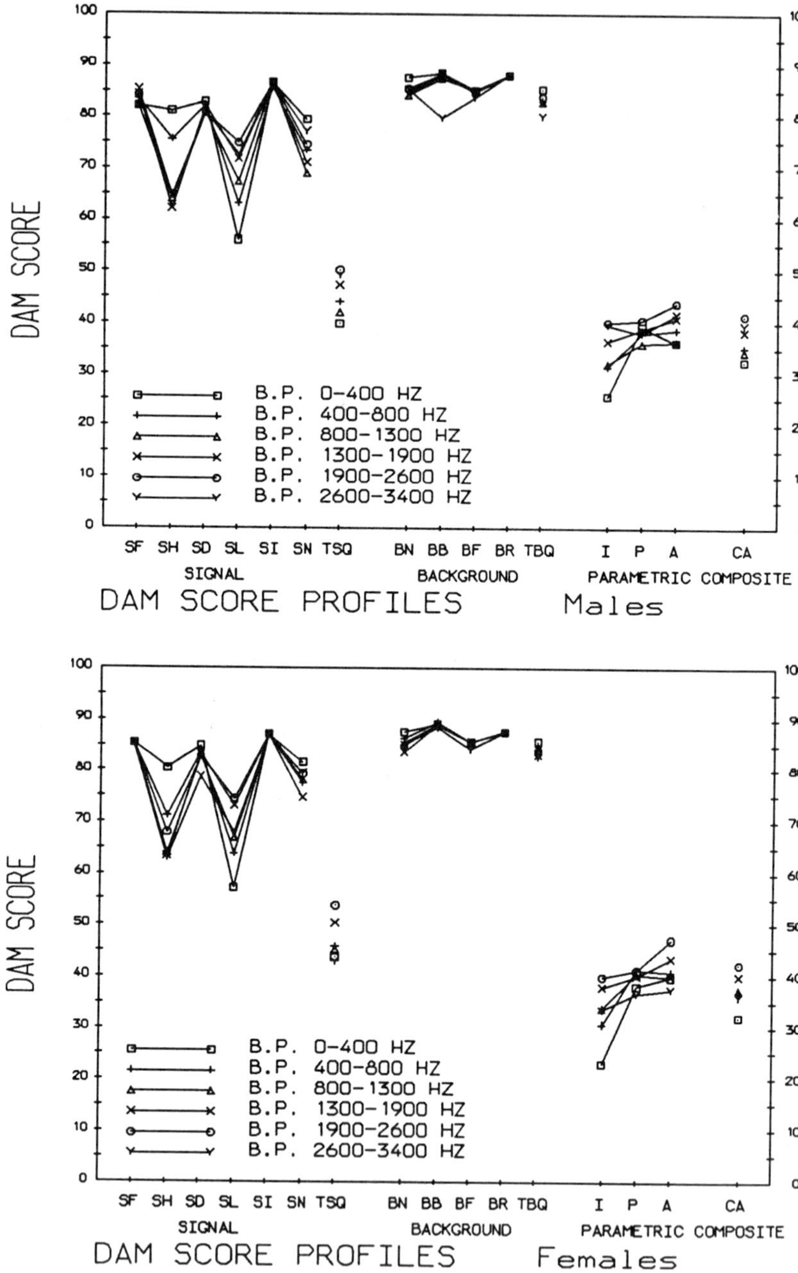

Figure 4.36: Effects of bandpass filtering.

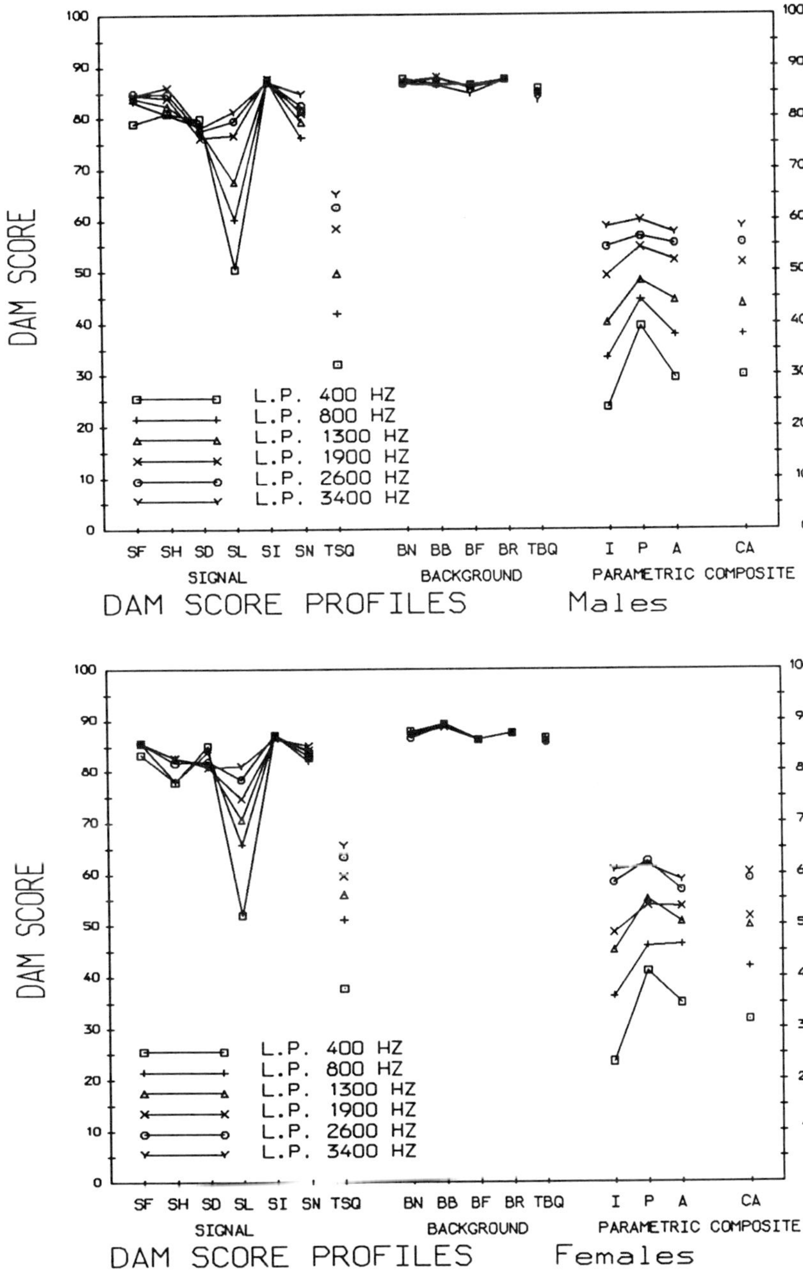

Figure 4.37: Effects of lowpass filtering.

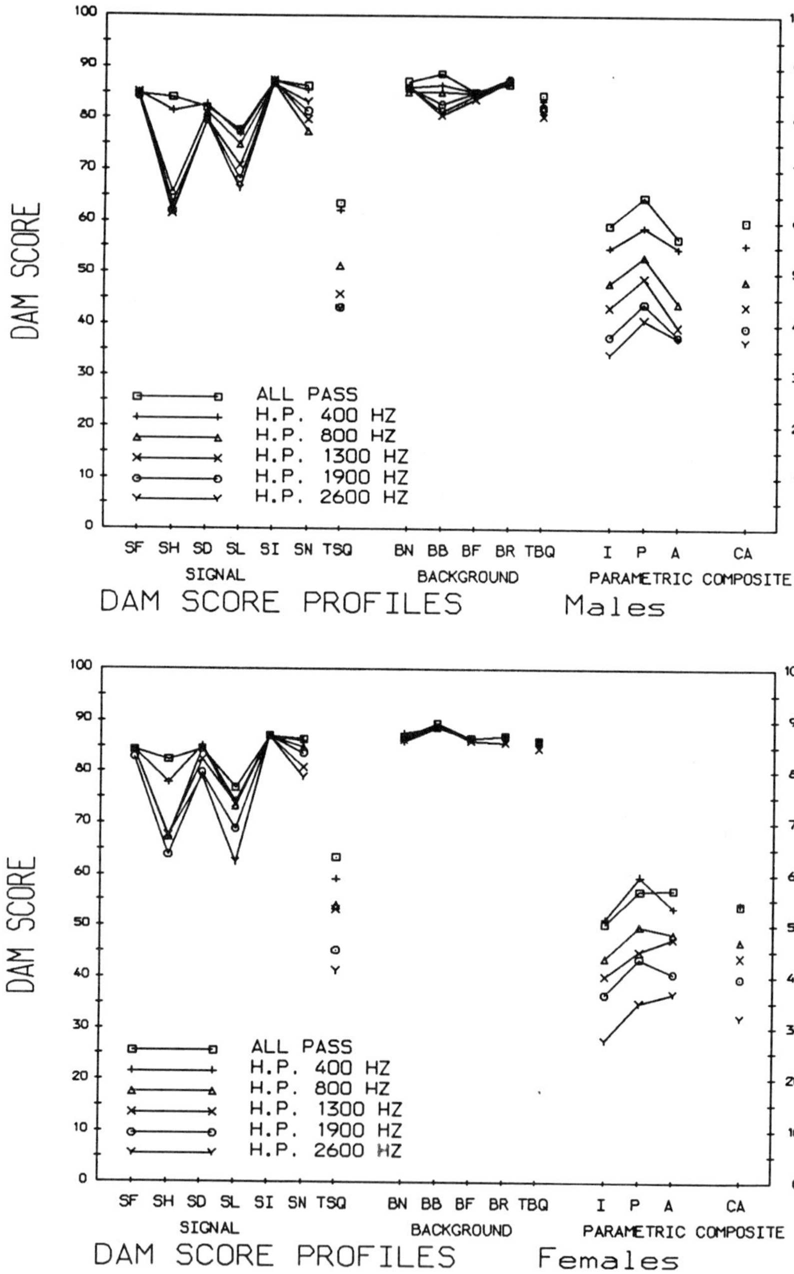

Figure 4.38: Effects of highpass filtering.

noise band is raised above 400 Hz, the quality "background rumble" decreases significantly.

As the frequency region of the noise band is increased beyond 1300 Hz, the perceptual consequences of the noise undergo several qualitative changes. The noise is perceived to have less "rushing-roaring" quality (BN) but more of a "buzzing-humming" quality as reflected in scores on the BB scale. At the highest noise levels listeners also tend to perceive an increasingly raspy (SD scale) quality which is most typical of amplitude distorted speech.

4.3.2.9 Pole Distortions - I

The first type of pole distortion considered involves distortion of pole frequencies within a given frequency band; the second, involves radial and, hence, bandwidth distortion. The final type involves both angular and radial distortions, and was adjusted to more fully excite the parametric scales of the DAM.

Angular Pole Distortion Figures 4.42–4.44 show the effects of pole distortion in each of the six frequency bands. The parameter which varies with each distortion level is the magnitude of frequency distortion, F, values of which are given in Table 4.26

It appears that the subjective effects of pole frequency distortion are expressed primarily via the SF (signal flutter) and BF (background flutter) scales. The remaining perceptual quality scales are virtually unaffected by this form of degradation. It appears, further, that the perceptual consequences of pole distortion are generally negligible in the upper and lower extremes of the 3.4 kHz band involved here.

Radial Pole Distortion Figures 4.45–4.47 show the effects of radial pole distortion. The parameter in all figures is relative "radius jitter," B, whose values are given in Table 4.26 From the figures it is evident pole bandwidth distortion has qualitatively different perceptual consequences than pole frequency distortion. The perceptual effects of radial or bandwidth distortion are confined primarily to the background and are quite pronounced in all but the highest frequency band.

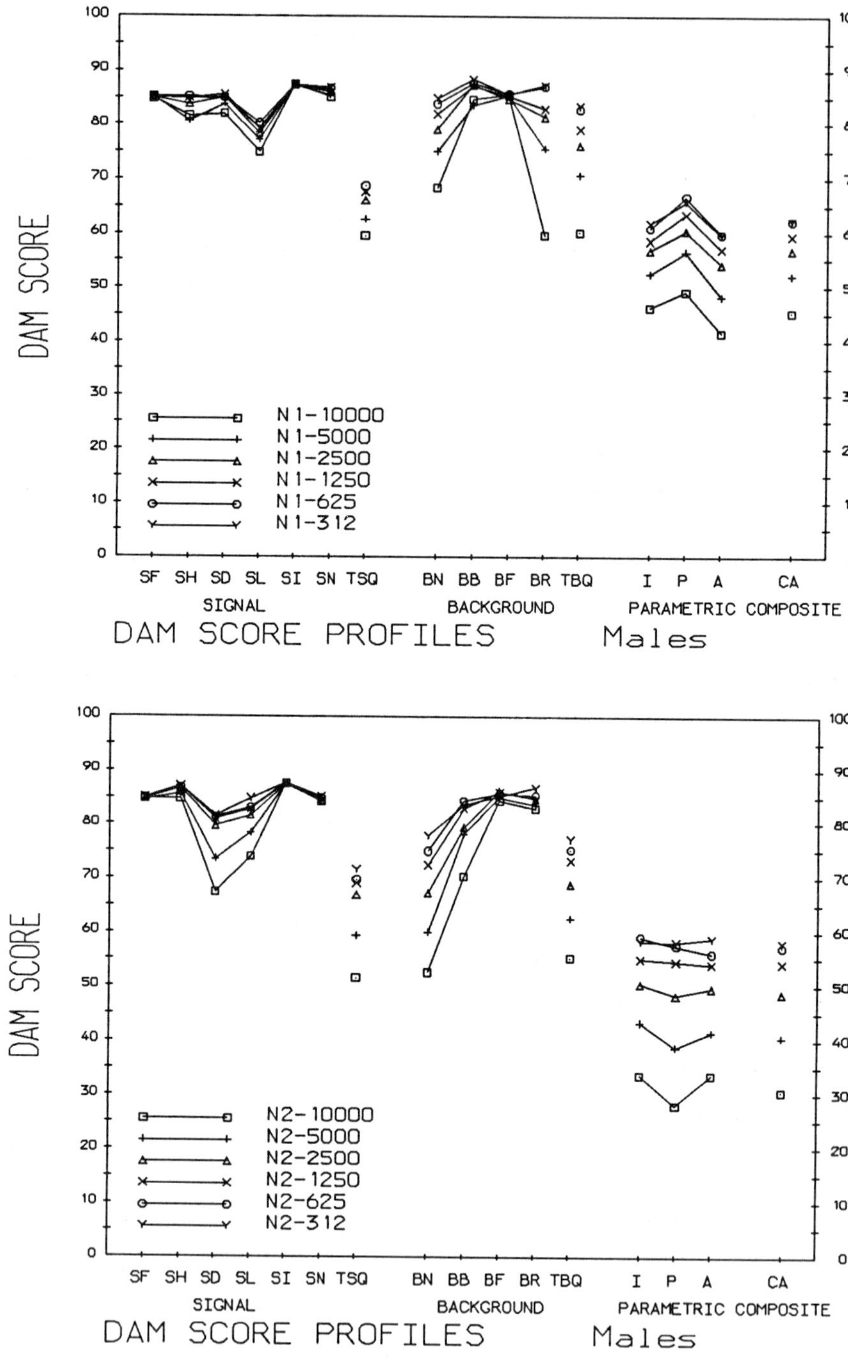

Figure 4.39: Effects of narrowband noise, males - 1.

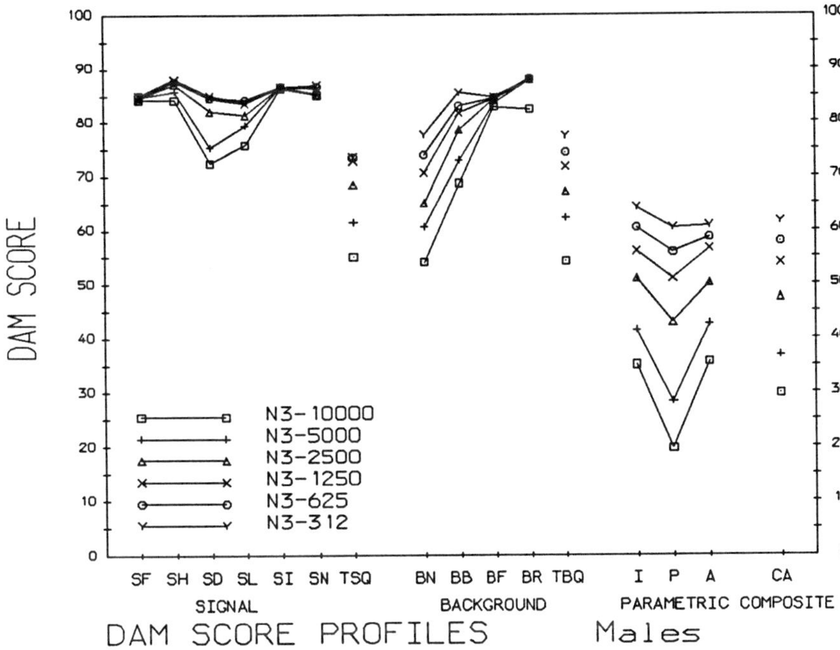

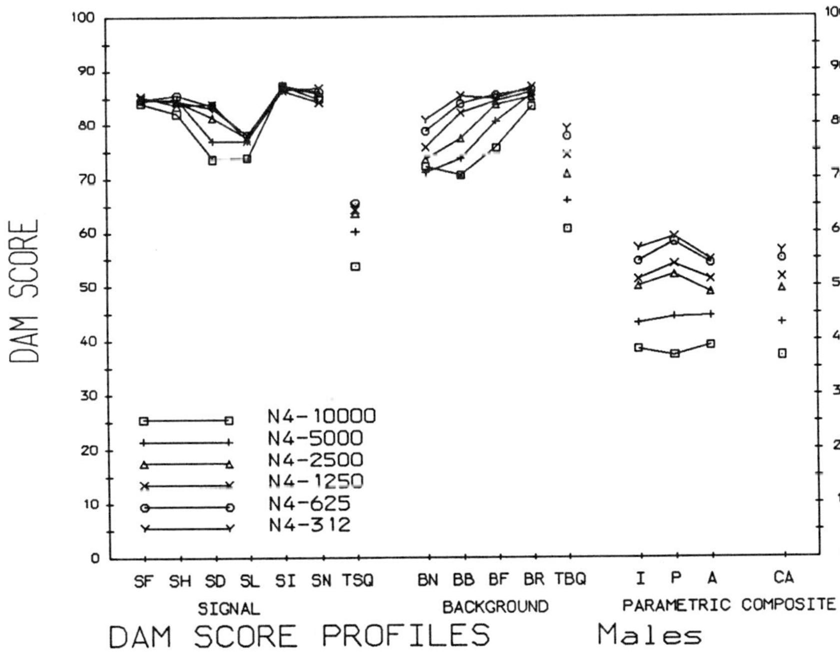

Figure 4.40: Effects of narrowband noise, males - 2.

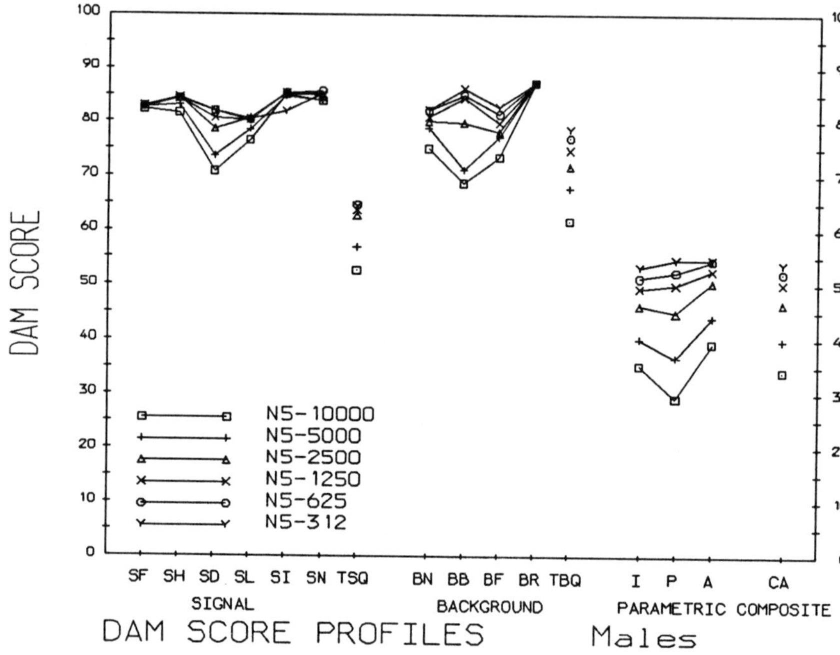

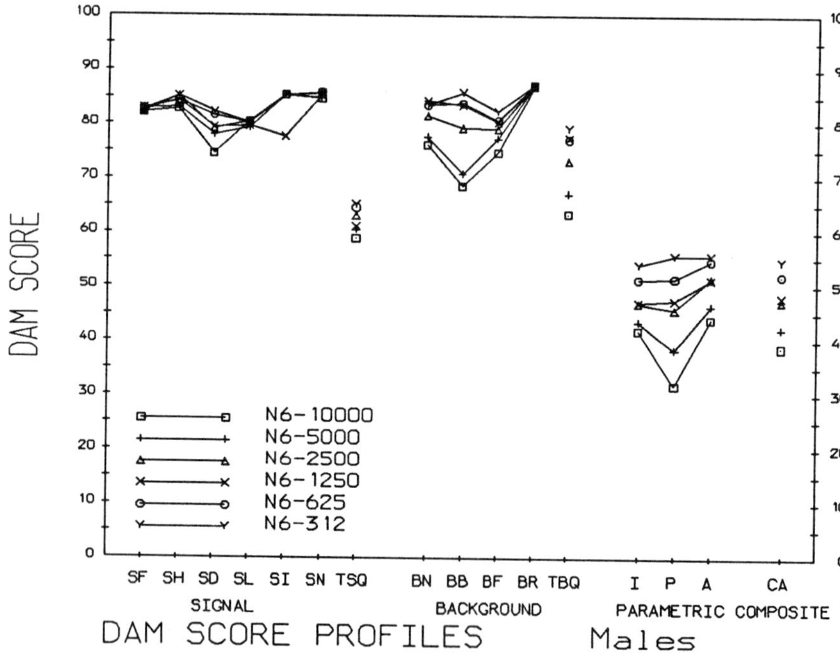

Figure 4.41: Effects of narrowband noise, males - 3.

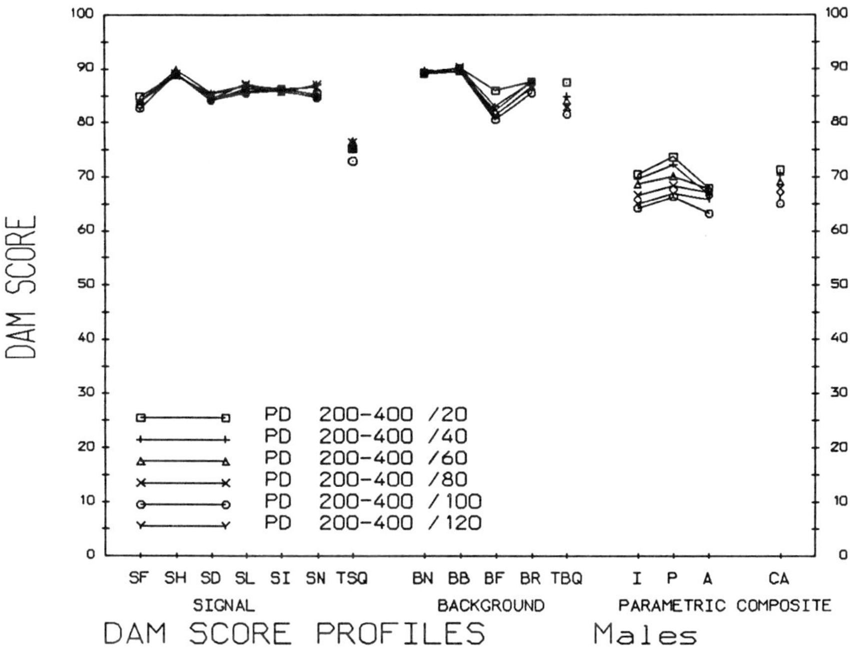

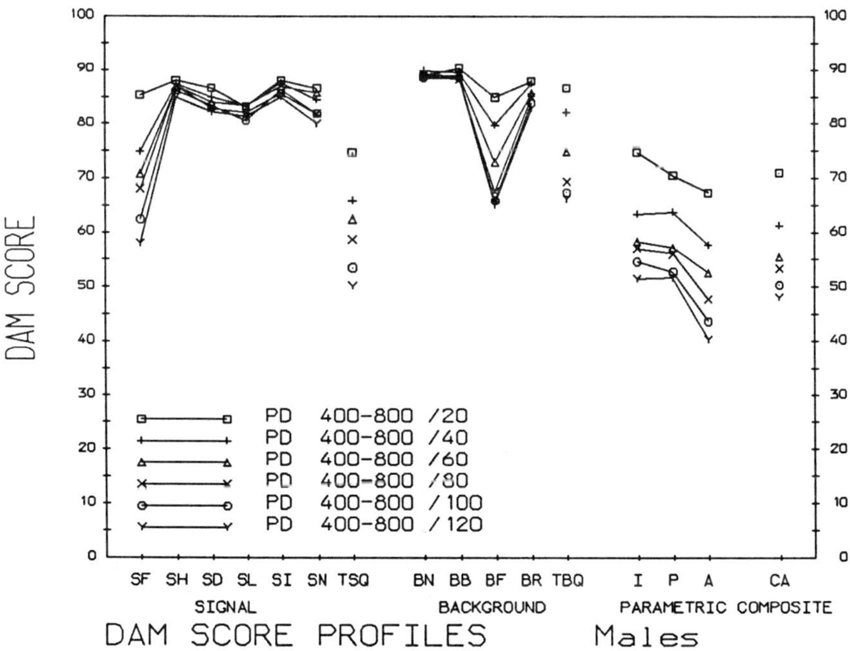

Figure 4.42: Effects of Angular Pole Distortion, Males - 1.

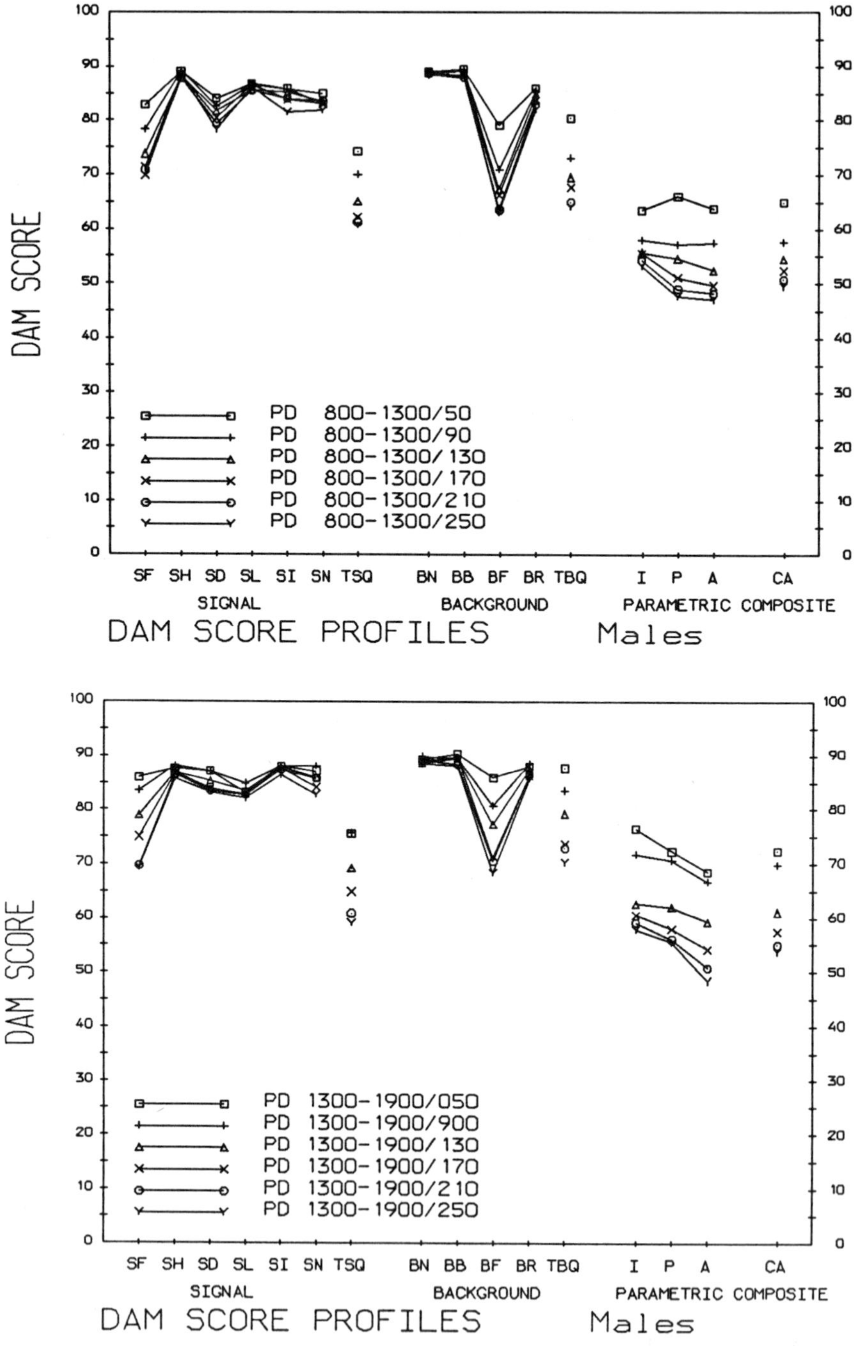

Figure 4.43: Effects of Angular Pole Distortion, Males - 2.

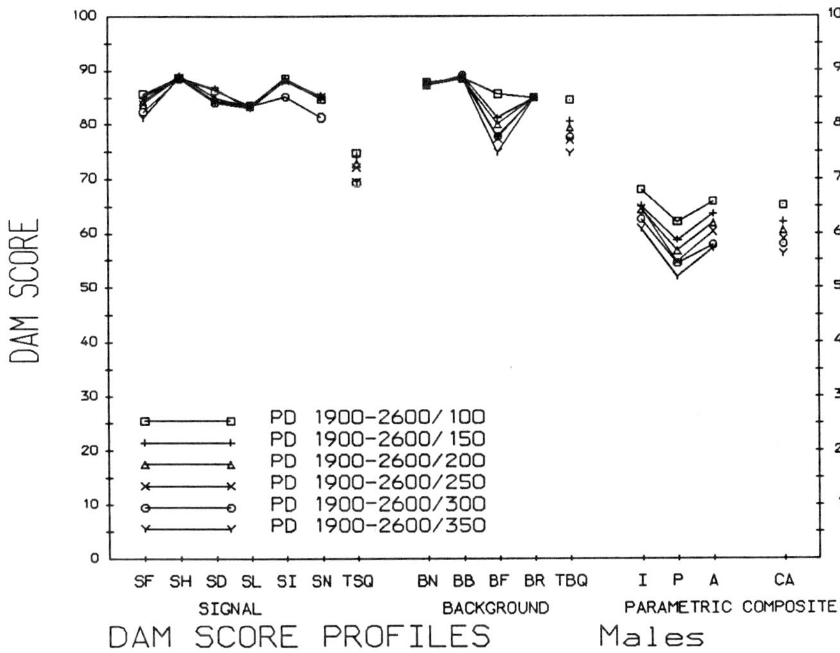

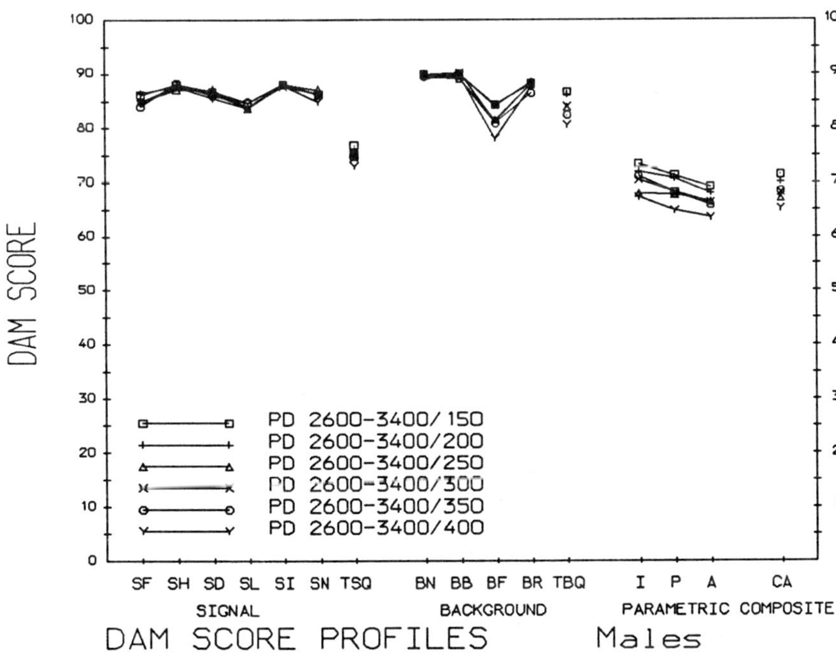

Figure 4.44: Effects of Angular Pole Distortion, Males - 3.

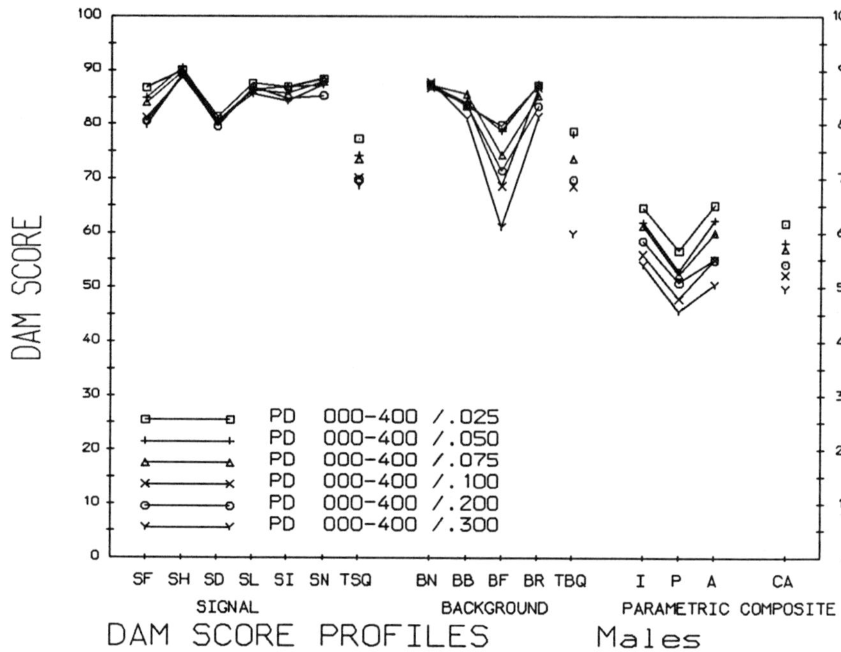

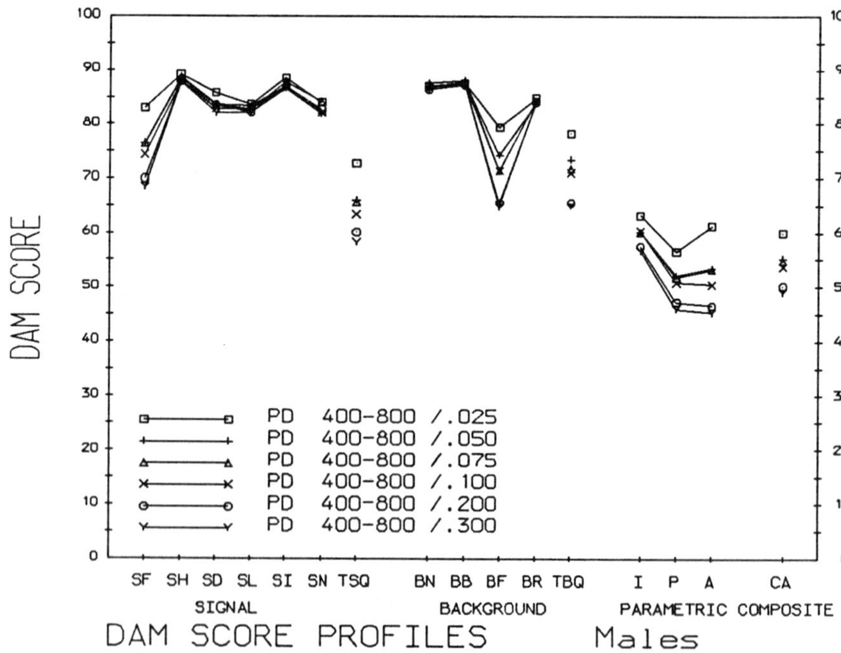

Figure 4.45: Effects of radial pole distortion, Males - 1.

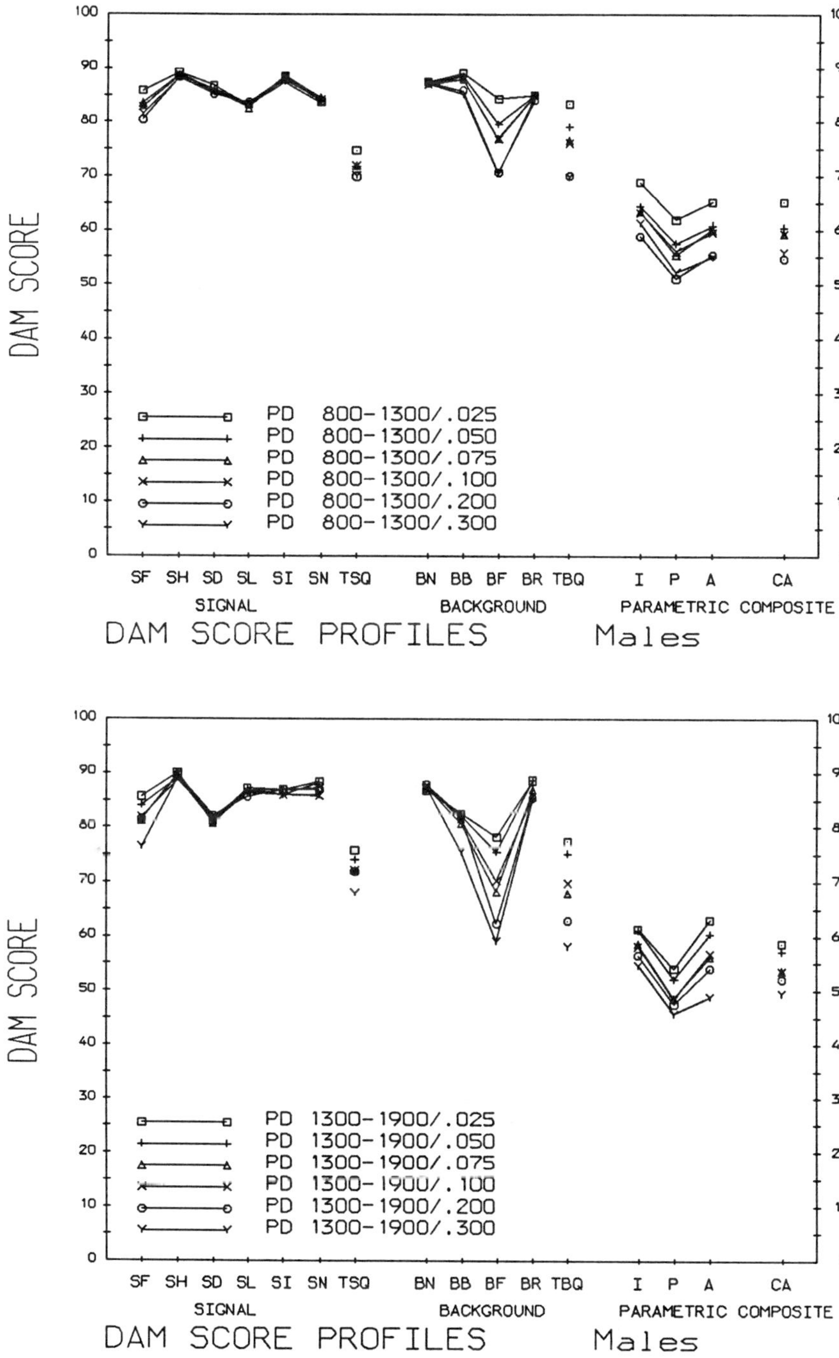

Figure 4.46: Effects of radial pole distortion, Males - 2.

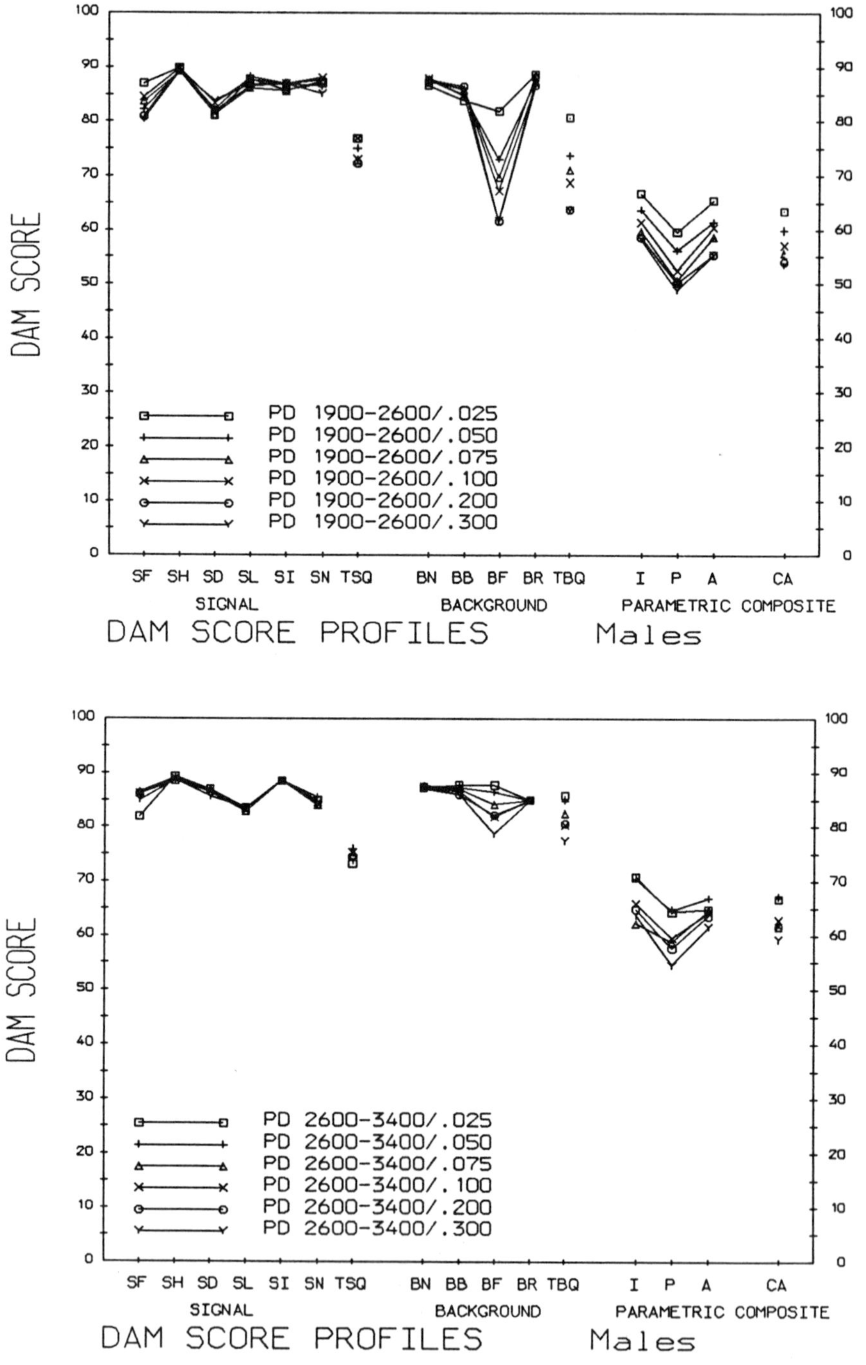

Figure 4.47: Effects of radial pole distortion, Males - 3.

4.3.2.10 Pole Distortions - II

Figures 4.48–4.49 show the effect of frequency variant pole distortion for the 0–420 Hz, 420–900 Hz, 900–1600 Hz, and 1600–3200 Hz bands. From these figures, it is clear that, for all frequency ranges, the scales which are most dramatically affected are SF (signal fluttering) and BF (background fluttering). Hence, the effect of quantizing the pole locations for LPC analysis can best be characterized as "fluttering" and "chirping." It is also clear that all frequency bands result in an acceptably wide range of perceived distortions.

4.3.2.11 Narrowband Frequency Distortion

Narrowband frequency distortion is of interest in relation to transform coding techniques, where noise may be produced in the power spectrum domain. To investigate such distortion, six levels of noise were produced in each of six spectral bands (See Sec. 4.2.2.10).

From Figs. 4.50–4.52 it appears that narrowband frequency distortion in the range treated here has relatively minor subjective consequences. In all but the lowest frequency band involved, perceived background flutter is the most pronounced effect.

Some amount of signal flutter (SF scale) and raspiness (SD scale) is evident in the 400–800 Hz and 800–1300 Hz bands, but these qualities are negligible in the remaining bands.

4.4 Testing Objective Measures

4.4.1 Testing

As mentioned in Chap. 1, speech perception as it applies to both speech understanding and speech quality evaluation is a poorly understood processes. Objective measures which attempt to model such largely unknown processes are therefore intrinsically limited in their ability to estimate speech quality. Simple models of perception as they relates to speech quality lead to objective measures which are typically inadequate for representing the true psychophysical dimensions of speech perception. However, despite such limitations there are still strong reasons to believe that good objective measures of speech quality can

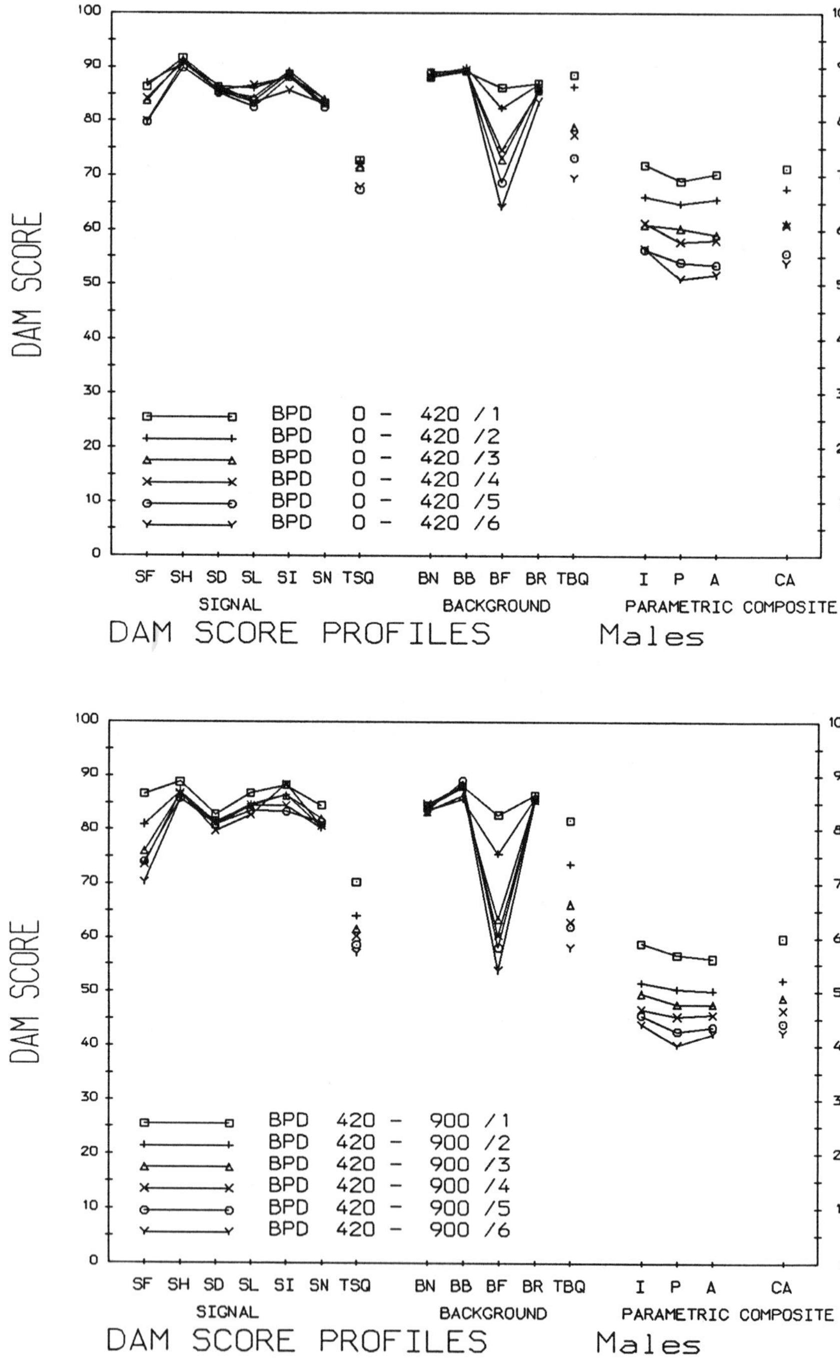

Figure 4.48: Effect of pole distortion - II, males - 1.

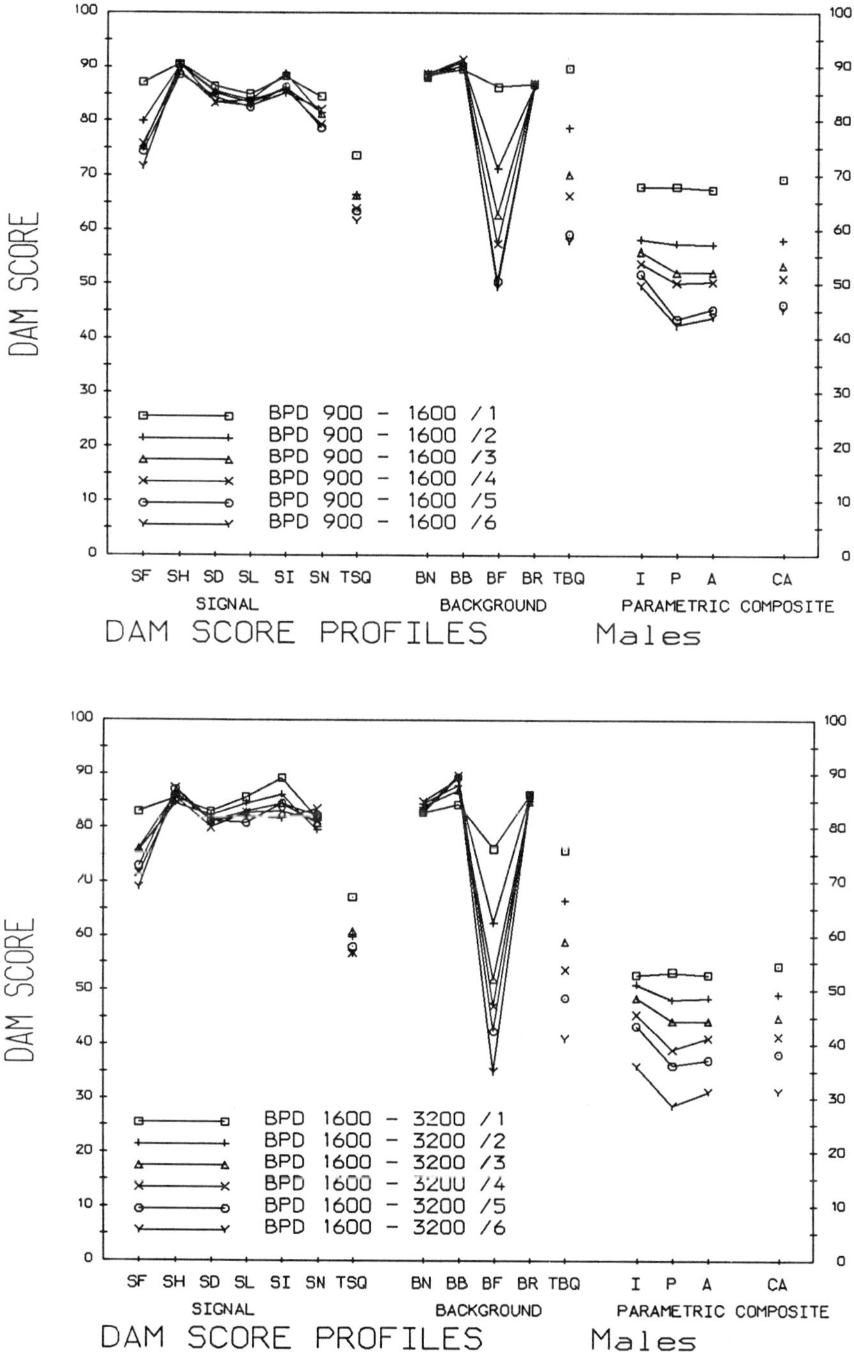

Figure 4.49: Effect of pole distortion - II, males - 2.

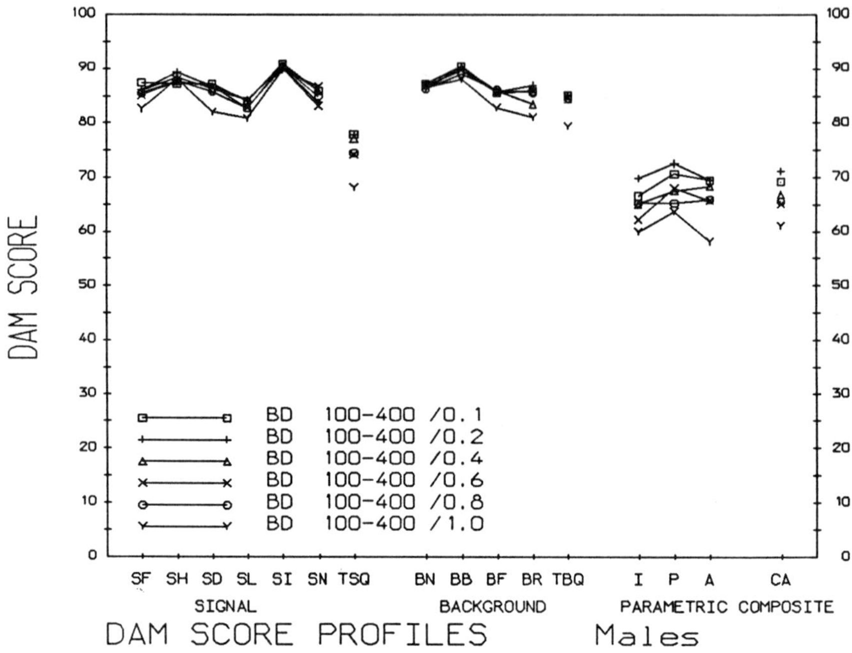

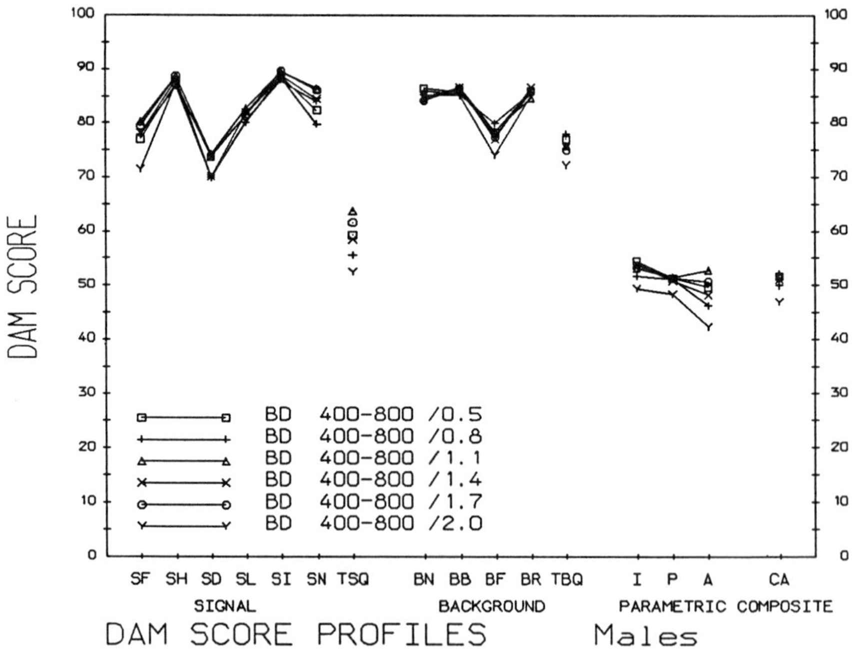

Figure 4.50: Effects of frequency distortion, males - 1.

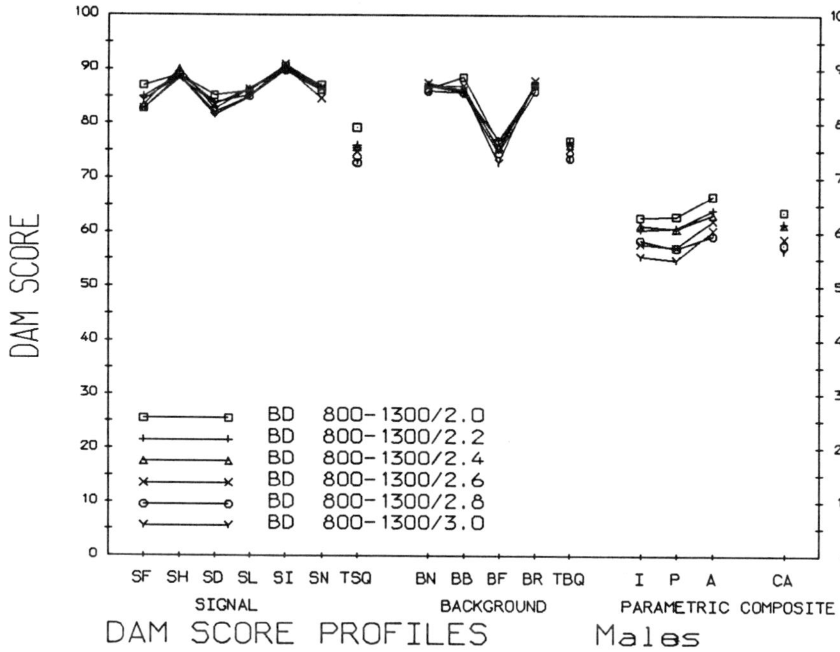

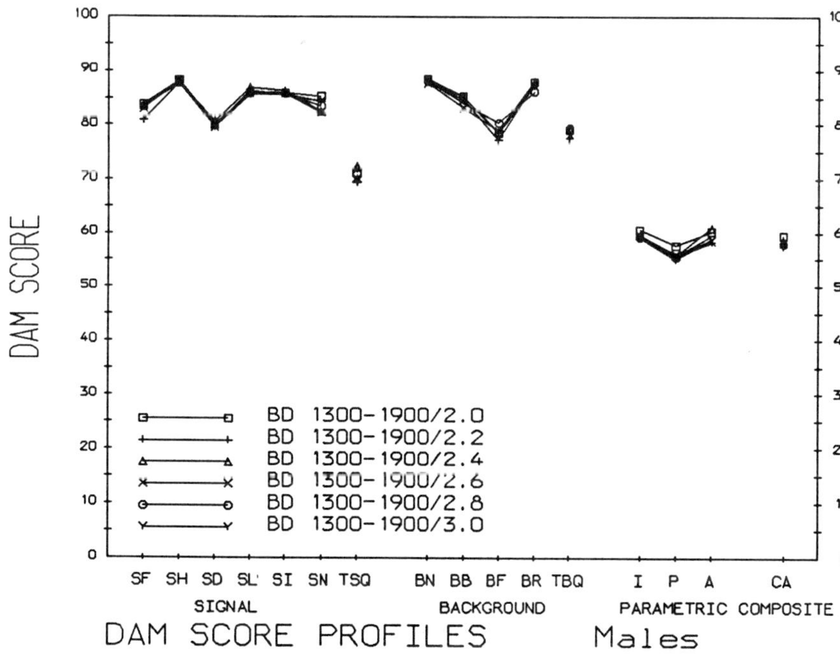

Figure 4.51: Effects of frequency distortion, males - 2.

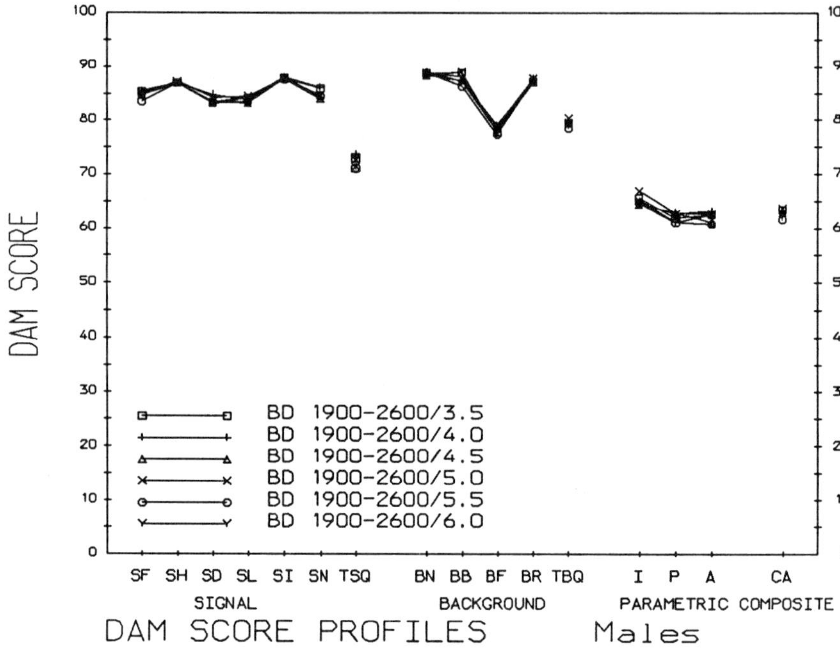

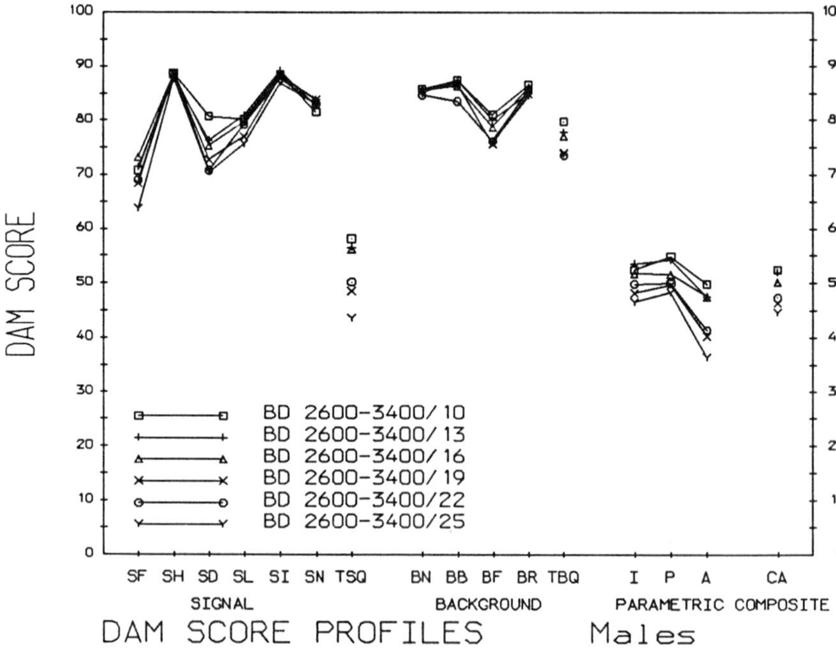

Figure 4.52: Effects of frequency distortion, males - 3.

be developed. This section presents a methodology for the design and testing of objective measures of speech quality.

Figure 4.53 shows a block diagram of a method for objective measure testing. In this diagram, the steps needed to produce the distorted speech and subjective quality data bases have been described earlier in this chapter. Designing an objective measure requires two additional steps, which must be repeated for each measure. The first step is to compute a set of objective measure variables which are some function of the original and the distorted speech signals. They are referred to as "variables" because they are used as the independent variables of a regression analysis. For each distorted speech signal, a set of objective measure variables is produced, where the set may have as few as one member. The challenging part of the objective measure design is that this set of variables must, in some sense, express the perceptual distance between the original speech and the distorted speech. The variables may be computed from the waveforms directly, or from some transformation of the waveforms, such as their linear prediction coefficients, or their short-time spectral representations. Furthermore, after their computation, the objective measure variables may themselves be transformed to yield new objective variables. Such transformations as simple as taking the log of a variable or raising it to a power, or as complex as taking products of variables to yield new variables.

The second step is to use multiple linear regression analysis to find the regression weights which maximize the correlation between the scores of a specified DAM subjective quality scale and the corresponding objective quality estimates. Because the research results reported here represent work done both over a considerable time span and by different individuals, several regression analysis programs were used. The results presented in Chaps. 5, 6, and the first section of Chap. 9 used in-house software running in the Georgia Tech Digital Signal Processing Laboratory. The results presented in Chaps. 7, 8 and 9 were obtained using the 1981 edition of the BMDP Statistical Software package, running on the Georgia Tech Control Data CYBER 180/855.

In the regression model, estimates of subjective quality are produced from a weighted sum of the objective measure variables, where the weighted sum is called an objective measure. This regression analysis is done across all distorted speech signals in either the original

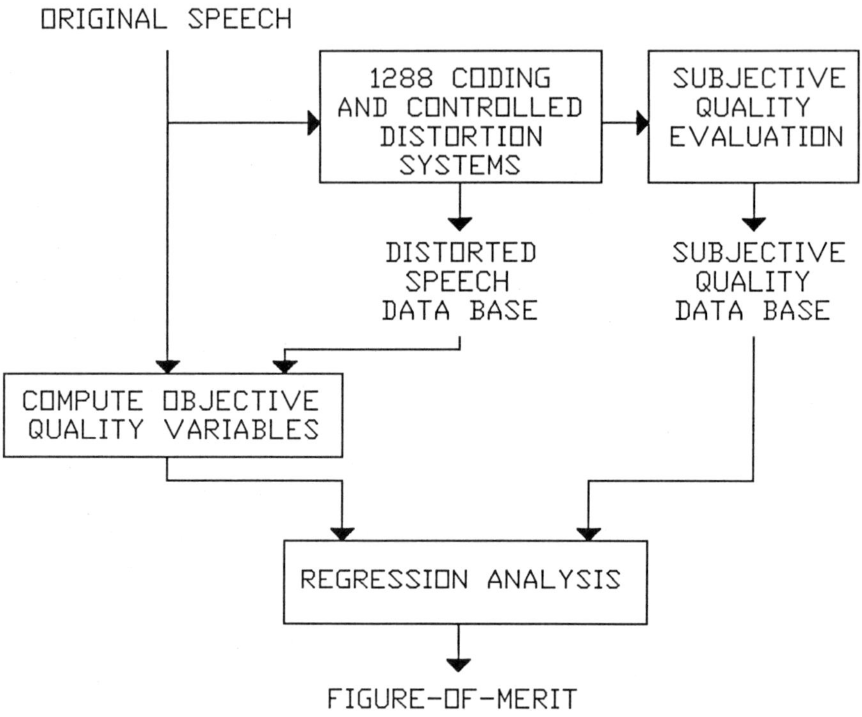

Figure 4.53: Technique for measure testing.

or the extended distorted data base. Chaps. 6, 7, 8, and 9 report results involving only the original data base, while the results of Chap. 5 involve the entire, extended data base. If the regression analysis produces valid results (that is, no assumptions of the analysis are violated), the magnitude of multiple R, or the correlation coefficient of the analysis, serves as a figure-of-merit for the objective measure. The figure-of-merit permits comparison of objective measures which estimate the same subjective quality and permits identification of the best of those measures. It also quantifies the absolute degree of success the objective measure has in estimating subjective quality.

4.4.2 Validation

Figure 4.54 shows a block diagram for the *cross-validation* method of testing an objective measure. The key point in this test procedure is that data used in estimating the measure's regression weights must not be the same data used in computing the measure's figure-of-merit. In this way the resulting figure-of-merit indicates the success the measure would have if it were applied to new samples of speech distortion. The data used to estimate the regression weights is called the training set, while the data used to to compute the figure-of-merit is called the validation set. The data is split nearly in half by randomly assigning each system to either the training data set or the validation data set. The random assignment assures an impartial division of the data, while the large size of the data set assures that the training and validation data sets will be statistically similar.

The *jackknife* method is another means of testing objective measures. In this method, the training data set is equal to the entire data set except for the removal of a single data point. The regression model is computed for the training data set and is used to predict subjective quality for the single data point not in the training set. This process is repeated N times, where N is the number of points in the data base. Each point therefore has associated with it a predicted value of subjective quality which is independent of the estimates of the regression coefficients. In this respect it is a "fair" test of the model's ability to predict speech quality. The objective measure's figure-of-merit is equal to the magnitude of the coefficient of correlation between the predicted subjective quality and the actual subjective quality. Since $N-1$ points in the data base are used to predict the single remaining point, the figure-of-merit in the jackknife method is generally higher than the that produced by the cross-validation method.

The jackknife validation procedure actually produces N different regression models and N sets of regression coefficients, since each prediction of subjective quality is associated with a slightly different training data set. However, for a large data set such as the speech quality data base, the difference between each of these models and the model obtained by using all the data (with no points removed) is usually statistically negligible. This assumption can be checked by computing a simultaneous 95 percent confidence interval for the regression coef-

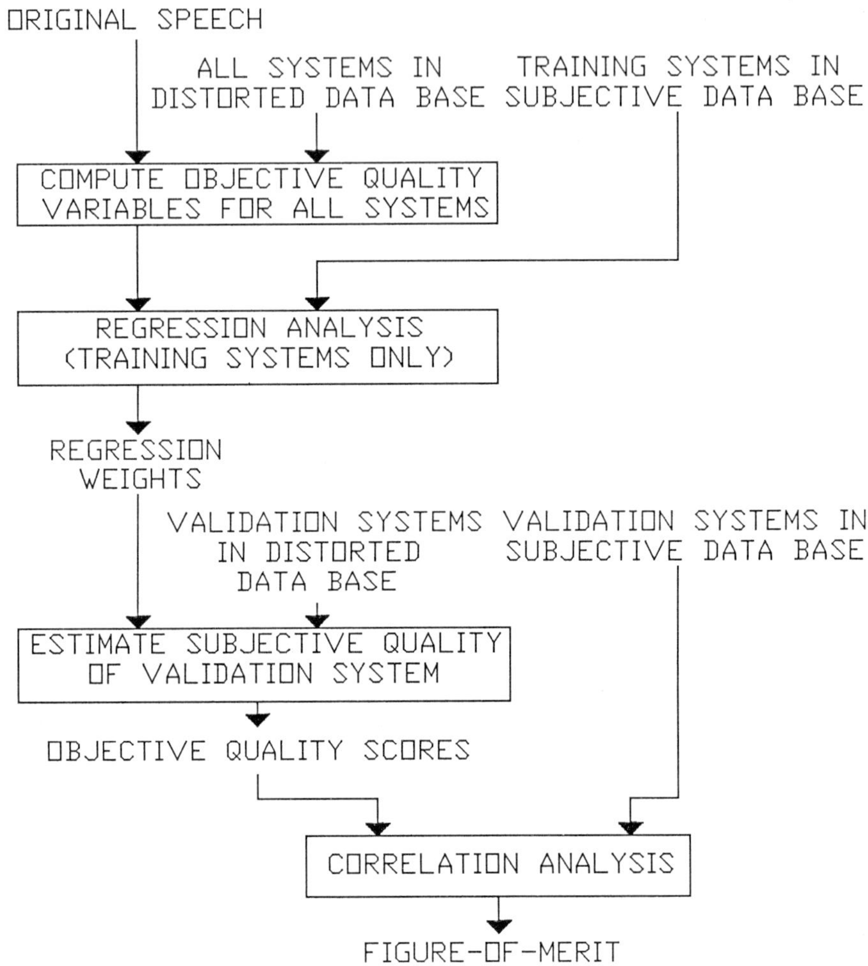

Figure 4.54: Technique for cross validation.

ficients obtained in modeling the entire data set. The assumption is violated if the removal of a data point, as is done in forming the training data sets, causes the training set regression coefficients to move outside of the confidence interval specified by the entire data set. If the assumption is valid, then a representative regression model for all data points is the model computed from the entire data set.

4.5 Analysis Tools

This section is a theoretical discussion of the analysis methods used in the design and testing of objective measures. The first topic discussed, regression analysis, is the most important of these methods because its model building capabilities are used to design the objective measures. The next three topics, t-tests, F-tests, and confidence intervals for ρ, relate to assessing the significance of the regression model and the performance of the associated objective measure. The fifth, and last, topic is multidimensional scaling, which is used to encapsulate the relationship between the objective and subjective quality measures. As applied here, multidimensional scaling is a qualitative rather than a quantitative tool, though it can still yield significant insights into the problem being addressed.

4.5.1 Multiple Linear Regression

4.5.1.1 The Regression Model

A linear model is used to estimate the Diagnostic Acceptability Measure's subjective quality scores based on the objective measure variables. The linear relationship between subjective and objective is hypothesized as

$$y_i = \beta_0 + \sum_{j=1}^{k} \beta_j x_{ij} + \epsilon_i \tag{4.56}$$

where y_i, the dependent variable, is the parametric subjective quality and the x_{ij}'s, the independent variables, are the objective measure variables. The β_j's are model parameters to be estimated and ϵ_i is a random error associated with each observation. Subscript j is the index of the independent, or objective measure, variable and subscript i is the index of the observation, or the index of the speaker and distortion system in the data base. In the speech quality data base an observation corresponds to the subjective and objective responses associated with a distorted speech signal. An observation can be considered to be an ordered n-tuple, in which the first number is the subjective quality estimate for a particular distorted speech signal, and the remaining $n-1$ numbers are the objective variables, as computed from the distorted and reference speech signals. Since observations in the distorted speech data base entail both a speaker and a distortion system, the observation index will more frequently indicate this explicitly as $y(s,d)$, where s indicates the speaker and d indicates the distortion system.

The β_j's are estimated in the classical manner by minimizing the mean square error over all distortion systems in the data base [4.35], [4.36], [4.37]. Let $\vec{Y}$ be a column vector composed of all observations of the dependent variable, and $\mathbf{X}$ be a matrix whose first column is all 1's and subsequent columns are made up of all observations of the independent variables. Let $\vec{B}$ be a column vector of length p, where $p = k + 1$ and k is the number of independent variables in the analysis. Vector $\vec{B}$ has entries b_j, which are the estimates of regression coefficients β_j. Column vector ϵ has entries equal to the random error associated with each observation. The vectors and matrix can be explicitly written as

$$\vec{Y} = \begin{bmatrix} y_1 \\ y_2 \\ \vdots \\ y_N \end{bmatrix} \quad \mathbf{X} = \begin{bmatrix} 1 & x_{11} & x_{12} & \cdots & x_{1k} \\ 1 & x_{21} & x_{22} & \cdots & x_{2k} \\ \vdots & \vdots & \vdots & \ddots & \vdots \\ 1 & x_{N1} & x_{N2} & \cdots & x_{Nk} \end{bmatrix}$$

$$\vec{\beta} = \begin{bmatrix} \beta_0 \\ \beta_1 \\ \vdots \\ \beta_k \end{bmatrix} \quad \vec{B} = \begin{bmatrix} b_0 \\ b_1 \\ \vdots \\ b_k \end{bmatrix} \quad \vec{\epsilon} = \begin{bmatrix} \epsilon_1 \\ \epsilon_2 \\ \vdots \\ \epsilon_N \end{bmatrix}$$

and the regression model can now be expressed as

$$\vec{Y} = \mathbf{X}\vec{\beta} + \vec{\epsilon} \tag{4.57}$$

The solution to the normal equations, which specify the b_j's is

$$\vec{B} = (\mathbf{X}^{\mathrm{T}}\mathbf{X})^{-1}\mathbf{X}^{\mathrm{T}}\vec{Y} \tag{4.58}$$

where $\mathbf{X}^{\mathrm{T}}$ is the transpose of the $\mathbf{X}$ matrix. This leads to the following regression model, which is the desired objective measure

$$\hat{y}_i = b_0 + \sum_{j=1}^{k} b_j x_{ij} \tag{4.59}$$

where $\hat{y}_i$ is the estimate of y_i.

4.5.1.2 Measuring Model Adequacy

A measure of model adequacy is the coefficient of multiple determination, or multiple R-squared, which is an estimate of the square of the coefficient of correlation between y_i and $\hat{y}_i$. It is computed as

$$R^2 = \frac{SSR}{SSY} = \left(1 - \frac{SSE}{SSY}\right) \tag{4.60}$$

where SSR is the regression sum of squares, SSE is the residual sum of squares and SSY is the corrected sum of squares of the observed dependent variable. These terms can be computed as

$$\begin{aligned} SSY &= \sum_{i=1}^{N}(y_i - \bar{y})^2 \\ SSR &= \sum_{i=1}^{N}(\hat{y}_i - \bar{y})^2 \\ SSE &= \sum_{i=1}^{N}(y_i - \hat{y}_i)^2 = \sum_{i=1}^{N}(e_i)^2 \end{aligned} \tag{4.61}$$

where $\hat{y}_i$ is the estimated independent variable, $\bar{y}$ is the average value of all of the y_i's, e_i is the residual associated with observation i, and N is the number of observations in the data base. Note that the residual e is not the same as the error ϵ, though the residual is often used as an estimate of model error when checking for model adequacy and verifying assumptions about the error distribution. The error term is a random component in the regression model, usually associated with errors in observing y, while the residual is the total error in the model for each observation, or $y_i - \hat{y}_i$, and may be composed of both random and non-random components (the non-random part arises from the model not fitting the data well.) Note that $SSE/(N-1)$ is an estimate of the regression error variance, σ^2.

4.5.1.3 Assumptions of Regression Analysis

In order for the model in Eq. (4.59) to be valid, the following assumptions must be satisfied:

1. The model errors ϵ_i, are uncorrelated.

2. The model error ϵ has zero mean.
3. The model error ϵ has constant variance equal to σ^2.
4. The relationship between y_i and the x_{ij}'s is, in fact, approximately linear.

To assess the validity of these assumptions, the source of the error term must be investigated. The basic source which determines the quality responses in the subjective data base is the nature of the distortions in the distorted speech data base. Therefore the distorted sentences themselves are, fundamentally, the independent variables. The x_{ij}'s, which are the objective measure variables, can be thought of as complex transformations of the distorted speech waveforms. Once the transformation is fixed, the x_{ij}'s are determined without uncertainty, as functions of the distorted and undistorted speech signals. Therefore the error terms, ϵ_i, should be interpreted as error in the subjective assessment of the quality of the distorted speech signals. Clearly, unless the model is perfect, the "observed error" (which is the residual e) will be composed of the random error ϵ plus any systematic errors due to deficiencies in the model itself. In this respect the model may violate the fourth assumption, and lack a nearly linear relationship between the y_i's and the x_{ij}'s. However, since the underlying statistics of such errors are unknown, it is reasonable to assume that these errors do satisfy all of the necessary conditions until proven otherwise by post analysis diagnostics. With this qualification, it is possible to evaluate the assumptions relative to the errors in subjective quality estimation.

First, the errors must be uncorrelated. In any subjective test this is ensured by randomizing the order in which the data is presented for evaluation. This prevents any evaluation bias based on previous subjective judgments of similar speech segments from occurring. Dynastat is fully aware of the need for unbiased evaluation, and carefully adjusts all aspects of the Diagnostic Acceptability Measure, including randomized presentation order, to ensure the validity of this assumption.

Second, and third, the error must have zero mean and constant variance. These two assumptions must be examined together. The subjective assessments of speech quality provided by the DAM are all mean opinion scores, that is, they are based on the judgments of several individuals. Before individual opinions are averaged together, they are adjusted to eliminate the effects of each individual's biases. This means that each listener's assessment error is transformed to have zero mean and constant variance relative to Dynastat's "average listener response." Furthermore, new listeners undergo a training period prior to the actual test, and can only proceed if they show a relatively small judgment error, relative to the average response, across a variety of distorted speech signals. Therefore, because individual judgments are adjusted to conform to a group norm and because listeners are carefully trained, assumptions two and three should be valid.

Fourth, the relationship between dependent and independent variables should be approximately linear. If this is not true, then the assumption of constant error variance will most likely be violated. In practice, the relationship is assumed to be linear, and then, after the regression analysis has been performed, the error is checked for constant variance. This check is most easily done by looking at a plot of the residual for each observation versus the predicted value for each observation. Here, the initial assumption is that the model fits the data, so that the residuals are a good estimate of the model error. If the model does exhibit non-constant variance, then a transformation of some or all of the x_{ij}'s may mitigate this problem. Using transformations, the relationship between independent and dependent variables can be linearized. If the residuals indicate that higher order terms in x_{ij} are needed, then powers of x_{ij} can be added to the model as additional independent variables. In this way a polynomial model can be built within the framework of the original regression model.

4.5.1.4 Outliers

Another violation of the approximate linearity assumption is caused by *outliers*, or data which do not lie anywhere near the regression curve and hence are associated with very large error terms. These points can greatly decrease the correlation of the objective estimates to the subjective quality, and hence make an otherwise good objective model appear quite poor. Residual plots can help identify cases in which outliers are decreasing the performance of the regression model. Correcting the problem consists of understanding why the outlier deviates from the regression model and then either correcting the model or eliminating the outlying points from the data.

Outliers can radically alter a regression model when they are *in-*

fluential observations. These are observations which disproportionately affect the regression coefficients, so that the regression model depends more on these points more than on the majority of the data. An example of such points in simple linear regression is a pair of outliers which are on either extreme of the independent variable axis and on opposite sides of the fitted regression line. This pair of points exerts an undue influence on the slope of the regression line. Because of this, all outliers should be checked to determine the extent to which they influence the regression model.

4.5.1.5 Normality of Residuals

Though not required for regression analysis to be valid, it is highly desirable for the error in the regression model to conform to a normal distribution. This permits the use of the t and F parametric statistical tests to check the validity of the linear model. Assessing error normality is most often done by making a normal probability plot of the residual. Let the model residuals be rank ordered such that $e_{(1)} < e_{(2)} < \cdots < e_{(N)}$, where N is the number of observations. The plot has cumulative probability $P_i = (i - 1/2)/N$, $i = 1, 2, \ldots, N$ on the ordinate and the value of $e_{(i)}$ on the abscissa axis. The ordinate scale is nonlinear in such a way that a truly Gaussian error plots as a straight line. It is most important that the center of the error distribution be most nearly normal, though small deviations from normality do not severely affect the t or F-statistics. In reference to the discussion of outliers, a heavy-tailed error distribution may indicate that an abnormal number of outliers are present in the data.

The quality scores in the Diagnostic Acceptability Test are computed from averages of the corrected responses of a relatively large listening crew. From central limit theorems, it is expected that this averaging will cause the model error to be approximately normal, without making any assumption as to the distribution of the error in the listener responses. Normal probability plots of several regression analyses confirm that the model errors are, in fact, close to normal and therefore make parametric statistical tests on regression model parameters valid.

4.5.2 t-Test

One of the parametric statistical tests used in validating the regression model is the partial t-test, which is used to test hypotheses concerning individual regression coefficients. The most useful hypothesis to test is whether a given regression coefficient is equal to zero, or

$$H_0 : \beta_j = 0 \tag{4.62}$$

This is the "null hypothesis," with the alternate hypothesis being that the regression coefficient is not equal to zero. Rejecting H_0 indicates that the jth independent variable has a statistically significant contribution to the regression model. If this is the case, then the jth independent variable explains a significant fraction of the variance of the dependent variable and should be included in the regression model. On the other hand, failing to reject H_0 means that the jth independent variable is not statistically significant, and should not be included in the regression model. The test statistic for the null hypothesis is

$$t_0 = \frac{b_j}{(s^2 C_{jj})^{1/2}} \tag{4.63}$$

where b_j is the estimated regression coefficient, s^2 is the estimated variance of the regression error ϵ, and C_{jj} is the jth diagonal element of $(\mathbf{X}^{\mathrm{T}}\mathbf{X})^{-1}$ (see Eq. (4.58)) [4.37]. It is a partial, or marginal, t-test because a given b_j depends on all other independent variables in the model. Therefore the statistic in Eq. (4.63) is testing the importance of one independent variable given that the rest of the independent variables are included in the regression model. H_0 is rejected in a two-sided test if

$$|t_0| > t_{N-k-1;\alpha/2} \tag{4.64}$$

where α is the level of significance of the test, N is the number of observations, k is the number of independent variables in the regression model, and $t_{N-k-1;\alpha/2}$ is the appropriate value from the Student-t distribution. The significance level of the test, α, is the probability of H_0 being true when t_0 indicates that H_0 should be rejected.

4.5.3 F-Test

A test for the significance of the entire regression model can be formulated using using the F-statistic. The null hypothesis for the test is

$$H_0 : \beta_1 = \beta_2 = \cdots = \beta_k = 0 \tag{4.65}$$

The alternate hypothesis is that at least one of the β_j's is not equal to zero. This determines whether there is a linear relationship between the dependent variable and *any* of the independent variables. The test statistic is

$$F_0 = \left[\frac{SSR/k}{SSE/(N-k-1)}\right] = \frac{MSR}{MSE} \tag{4.66}$$

where SSR is the regression sum of squares, SSE is the error sum of squares (see Eq. (4.61)), N is the number of observations, and k is the number of independent variables in the regression model. In the term to the right of the second equals sign, MSR is the mean square for regression and MSE is the mean squared error. The null hypothesis is rejected if $F_0 > F_{k,N-k-1;\alpha}$. If rejected, the regression model explains a statistically significant fraction of the variance of the dependent variable. The test for significance of regression can be summarized in an analysis of variance table as follows:

Source of Variation	Sum of Squares	Degrees of Freedom	Mean Square	F_0
Regression	SSR	k	MSR	MSR/MSE
Residual	SSE	$N-k-1$	MSE	
Total	SSY	$N-1$		

The fundamental idea in the test for significance of regression is that the corrected sum of squares of the dependent variable, SSY, can be partitioned into a regression sum of squares and an error sum of squares. Referring back to Eq. (4.61), it can be shown that

$$SSY = SSR + SSE \tag{4.67}$$

where the expected value of SSE is σ_ϵ^2. It can be shown that the expected value of the MSR is

$$E\{MSR\} = \sigma_\epsilon^2 + (\vec{\beta}^{\,\mathrm{T}}\mathbf{X}^{\mathrm{T}})[\mathbf{X}(\mathbf{X}^{\mathrm{T}}\mathbf{X})^{-1}\mathbf{X}^{\mathrm{T}} - \mathbf{A}/N](\mathbf{X}\vec{\beta}/k) \tag{4.68}$$

where σ_ϵ^2 is the regression error variance, $\mathbf{A}$ is an N by N matrix composed of all 1's, and E is the expectation operator [4.38]. If all the β_j's are not significantly different from zero, so that H_0 is not rejected, then the second term in Eq. (4.68) will be small. This makes the F_0 statistic nearly equal to 1, which will be less than $F_{k,N-k-1;\alpha}$.

4.5.4 Confidence Interval on ρ

A final statistic that will be of use in analyzing objective measure models is a confidence interval for the correlation coefficient ρ. Though not exact, the expression is asymptotically correct as the number of observations increases. Since there are more than twelve hundred observations in the speech quality data base, this formula is very close to the actual confidence interval [4.37]. A $100(1-\alpha)$ percent confidence interval for the correlation coefficient can be expressed as

$$\begin{aligned} \tanh\left(\operatorname{arctanh}(R) - \frac{Z_{\alpha/2}}{(N-3)^{1/2}}\right) \\ \leq \rho \leq \\ \tanh\left(\operatorname{arctanh}(R) + \frac{Z_{\alpha/2}}{(N-3)^{1/2}}\right) \end{aligned} \tag{4.69}$$

where ρ is the correlation coefficient, R is the Multiple R of the objective measure regression model, $Z_{\alpha/2}$ is the normal deviate and tanh( ) is the hyperbolic tangent function.

The confidence interval enables meaningful comparison of the multiple R associated with different objective measures. If repeated and independent trials were executed in generating the distorted data base, estimating subjective quality, and computing the multiple R figure-of-merit for a given objective measure, $100(1-\alpha)$ percent of the multiple R's associated with each trial would fall within the confidence interval. Therefore objective measures whose associated multiple R's are different, but still within each other's $100(1-\alpha)$ percent confidence interval do not have statistically different performance.

4.5.5 Multidimensional Scaling

In the context of this research, multidimensional scaling (MDS) is a tool used to show the relationship between several speech quality measures graphically. It maps similarity between quality measures, as measured by the magnitude of the coefficient of correlation, into distances between quality measures as measured in an N dimensional space. Using this technique the relationship between many measures can be easily comprehended by graphing the space, as opposed to scanning a large table of correlation coefficients. The principles of multidimensional scaling are best set forth by R. N. Shepard [4.39], [4.40] and J. B. Kruskal [4.41], [4.42], [4.43], and their KYST computer program, 1973 release, was used to perform the scaling. Before a discussion of the theory of multidimensional scaling can be presented, several terms must first be defined:

OBJECT – the thing or event to be investigated. In this research objects are speech quality measures.

PROXIMITY – also referred to as similarity. This is a measure of the distance between objects as quantified by the magnitude of a correlation coefficient or some other distance measure.

DATA MATRIX – MDS operates on proximities associated with pairs of objects. It is convenient to think of proximities among N objects as entries in an N by N data matrix, where the entry in row i column j, m_{ij}, is the proximity of object i to object j. It generally can be assumed that m_{ij} is equal to m_{ji} so that the data matrix is symmetric. Furthermore if it is assumed that the proximity of an object to itself is constant for all objects, zero for example, then the data matrix contains only $(N)(N-1)/2$ unique entries. For the applications in this study, the data matrix can effectively be reduced to a triangular matrix of $(N)(N-1)/2$ proximities.

REALIZATION SPACE – the output of a multidimensional scaling is a table of coordinates which locate each object in the realization space. In all analyses presented here, this is a Euclidean space, with dimensionality specified by the user. The distance between objects in the realization space is a function of the proximity associated with the two objects. The distance between object i and object j in the realization space is denoted as $\hat{d}_{ij}$. The dimensionality of the realization space is an important issue. For N objects it can be shown that the realization space spans at most $N-1$ dimensions for metric scaling and $N-2$ dimensions for non-metric scaling [4.43]. If the data are error free, then this dimensionality may be appropriate, though with the more common case of noisy data, higher dimensions may be accounting for noise only. Lower dimensional spaces tend to smooth out noise since, with fewer object coordinates to estimate from the data, the coordinates have greater statistical reliability. If the scaling is non-metric, Kruskal [4.43] recommends that there be twice as many stimulus pairs as there are parameters to be estimated, to assure a statistically stable configuration. For a symmetric proximity matrix, this means that $(N)(N-1)/2 > 2NR$, where N is the number of objects and R is the dimension of the realization space.

METRIC and NON-METRIC SCALING – scaling can be divided into these two broad categories. Mapping the proximities of the data matrix onto distances in the realization space in general requires a transformation of the proximities. If the function which transforms proximities to distances in the realization space is linear, then the scaling is metric. If the function is merely monotonic then the scaling is non-metric. Transformed proximities can be thought of as estimates of inter-object distances in the realization space. The transformed proximity associated with object i and object j is denoted as d_{ij}.

STRESS – points are placed in the realization space such that they minimize an error function, defined as

$$Stress = \left[\frac{\sum_{i=1}^{N} \sum_{j=1}^{N} (d_{ij} - \hat{d}_{ij})^2}{\sum_{i=1}^{N} \sum_{j=1}^{N} (d_{ij})^2} \right]^{1/2} \tag{4.70}$$

"Stress" measures the differences between the distance between points in the realization space and the estimated distance between points as specified by the transformed proximities. In another sense it measures how well the dimension of the realization space suits the data. The value of stress should guide the experimenter in choosing the appropriate dimensionality for the realization space. For a realization space of two or three dimensions, a rough interpretation of stress is as follows:

0%	perfect
5%	good
10%	fair
20%	poor

As an example of non-metric MDS, consider the data in Table 4.29 in which proximities are actually distances, in miles, between 11 cities in the United States. MDS can be used to construct a "map" of the U.S. as in Fig. 4.55. This scaling is non-metric because the transformation of m_{ij}, the distances in Table 4.29, to yield d_{ij} is monotonic. That is, if you were to construct ordered pairs (m_{ij}, d_{ij}), and then rank the m_{ij}'s in descending order, their corresponding d_{ij}'s would also be ranked in descending order. This is the only restriction on the transformation. An important point in this scaling is that a nearly perfect reconstruction was achieved *with only inequality constraints.* If there are sufficient objects in a scaling of a given dimensionality, a monotonic

Table 4.29: Distances between cities.

	Atlanta									
Boston	1037	Boston								
Chicago	674	963	Chicago							
Denver	1398	1949	996	Denver						
Houston	789	1804	1067	1019	Houston					
Los Angeles	2182	2779	2054	1059	1538	Los Angeles				
Miami	655	1504	1329	2037	1190	2687	Miami			
New York	841	206	802	1771	1608	2786	1308	New York		
San Francisco	2496	3095	2142	1235	1912	379	3053	2934	San Fran.	
Seattle	2618	2976	2013	1307	2274	1131	3273	2815	808	Seattle
Washington	608	429	671	1616	1375	2631	1075	233	2799	2684

transformation of proximities produces a set of d_{ij} which is nearly the same as those derived from a linear transformation.

Since this data was measured from a very nearly two dimensional space (the surface of the United States land mass) the realization space need not have more than two dimensions. In this example the stress, or error of fit in the realization space, is extremely small (one percent), indicating an excellent fit to the data. Figure 4.55 illustrates another important aspect of MDS: the Euclidean distance measure used in the realization space is rotation and reflection invariant, which means that the resultant configuration of points can have any angular orientation in the realization space, or undergo a reflection transformation, as is the case in this scaling. MDS produces a configuration of points, but it is up to the researcher to identify the orientation and meaning of spatial dimensions in that configuration.

4.6 Summary

This chapter has described the tools and methodology used in designing and testing objective measures. The first three sections described the speech quality data base, which is the essential ingredient for meaningful research in objective measures. The fourth section described the procedures used in the design and performance evaluation of objective measures, and outlined the basis of a research method by which objective measures can be reliably compared. The fifth section described statistical tools that permit the data bases to be effectively utilized to design and test objective measures of speech quality.

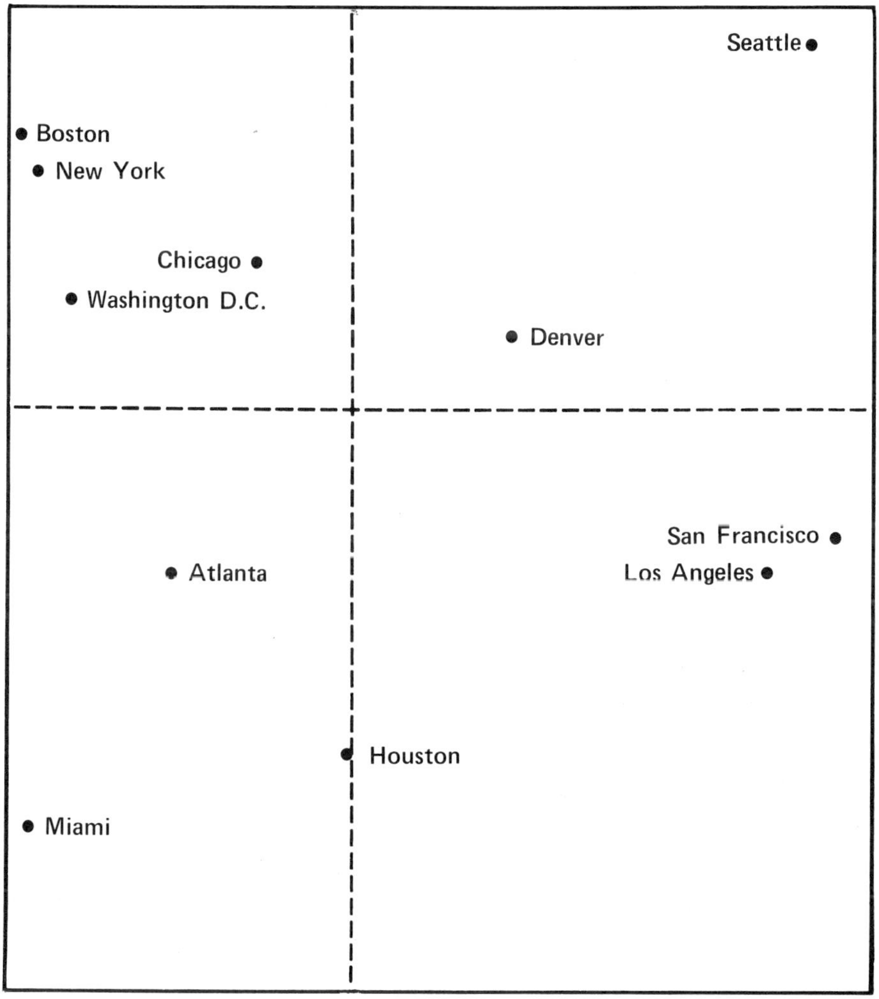

Figure 4.55: Non-metric scaling of data in table 4.29.

References

[4.1] T. P. Barnwell and A. M. Bush, "A Minicomputer Based Digital Signal Processing Laboratory," EASCON '74, Washington, DC, pp. 555–8, Oct. 1974.

[4.2] N. S. Jayant, "Adaptive Delta Modulator with a One Bit Memory," *Bell Systems Tech. J., vol. 49,* pp. 321–42, Mar. 1970.

[4.3] A. H. Gray, Jr., and J. D. Markel, "Quantization and Bit Allocation in Speech Processing," *IEEE Trans. Acoust., Speech, and Signal Processing, vol. ASSP-24* no. 6, pp. 459–72, Dec. 1976.

[4.4] A. M. Noll, "Cepstrum Pitch Determination," *J. Acoust. Soc. Amer., vol. 41*, no. 2, pp. 293–309, Feb. 1967.

[4.5] A. M. Noll, "Short-Time Spectrum and Cepstrum Techniques for Vocal Pitch Detection," *J. Acoust. Soc. Amer., vol. 36*, no. 2, pp. 296–302, Feb. 1964.

[4.6] R. Zelinski and P. Noll, "Adaptive Transform Coding of Speech Signals," *IEEE Trans. Acoust., Speech, and Signal Processing, vol. ASSP-25,* no. 4, pp. 300–309, Aug. 1977.

[4.7] J. M. Tribolet and R. E. Crochiere, "An Analysis/Synthesis Framework for Transform Coding of Speech," *Proc. 1979 IEEE ICASSP*, Washington, DC, pp. 81–4, Apr. 1979.

[4.8] T .P. Barnwell III and W. D. Voiers, "An Analysis of Objective Measures for User Acceptance of Voice Communications Systems," Final Report to the Defense Communications Agency, DCA100-78-C-0003, Sep. 1979.

[4.9] T. P. Barnwell III, A. M. Bush, R. M. Mersereau, and R. W. Schafer, "Speech Quality Measurement," Final Report RADC-TR-78-122, May, 1978.

[4.10] T. P. Barnwell III, R. W. Schafer, and A. M. Bush, "Tandem Interconnections of LPC and CVSD Digital Speech Coders," Final Report, DCA100-76-6-0073, Nov. 1977.

[4.11] T. P. Barnwell III and A. M. Bush, "Statistical Correlation Between Objective and Subjective Measures for Speech Quality," *Proc. 1978 IEEE ICASSP*, Tulsa, OK, pp. 595–8, Apr. 1978.

[4.12] T. P. Barnwell III and W. D. Voiers, "Objective Fidelity Measures for Speech Coding Systems," presented at the meeting of the Acoustical Society of America, Honolulu, Dec. 1978.

[4.13] T. P. Barnwell III, "Objective Fidelity Measures for Speech Coding Systems," *J. Acoust. Soc. Amer., vol. 65*, no. 6, pp. 1658–63, Dec. 1979.

[4.14] T. P. Barnwell III, "Correlation Analysis of Subjective and Objective Measures for Speech Quality," *Proc. 1980 IEEE ICASSP*, Denver, Colorado, pp. 706–9, Apr. 1980.

[4.15] T. P. Barnwell III, "A Comparison of Parametrically Different Objective Speech Quality Measures Using Analysis with Subjective Quality Results," *Proc. 1980 IEEE ICASSP*, Denver, Colorado, pp. 710–3, Apr. 1980.

[4.16] P. Breitkopf and T. P. Barnwell III, "Segmental Preclassification for Improved Objective Speech Quality Measures," *Proc. 1981 IEEE ICASSP*, Atlanta, Georgia, pp. 1101–4, Mar. 1981.

[4.17] T. P. Barnwell III, "On the Standardization of Objective Measures for Speech Quality Testing," *Proceedings of 1982 NBS Workshop on Standards for Speech Recognition and Synthesis*, Washington, D.C., pp, 193–200, Mar. 1982.

[4.18] T. P. Barnwell III and S. R. Quackenbush, "An Analysis of Objectively Computable Measures for Speech Quality Testing," *Proc. 1982 IEEE ICASSP*, Paris, France, pp. 996–9, May 1982.

[4.19] S. R. Quackenbush and T. P. Barnwell III, "An Approach to Formulating Objective Speech Quality Measures," *Proc. 15th Southeastern Symposium on System Theory*, Huntsville, Alabama, pp. 10–13, Mar. 1983.

[4.20] S. R. Quackenbush and T. P. Barnwell III, "The Estimation and Evaluation of Pointwise Nonlinearities for Improving the Performance of Objective Speech Quality Measures," *Proc. 1983 IEEE ICASSP*, Boston, pp. 547–50, Apr. 1983.

[4.21] B. S. Atal and J. R. Remde, Jr., "A New Model of LPC Excitation for Producing Natural Sounding Speech at Low Bit Rates," *Proc. 1982 IEEE ICASSP*, Paris, France, pp. 614–7, May 1982.

[4.22] B. S. Atal and M. R. Schroeder, "Improved Quantizer for Adaptive Coding of Speech Signals at Low Rates," *Proc. 1980 IEEE ICASSP*, Denver, Colorado, pp. 535–8, Apr. 1980.

[4.23] S. Singhal and B. S. Atal, "Improving Performance of Multi-Pulse LPC Coders at Low Rates", *Proc. 1984 IEEE ICASSP*, San Diego, pp. 1.3.1–4, Apr. 1984.

[4.24] V. K. Jain and R. Hangartner, "Efficient Algorithm for Multi-Pulse LPC Analysis of Speech," *Proc. 1984 IEEE ICASSP*, San Diego, pp. 1.5.1–4, Apr. 1984.

[4.25] J. M. Tribolet and R. E. Crochiere, "Frequency Domain Coding of Speech," *IEEE Trans. on Acoust., Speech, and Signal Processing, vol. ASSP-27*, no. 5, pp. 512–30, Oct. 1979.

[4.26] J. R. Crosmer and T. P. Barnwell III, "An Algorithm for Designing Optimum Quantizers Subject to a Multiclass Distortion Criterion," *Proc. 1983 IEEE ICASSP*, Boston, pp. 539–42, Apr. 1983.

[4.27] R. E. Crochiere, "On the Design of Subband Coders for Low-Bit-Rate Speech Communications," *Bell System Tech. J., vol. 56*, pp. 747–70, May–June, 1977.

[4.28] T. P. Barnwell III, "Subband Coder Design Incorporating Recursive Quadrature Filters and Optimum ADPCM Coders," *IEEE Trans. on ASSP, vol. ASSP-30*, no. 5, pp. 751–65, Oct. 1982.

[4.29] M. Krasner, "The Critical Band Coder-Digital Encoding of Speech Signals Based on the Perceptual Requirements of the Auditory System," *Proc. 1980 IEEE ICASSP*, Denver, Colorado, pp. 327–30, Apr. 1980.

[4.30] J. D. Johnston, A Filter Family for Use in Quadrature Mirror Filter Banks," *Proc. 1980 IEEE ICASSP*, Denver, Colorado, pp. 291–4, Apr. 1980.

[4.31] B. S. Atal and M. R. Schroeder, "Predictive Coding of Speech Signals and Subjective Error Criteria," *IEEE Trans. on Acoust.,*

Speech, and Signal Processing, vol. ASSP-27, no. 3, Jun. 1979, pp 247–54.

[4.32] W. D. Voiers, "Diagnostic Acceptability Measure for Speech Communications System," *Proc. 1977 IEEE ICASSP*, Hartford, CN, pp. 204–7, May 1977.

[4.33] J. C. R. Licklider, "Effects of Amplitude Distortion Upon the Intelligibility of Speech," *J. Acoust. Soc. Amer., vol. 18*, no. 2, 1946.

[4.34] W. D. Voiers, "Methods of Predicting User Acceptance of Voice Communication Systems," Final Report, DCA100-74-C-0056, Jul. 1976.

[4.35] B. Ostle and R. W. Mensing, *Statistics in Research: Basic Concepts and Techniques for Research Workers* (3rd ed.). Ames: Iowa State University Press, 1975.

[4.36] N. R. Draper and H. Smith, *Applied Regression Analysis* (2nd ed.). New York: John Wiley, 1981.

[4.37] D. C. Montgomery and E. A. Peck, *Introduction to Linear Regression Analysis*, New York, John Wiley, 1982.

[4.38] O. J. Dunn and V. A. Clark, *Applied Statistics: Analysis of Variance and Statistics.* New York: John Wiley, 1974.

[4.39] R. N. Shepard, "The Analysis of Proximities: Multidimensional Scaling with an Unknown Distance Function," *Psychometrika, vol. 27*, pp. 125–39, pp. 219–46, 1962.

[4.40] R. N. Shepard, *Multidimensional Scaling: Theory and Applications in the Behavioral Sciences, vol. 1*, New York, Academic Press, 1972.

[4.41] J. B. Kruskal, "Multidimensional Scaling by Optimizing Goodness of Fit to a Numerical Hypothesis," *Psychometrika, vol. 29*, pp. 1–27, 1964.

[4.42] J. B. Kruskal, "Nonmetric Multidimensional Scaling: a Numerical Method," *Psychometrika, vol. 29*, pp. 115–29, 1964.

[4.43] J. B. Kruskal, *Multidimensional Scaling*, Beverly Hills; Sage Press, 1978.

Chapter 5

Basic Objective Measures

5.1 Introduction

The primary purpose of this chapter is to investigate the performance of a large number of basic objective measures for speech quality testing. In this context, a *basic* objective measure is defined as a compactly computable function which can be applied uniformly across any form of speech distortion in order to estimate subjective quality. By this definition, most objective speech quality measures which have been widely proposed and widely used are basic measures. Commonly used basic objective measures include signal-to-noise ratios, segmental signal-to-noise ratios, log likelihood ratios, spectral distance measures, and many more (see Chap. 2).

It is unlikely that any single, simple basic objective measure will ever be found that can accurately predict subjective quality across all of the possible distortions which occur in the coding and processing of speech. Speech perception is too complex a process to be modeled so simply. And yet, basic objective measures have traditionally been used to great advantage in designing and measuring the performance of speech processing systems. There are two reasons for this apparent paradox. First, most basic measures have traditionally been used with fairly narrow classes of distortions. A great many objective measures which do not perform well when comparing dissimilar speech distortions are still capable of rank ordering similar distortions. A second, and perhaps more important, reason is that even an objective measure with modest performance is better than no measure at all. Many objective measures are capable of measuring gross distinctions between dissimilar systems and moderate distinctions between similar systems. However, they simply are not capable of providing the fine distinctions between dissimilar distortions that are necessary if one wishes truly to optimize the quality of a speech coding or processing system.

The purpose of this chapter is to quantify the performance of a large class of basic objective measures across a very broad class of speech distortions. By any standard, this is a huge study, involving hundreds of distortions, thousands of objective measures, and millions of correlation analyses which was, in fact, performed over a period of more than ten years. Regardless of its size and scope, however, this study cannot claim to include all the distortions and all the objective measures of interest. It is large enough, and sufficiently well planned, however, to lend credibility to the trends which are observable in the data.

This chapter describes in detail the basic objective measures investigated in this research, and discusses their performance. These measures are chiefly intended to test three classes of digital coding systems: waveform coders, in which the coding system tries to duplicate the input signal at the output; vocoders, in which speech is decomposed into an excitation function and a model of the upper vocal tract filter; and frequency-domain coders, in which a time-frequency representation of the speech waveform is coded instead of the waveform itself. This bias toward digital coding systems is mainly motivated by current trends in technology. This does not mean that the results here are not applicable to analog systems, but such systems do pose greater problems in the area of synchronization and phase normalization.

The objective measures treated in this chapter can be divided into seven classes. These classes are referred to as simple spectral distance, simple noise, simple LPC-based parametric, frequency variant spectral distance, frequency variant noise, time variant spectral distance, and time variant LPC-based parametric. Simple spectral distance measures include all measures in which the distortion is computed in the frequency domain and in which the spectral weighting of the measure is either unity or is derived from the original speech signal. Simple noise measures include all measures in which the main component is the "noise" between the original speech signal and a time-aligned version of the distorted speech signal. LPC-based parametric measures include all measures based on a parameter set derived from the short-time LPC representation of the speech signals under test. Frequency variant spectral distance measures are those in which the measurements are applied in distinct bands in the frequency domain, rather than applied uniformly across the entire frequency range. Frequency variant noise measures are those in which noise is measured in frequency bands by appropriate prefiltering. Time variant spectral distance measures and time variant LPC-based parametric measures are segmental versions of their time invariant counterparts. In time variant measures, the speech waveform is segmented based on an objectively derived "phonemic" classification in which separate measures are computed for each phonemic class.

The three types of "simple" measures are being studied for three reasons: first, to quantify the effectiveness of many of the measures currently in common use for speech quality prediction; second, to test the effect of parametrically different forms of the various measures; and third, to test the utility of such measures as compared with more complex measures.

The two classes of frequency variant measures are being studied largely because it is well known [5.3] that auditory perception is itself frequency variant. Although auditory perception has been studied psychophysically, the measurement of precise physical parameters is very difficult. The frequency variant measures are a vehicle by which indirect measurement of frequency variant auditory effects can be made using regression analysis [5.1]. In addition, it is well known that many of the parametric subjective measures from the DAM (see Chap. 3) are frequency dependent. The frequency variant objective measures have the ability to be "tuned" to predict such parametric subjective quality scores.

The time variant measures are being studied to assess the possible gains that could be realized by measuring the distortion in different classes of speech sounds independently. For example, it has been observed that distortions which may be easily perceived in a vocalic segment may not be perceived at all in a stop segment, so that the measured distortion associated with these two segments should not be given equal weight. The time variant measures investigated here have even greater significance because they represent a general concept that can be used to improve the performance of any type of objective quality measure.

5.2 Methodology and Notation

Objective measures are made using an undistorted speech data set, which is labeled as ϕ, and a distorted speech data set, which is labeled with a distortion index d. The undistorted speech data set in this study is made up of four speakers, indexed by s.

In computing objective measures, the estimate is generally formed by averaging the results from a number of "frames" of the undistorted and distorted speech. In order for the measures to be unbiased, precise synchronization between the distorted and undistorted speech signals must be maintained. Since all of the distortions in this study were produced digitally, synchronization was not a great problem. However for the testing of non-simulated coding systems, the synchronization problem would have to be considered carefully.

Measures are computed from a set of undistorted speech frames, $x_n(m, s, \phi)$, where n is the frame index, m is the sample index, s is the speaker index, and ϕ indicates no distortion; and a distorted speech set, $x_n(m, s, d)$, where d is the distortion index. Each frame is composed of a set of samples from the distorted or undistorted speech signals. Each distortion measure is characterized at the frame level by a specific function, F, with the objective measures, $D(d)$, computed from

$$D(d) = \frac{\sum_{s=1}^{4} \sum_{n=1}^{N} W(n,s) F[x_n(m,s,\phi), x_n(m,s,d)]}{\sum_{s=1}^{4} \sum_{n=1}^{N} W(n,s)} \tag{5.1}$$

where N is the number of frames in the analysis, and $W(n, s)$ is a time domain weighting function for the nth frame and the sth speaker. Note that $W(n, s)$ may also be a function of $x_n(m, s, \phi)$, $x_n(m, s, d)$, or both. In this environment, therefore, describing the objective measures reduces to describing the functions W and F used for each measure.

Recall from Chap. 4 that the data base contains coding distortions, wideband controlled distortions, and frequency variant controlled distortions. The distorted data base was relatively heavily loaded with frequency variant distortions because it was felt that considerable improvement in objective quality measures might be achieved by better

Table 5.1: Distortion classes.

Coding Distortion	No. of Cases	WBD	WFC	CODE	CON	WBN	NBN	BD	PD	ND
CVSD	6	X	X	X						
ADM	6	X	X	X						
APCM	6	X	X	X						
ADPCM	6	X	X	X						
LPC	6	X		X						
APC	6	X	X	X						
VEV	12	X		X						
MPLPC	6	X		X						
ATC-I	6	X	X	X						
ATC-II	6	X	X	X						
Subband coder	6	X	X	X						
Channel vocoder	6		X	X						
Controlled Distortion										
Additive noise	6	X	X		X	X				X
Interruption	12	X			X					
Clipping	6	X			X					
Center clipping	6	X			X					
Quantization	6	X	X	X	X					
Echo	6				X					
Lowpass filter	6	X			X					
Highpass filter	6	X			X					
Bandpass filter	6	X			X					
Frequency Variant										
Narrowband noise	36				X		X			X
Pole distortion-I	78				X				X	
Pole distortion-II	24									
Frequency distortion	36				X			X		

understanding of the frequency variant perceptual effects. Hence, measures tested over the set of all distortions (called ALL) are of considerable interest. However, an ensemble of distortions which contains as many frequency variant distortions as are in the ALL distortion set does not represent a good estimate of a true coding environment. Hence, a second major distortion set was identified, called WBD (wideband distortions), which was intended to give a better estimate of the true behavior of the measures in a coding environment. A description of the distortion set WBD is given in Table 5.1.

In addition to WBD, eight other data subsets are used. These are WFC (waveform coders), CODE (coding distortions), CON (controlled distortions), WBN (wideband noise), NBN (narrowband noise), BD (band distortions), PD (pole distortions), and ND (noise distortions). The contents of these various sets are also shown in Table 5.1.

Although the subjective data base contains nineteen subjective scores per distortion, only a few of these are emphasized. These include CA (composite acceptability), TBQ (total background quality), and TSQ (total signal quality) for the isometric measures, and the entire set of parametric measures. Of these, CA was considered most important, and most performance results are based on this measure.

During the period from 1976 to 1987, roughly 2000 basic objective measures and parametric variations of basic measures were studied. The analyses performed on the 2000 objective measures included linear, 3rd order polynomial, and 6th order polynomial regressions. All were performed across the nine distortion subsets (ALL, WBD, WFC, CODE, WBN, NBN, BD, PD, ND) for sixteen subjective measures of the DAM. Since the total number of analyses was $2000 \times 3 \times 9 \times 16 = 864,000$, it is obviously unreasonable even to print the results. Instead, this chapter takes the approach of illustrating the important results by addressing specific questions about the performance of various classes of objective measures using the results of the correlation analyses. These questions largely concern the utility of various objective measures and the optimality of their various controlling parameters.

5.3 Simple Measures

Simple measures are the set of measures which produce an estimate of composite acceptability from a single compact computational algorithm. These include such traditional measures as the SNR and spectral distance. This section describes measures of this type and compares their performance.

5.3.1 Spectral Distance Measures

All of the simple spectral distance measures used in this study are based on a function $V(n, s, d, \theta)$, which is the magnitude spectrum for frame n, speaker s, distortion d, and the frequency variable θ. In general, the spectral estimates are always derived from the non-overlapping speech frames, $x_n(m, s, d)$. Let $x(m, s, d)$ be the sampled digital representation (sampled at 8 kHz) of the distorted signal for the sth speaker and the dth

distortion. Then the "framed" speech time sample for the nth frame, $x_n(m, s, d)$, is given by

$$x_n(m, s, d) = x(m, s, d)w(m - nI) \tag{5.2}$$

where $w(m)$ is a finite length window function and I is the frame interval in samples. The discrete Fourier spectrum (DFS) for this signal is given by

$$V(n, s, d, \theta) = \left| \sum_{m=-\infty}^{+\infty} x_n(m, s, d)e^{-j\theta m} \right| \tag{5.3}$$

where the limits on the sum are finite in practice because of the finite length of $x_n(m, s, d)$. The short-time stationarity of the speech signal [5.3] suggests that a good window length is 10–30 ms. Although the DFS is a very natural function to consider, it is not particularly well suited to this application. The primary reason for this is that for voiced sounds and reasonable window lengths, $x_n(m, s, d)$ often includes several pitch periods. This causes the DFS to approach a line spectrum, as shown in Fig. 5.1. Because pitch is a coded parameter in many speech coding systems, and because small variations in pitch, which have little impact on quality, would cause great differences between such spectra, the DFS is not a good candidate for a spectral distance measure. A better representation is the spectral envelope of the DFS, which can be approximated in several ways. First, the analysis window in a DFS can be constrained always to contain exactly one pitch period. This method, however, requires the use of a pitch detector plus additional synchronization logic which makes this approach unattractive. Second, the spectral envelope can be estimated using the parametric LPC analysis technique [5.4], [5.5], [5.6]. The LPC technique's advantages include its computational simplicity and its ability to represent the spectral envelope compactly. However, like all parametric approaches, LPC analysis is subject to modeling errors. Finally, the spectral envelope can be extracted using cepstral deconvolution techniques [5.7], [5.9]. Previous research has shown [5.8], [5.10], however, that cepstral estimate is closely related to the corresponding LPC technique, and is more computationally intense.

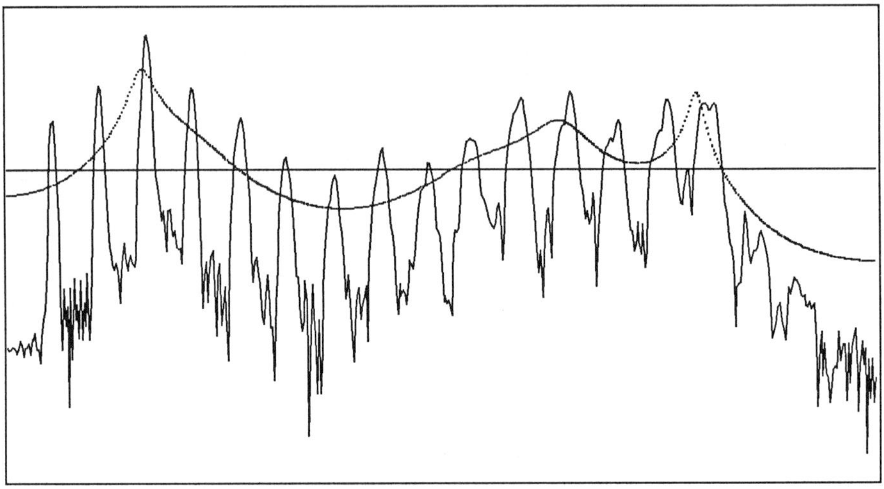

Figure 5.1: Comparison of Fourier and LPC log spectra for a vowel.

5.3.1.1 The LPC Analysis Technique

In the measures presented in this chapter (but not Chap. 6), the basis for the spectral envelope approximations is always the LPC parametric technique. In this technique, a set of autocorrelation functions, given by

$$R_n(k) = \sum_{m=-\infty}^{+\infty} x_n(m,s,d)x_n(m+k,s,d) \tag{5.4}$$

for the nth frame and $0 \leq k \leq 10$, are computed. A set of 10 feedback coefficients, $a(k)$, are computed from Durbin's recursion, given by the initial conditions

$$E(0) = R(0), \qquad K(1) = \left(\frac{-R(1)}{R(0)}\right), \qquad a^{(1)}(1) = -K(1)$$

and the recursion equations

$$\begin{aligned} K(n) &= \left[\sum_{i=1}^{n-1} a^{(n-1)}(i)R(n-i) - R(n)\right] / E(n-1) \\ a^{(n)}(n) &= -K(n) \end{aligned} \tag{5.5}$$

$$\begin{aligned} a^{(n)}(i) &= a^{(n-1)}(i) + K(n)a^{(n-1)}(n-i) \\ E(n) &= [1 - K^2(n)]E(n-1) \end{aligned}$$

where the autocorrelation subscript has been dropped, and $a(i) = a^{(n)}(i)$ when the desired model order has been obtained. In this recursion, the $K(n)$ parameters are the well-known PARCOR (partial correlation) coefficients first used by Itakura [5.11]. From the feedback coefficients, the magnitude spectrum for frequency θ can be computed by

$$V(n,s,d,\theta) = \left| \frac{G}{1 - \sum_{k=1}^{10} a(k)e^{-j\theta k}} \right| \tag{5.6}$$

where G is the gain term, given by

$$G = [R(0) - \sum_{k=1}^{10} a(k)R(k)]^{1/2} \tag{5.7}$$

The LPC approach has several specific advantages when used for spectral analysis. First, the entire analysis for a frame results in only eleven numbers, $a(1), \ldots, a(10)$, and G. This means that a large number of spectral analysis results may be stored relatively compactly. Second, the gain analysis is separate from the spectral analysis. Since overall level does not have a great impact on perception, it is often desirable to remove gain effects from the spectral distance measure. If the gain term in Eq. (5.6) is forced to be 1, i.e.,

$$V(n,s,d,\theta) = \left| \frac{1}{1 - \sum_{k=1}^{10} a(k)e^{-j\theta k}} \right| \tag{5.8}$$

the spectrum $V(n,s,d,\theta)$ is in some sense normalized. Finally, the LPC method results in a relatively compact computation of $V(n,s,d,\theta)$ from $a(1), \ldots, a(10)$. $V(n,s,d,\theta)$ may be thought of as the magnitude of the discrete time Fourier transform (DTFT) of the impulse response of an infinite impulse response (IIR) filter whose z-transform is given by

$$V(z) = \frac{1}{1 - \sum_{k=1}^{10} a(k)z^{-k}} \tag{5.9}$$

The inverse of this filter is an finite impulse response (FIR) filter whose z-transform $A(z)$ is given by

$$A(z) = \frac{1}{V(z)} = 1 - \sum_{k=1}^{10} a(k)z^{-k} \tag{5.10}$$

The inverse of $V(n,s,d,\theta)$ can hence be computed from

$$|A(n,s,d,\theta)| = \left|1 - \sum_{k=1}^{10} a(k)e^{-j\theta k}\right| \tag{5.11}$$

Since this sum has only eleven terms, it can be represented compactly and computed efficiently. Once $A(n,s,d,\theta)$ is known, $V(n,s,d,\theta)$ may be simply obtained from

$$V(n,s,d,\theta) = \frac{1}{|A(n,s,d,\theta)|} \tag{5.12}$$

5.3.1.2 The Formulation of the Spectral Distance Measures

Six variations of the distance function for spectral distance analysis (function F in Eq. (5.1)), were studied. The first, called the *linear unweighted* spectral distance, is given by

$$F = \left\{\frac{1}{L}\sum_{l=0}^{L-1} |V(n,s,\phi,\theta_l) - V(n,s,d,\theta_l)|^p\right\}^{1/p} \tag{5.13}$$

This measure is simply the L_p norm of the magnitude of the differences of the spectral samples. In general, $L = 128$ and

$$\theta_l = \frac{\pi l}{L} \text{ for } l = 0,\ldots,L-1 \tag{5.14}$$

The second form, called the *linear frequency weighted* spectral distance, is given by

$$F = \left\{\frac{\sum_{l=0}^{L-1} |V(n,s,\phi,\theta_l)|^\gamma |V(n,s,\phi,\theta_l) - V(n,s,d,\theta_l)|^p}{\sum_{l=0}^{L-1} |V(n,s,\phi,\theta_l)|^\gamma}\right\}^{1/p} \tag{5.15}$$

In this form, the frequency average is weighted by the undistorted spectrum raised to the γ power. For γ equal to 0, Eq. (5.15) reduces to Eq. (5.13). The third form, called the *log unweighted* spectral distance, is given by

$$F = \left\{\frac{1}{L}\sum_{l=0}^{L-1} \left|20\log_{10}\frac{V(n,s,\phi,\theta_l)}{V(n,s,d,\theta_l)}\right|^p\right\}^{1/p} \tag{5.16}$$

Here the constant 20 is used to produce results in dB. The fourth form, called the *frequency weighted log* spectral distance, is given by

$$F = \left\{ \frac{\sum_{l=0}^{L-1} |V(n,s,\phi,\theta_l)|^\gamma \, |20 \log_{10}\{V(n,s,\phi,\theta_l)/V(n,s,d,\theta_l)\}|^p}{\sum_{l=0}^{L-1} |V(n,s,\phi,\theta_l)|^\gamma} \right\}^{1/p} \tag{5.17}$$

The fifth form of the spectral distance, called the *unweighted δ* form, is given by

$$F = \left\{ \frac{1}{L} \sum_{l=0}^{L-1} |V(n,s,\phi,\theta_l)^\delta - V(n,s,d,\theta_l)^\delta|^p \right\}^{1/p} \tag{5.18}$$

Finally, the *frequency weighted δ* form is given by

$$F = \left\{ \frac{\sum_{l=0}^{L-1} |V(n,s,\phi,\theta_l)|^\gamma \left|V(n,s,\phi,\theta_l)^\delta - V(n,s,d,\theta_l)^\delta\right|^p}{\sum_{l=0}^{L-1} |V(n,s,\phi,\theta_l)|^\gamma} \right\}^{1/p} \tag{5.19}$$

All of the measures described above could be applied in either a direct or an inverted form. To understand the inverted form, first notice that all of the functions F can be expressed as

$$F = \left\{ \frac{\sum_{l=0}^{L-1} W(l) Q[V(n,s,\phi,\theta_l), V(n,s,d,\theta_l]^p}{\sum_{l=0}^{L-1} W(l)} \right\}^{1/p} \tag{5.20}$$

for some value of the function Q. If the inverted form was used, then the transformation which was applied is given by

$$Q \rightarrow \frac{1}{b+Q} \tag{5.21}$$

where the constant b is required for the case where $Q = 0$. Using this transformation, the spectral distance measures are generally transformed from distortion measures to fidelity measures, and the correlation with subjective quality scores goes from negative to positive. The log, $|\ |^\delta$, and inverted transformations are normally referred to as "pointwise" nonlinearities because they apply to each of the points in the averaging operations which define F.

Implicit in the definitions of the spectral distances above are four major questions. First, what (if any) pointwise nonlinearity should be applied to the estimated spectra before computing the differences? The three candidates here are none (linear), log, and raising the spectrum to the δ power. The $|\ |^{\delta}$ form can be thought of as an approximate bridge between the other two forms because it includes the linear form as a special case ($\delta = 1$) and for values of $|\ |^{\delta}$ close to about 0.2, it is very similar to the log function over a large portion of its range. Second, should the spectrum be weighted by a function of the undistorted spectrum, and, if so, by how much? The control parameter for this case is γ. Third, what value of p for the L_p norm should be used? For this case, as $p \to \infty$, a maximum deviation criterion is approached. Finally, what is the effect of the inversion transformation, and what value of b gives the highest correlation scores?

A total of 1574 variations of the spectral distance measures were investigated in this phase of the study. Each of the objective measures studied can be uniquely described by five parameters. First, the spectral distance measure may be between linear spectra, log spectra, or a spectrum taken to the δ power. If the $|\ |^{\delta}$ form is used, the value of δ must be specified. Second, between frames (see Sec. 5.1), the measures are weighted by $W(n, s)$ from Eq. (5.1). In the measures studied, the form of $W(n, s)$ was the energy of the original signal taken to the α power, leaving only α to be specified. If $\alpha = 0$, then there is no energy weighting. Third, all of the measures always involve an L_p norm, and the value of p must be specified. Fourth, within frames, the distance measure may be spectrally weighted by $V(n, s, \phi, \theta_l)^{\gamma}$. If $\gamma = 0$, there is no spectral weighting. Finally, the inverted form of the measure may be specified. For inverted measures, the value of b in Eq. (5.21) must be specified. In these terms, Table 5.2 summarizes the 1574 spectral distance measures considered in this research. The reader should note the inverted form was not implemented for the $|\ |^{\delta}$ measures.

5.3.1.3 The Best Spectral Distance Measures

The obvious first questions of interest are which of all these measures perform best, and how well do the best measures perform? Table 5.3 gives a summary of the best spectral distance measures for CA, TSQ, and TBQ over the ALL and WBD distortion subsets for each of the classes of spectral distance measures studied.

Several points can be seen in this data. First the best overall spec-

Table 5.2: Parameters for the spectral distance measures.

Linear Spectral Distance Measures	
Energy Weighting (α)	0, .05, .1, .2, .3, .5, .6, 1, 2
L_p Norm (p)	1, 2, 4, 8, 10, 16, 24, 32
Spectral Weighting (γ)	0, .1, .2, .3, .5, .6, 1, 2
Inversion Parameter (b)	.001, .002, .005, .01, .02, .05, .07, .1, .2, .5, 1, 2, 5, 10, 20, 50, 70, 80, 82–100, 110, 120, 150, 170, 200, 300 400, 500, 700, 1000

Log Spectral Distance Measures	
Energy Weighting (α)	0, .05, .1, .2, .3, .5, .6, 1, 2
L_p Norm (p)	1, 2, 4, 8, 10, 12, 14, 16, 24, 32
Spectral Weighting (γ)	0, .1, .2, .3, .5, .6, 1, 2
Inversion Parameter (b)	.001, .002, .005, .01, .02, .05, .07, .1, .2, .5, 1, 2, 5, 10, 20, 50, 70, 80, 82–100, 110, 120, 150, 170, 200, 300 400, 500, 700, 1000

Nonlinear Spectral Distance Measures	
Energy Weighting (α)	0, 0.5, 1, 2
L_p Norm (p)	1, 2, 4, 8, 10, 12, 14, 16, 24, 32
Spectral Weighting (γ)	0, .1, .2, .3, .5, .6, 1, 2
Nonlinearity (δ)	.03, .05, .1, .2, .3, .4, .5, .6, .7, .8, .9

tral distance measure over all distortions for CA is an inverted linear measure with no spectral weighting, an energy weighting parameter of 0.5, and a power for the averaging operation of 8. The correlation coefficient for this measure is positive, since the inversion operation makes this a fidelity measure rather than a distortion measure. It is noteworthy that without the inversion operation, the linear spectral distance measures perform quite badly. The performance of the best log measure and the best $|\ |^{\delta}$ measure are quite close to the best inverted linear measure, as well as to one another. Because of the similarities between the $|\ |^{0.2}$ and the log functions, this result not particularly surprising.

The maximum magnitude correlation coefficient is 0.6319, corre-

Table 5.3: Best spectral distance measures for CA, TSQ, and TBQ across ALL.

	$\hat{\rho}$	$\hat{\sigma}_e$	Nonlinearly (δ)	L_p (p)	Spectral (γ)	Energy (α)	Inversion (b)
CA(ALL)	-.61	7.8	$\|\ \|^{.2}$	1	1.0	0.0	–
	-.60	7.9	log	2	0.5	0.5	–
	-.38	9.1	linear	1	0.0	0.0	–
	-.59	8.0	log	1	0.6	1.0	2
	.63	7.6	linear	8	0.0	0.5	50
TSQ(ALL)	-.58	8.7	$\|\ \|^{.2}$	1	1.0	0.5	–
	-.57	8.8	log	1	0.8	1.0	–
	-.38	9.9	linear	1	0.0	0.0	–
	-.55	9.0	log	1	1.0	2.0	2
	.58	8.8	linear	8	0.0	0.5	70
TBQ(ALL)	.13	7.2	$\|\ \|^{.8}$	1	0.0	2.0	–
	.15	7.2	log	1	0.4	8.0	–
	.14	7.2	linear	1	0.0	2.0	–
	.17	7.1	log	1	2.0	0.5	1
	.17	7.1	linear	1	1.0	0.5	1
CA(WBD)	-.55	7.5	$\|\ \|^{.1}$	4	1.0	0.3	–
	-.53	7.6	log	2	1.0	1.0	–
	-.40	8.2	linear	1	0.2	1.0	–
	-.53	7.6	log	1	0.2	1.0	2
	.48	7.8	linear	8	0.0	0.5	50
TSQ(WBD)	-.53	8.7	$\|\ \|^{.1}$	4	1.0	0.3	–
	-.54	8.8	log	8	2.0	2.0	–
	-.40	9.9	linear	1	0.2	8.0	–
	-.53	9.0	log	1	0.2	2.0	2
	.45	8.8	linear	8	0.0	0.5	70
TBQ(WBD)	.23	6.0	$\|\ \|^{.6}$	1	0.0	2.0	–
	.23	6.0	log	1	0.4	8.0	–
	.24	6.0	linear	1	0.0	8.0	–
	.23	6.0	log	1	2.0	8.0	2
	.24	6.0	linear	1	1.0	1.0	3

sponding to a standard deviation of error of 7.63. Even though this is one of the better simple measures, it must be said that it does not perform very well over this broad class of distortions. As will become obvious as this chapter progresses, this is a general result which suggests that composite measures are necessary if truly broadly applicable objective measures are to be designed.

The results for TSQ are similar to, though slightly lower than, those for CA. By comparison, the results for TBQ are very poor, with a maximum magnitude correlation of only 0.149. Since all the spectral distance measures explicitly measure signal distortion, it is not surprising that they do a poor job of estimating background subjective qualities.

Although finding the very best measure from among all the measures tested is important and interesting, it gives very little insight into the relative importance of the various parameters tested. In the sections which follow, the relative effects of all the parameters tested will be illustrated using parametric graphs of the correlation results.

5.3.1.4 The Effects of the Pointwise Nonlinearities

A fundamental defining component of any spectral distance measure is the pointwise nonlinearity which is applied to the distorted and undistorted speech spectra. In this study, in effect, three types of nonlinearity were used: the logarithm nonlinearity, the $|\ |^{\delta}$ nonlinearity, and the inversion nonlinearity. As previously noted, the inversion nonlinearity was applied to both linear and log spectral distance measures, but not to $|\ |^{\delta}$ measures (see Table 5.2). These three classes of measures were not studied with equal resolution. In particular, the log and the $|\ |^{\delta}$ nonlinearities were broadly studied, while the study of inverted nonlinearities was much more directed. Stated another way, although there were many more control parameter values associated with the inverted nonlinearity, there were more actual combinations of parameters applied to the log and $|\ |^{\delta}$ measures.

A major goal of this study was to illustrate the important role of the pointwise nonlinearity in the performance of spectral distance measures. The fact that such nonlinearities play an important role is most easily seen by examining the results for the $|\ |^{\delta}$ measures. In all, twelve different values of δ were studied, ranging from 0.03 to 1.0, with

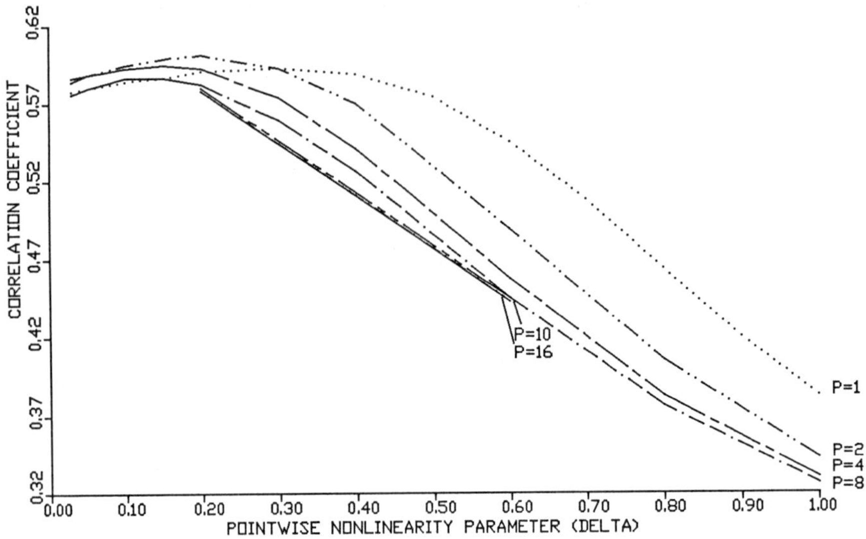

Figure 5.2: Correlation coefficients for CA across the ALL distortion subset for $|\ |^{\delta}$ spectral distance measures as a function of δ for $\gamma = 0$, $\alpha = 0$, and several values of the L_p parameter, p.

a δ of 1.0 corresponding to a linear spectral distance.

The effects of the $|\ |^{\delta}$ nonlinearity are illustrated in Figs. 5.2, 5.3, and 5.4. In Fig. 5.2, the correlation coefficients for the CA subjective measure over the ALL distortion subset are plotted as a function of δ for several different values of p in the L_p norm. Figs. 5.3 and 5.4 give similar plots for different values of the spectral weighting parameter, γ, and the energy weighting parameter, α, respectively.

It is clear from Figs. 5.2, 5.3, and 5.4 that the effect of varying δ is considerable, causing the correlation coefficient magnitude to vary from 0.61 to 0.32. In these results, the linear spectral distance alternative, $\delta = 1$, always had the worst performance. Varying the value of p in the L_p norm also can have a nontrivial effect, but only for values of δ greater than about 0.4. The very best performance is generally obtained around $\delta = .2$, where the value of p has very little impact on the correlation. For values of δ greater than about 0.3, $p = 1$ always gives the best performance.

In comparison, varying the values of γ and α has very little effect

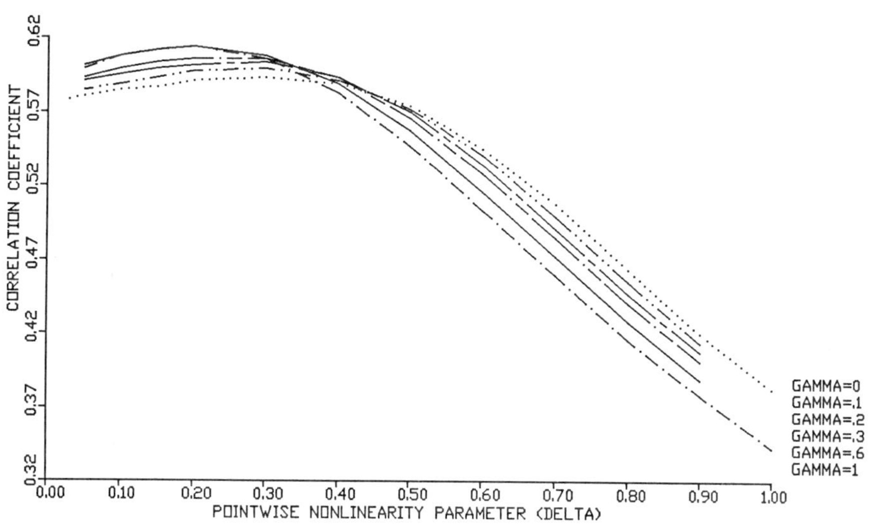

Figure 5.3: Correlation coefficients for CA across the ALL distortion subset for $|\ |^{\delta}$ spectral distance measures as a function of δ for $p = 1$, $\alpha = 0$, and several values of the spectral weighting parameter, γ.

on the correlation coefficients. In general, the best performance for energy weighting is obtained for $\alpha = 0$ (no energy weighting), while the best performance for the spectral weighting parameter is obtained with $\gamma = 0$ for values of δ greater than about 0.4, and $\gamma = 1$ for values below 0.4. The primary result here remains, however, that neither spectral weighting nor energy weighting has much effect on the performance of the spectral distance measures.

As expected, the effect of the logarithm pointwise nonlinearity is very similar to the effect of the $|\ |^{\delta}$ nonlinearity for values of δ around 0.2. This effect is observable below in Figs. 5.6, 5.7, and 5.8.

The effects of the inverted pointwise nonlinearity were studied in a different way from the other nonlinearities. This was because the large number of parameters involved made an extensive sampling of the total parameter space impractical. To begin with, the study was limited to only the linear and log spectral distance measures. Then, the parameter space was sampled very coarsely, and correlation analyses were performed. Based on the results of these experiments, the most

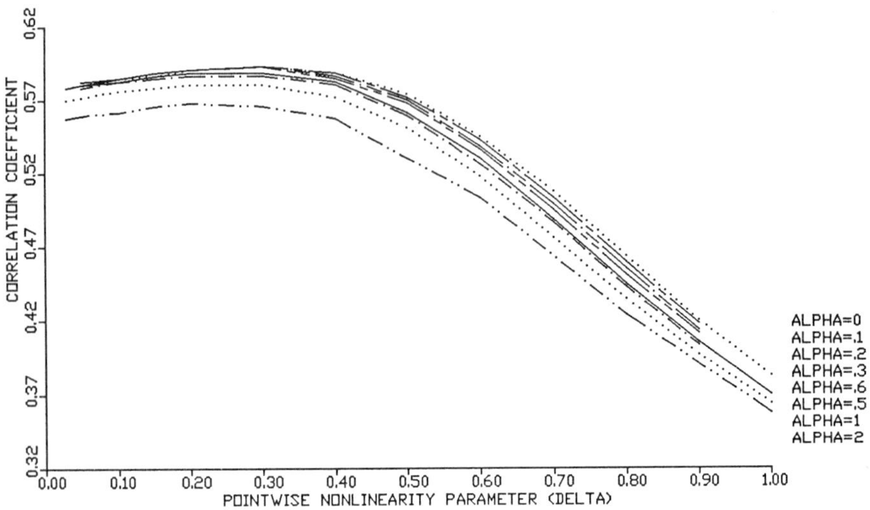

Figure 5.4: Correlation coefficients for CA across the ALL distortion subset for $|\ |^{\delta}$ spectral distance measures as a function of δ for $p = 1$, $\gamma = 0$, and several values of the energy weighting parameter, α.

promising regions of parameter space were examined more closely by sampling the parameters more finely. The procedure was continued until no further improvement in correlation was observable.

The effect of the experimental procedure outlined above is to search the parameter space sparsely and suboptimally. If the multidimensional correlation surface is smooth and convex, then this procedure should result in the optimal parameter values. If it is not, then a suboptimal parameter set may well result.

During the optimization procedure, about four hundred different inverted spectral distance measures were tested. To the extent they were observable, the general results for the inverted pointwise nonlinearity matched the results for the $|\ |^{\delta}$ and the log nonlinearities. It was observed that the impact of energy weighting and spectral weighting was very small, while the impact of the L_p norm was slightly greater. It was also observed that for the highest performance measures studied, good performance was obtained for a very wide range of the inverted control parameter, b. This point is clearly illustrated in Fig. 5.5, in

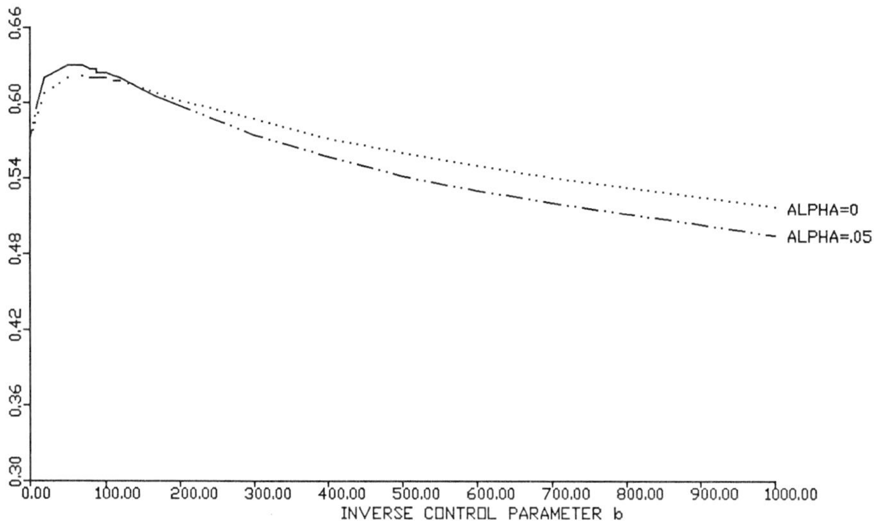

Figure 5.5: Correlation coefficients for CA across the ALL distortion subset for the inverted pointwise nonlinearity applied to the linear spectral distance measure as a function of b for $p = 1$, $\gamma = 0$, and two different values of the energy weighting parameter, α.

which b is varied from 0.01 to 1000 for two different parameter settings. One point which is quite clear is that the application of this nonlinearity can greatly improve performance in some cases. This is evidenced by the dramatic improvement in performance for the linear spectral distance when the inversions were made.

5.3.1.5 The Effects of the L_p Averaging

One of the primary goals of this study was to measure the effect of the L_p averaging operation on the correlation between the subjective and objective measures. The data in Fig. 5.2 above provides a partial answer to the question. In particular, it was observed above that variations in p had, at most, a moderate effect on the performance of the spectral distance measures tested.

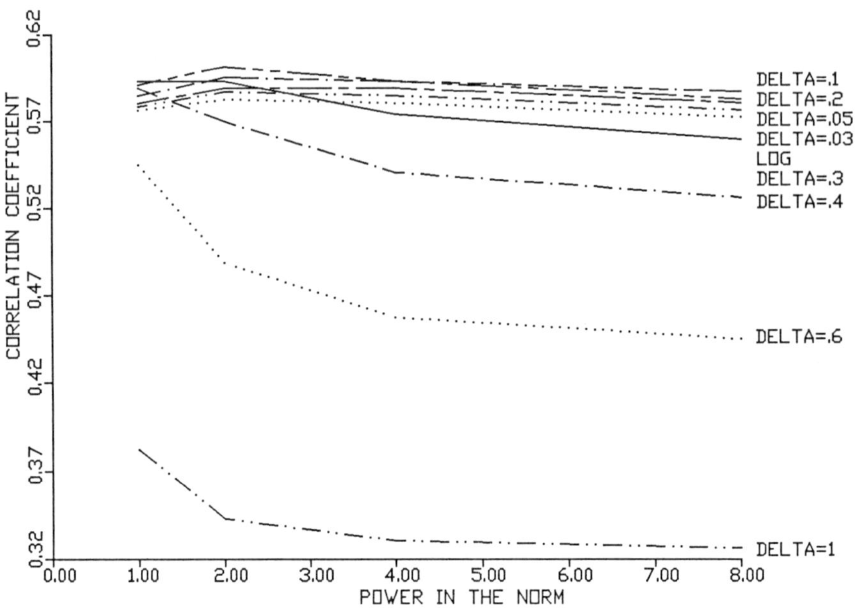

Figure 5.6: Correlation coefficients for CA across the ALL distortion subset for the value of p in the L_p averaging operation for $\alpha = 0$, $\gamma = 0$ and several different $| \ |^\delta$, log, and linear spectral distance measures.

A more complete picture of the impact of p is presented in Fig. 5.6, which plots the correlation coefficients as a function of p for a variety of $| \ |^\delta$, log, and linear spectral distance measures. The results show that for the highest performance spectral distance measures ($| \ |^\delta$ for $\delta = 0.05$, 0.1, 0.2 and log), the value of p has very little impact on the correlation. For the lower performance measures ($| \ |^\delta$ for $\delta = 0.4$, 0.6, and linear), higher correlations are consistently obtained for lower values of p.

Figure 5.6 only shows results for values of p from 1 to 8. In the broad study, p was allowed to vary from 1 to 32. For all measures tested, it was observed that all values of p greater than 8 resulted in essentially the same performance as for $p = 8$.

5.3.1.6 The Effects of Spectral Weighting

The fundamental question for both spectral weighting and energy weighting is whether or not there is any advantage to be gained by the use of these functions in the spectral distance measures at all. On the one hand, it was observed above (Table 5.3) that several of the best spectral distance measures have non-zero values of γ. On the other hand, it was also observed in the discussion of pointwise nonlinearities that large variations in γ did not result in correspondingly large variations in the correlation coefficients.

Figure 5.7 shows a plot of the correlation coefficients as a function of γ for a number of linear and nonlinear spectral distance measures. Clearly, the primary result is that spectral weighting has very little effect on the performance of any of the spectral distance measures. There is, however, a small but clear secondary effect observable in Fig. 5.7. This is that for the higher performance measures, there is a small but consistent gain in performance with larger values of γ. Similarly, for lower performance measures, there is a small degradation associated with larger values of γ. Thus, the results of these experiments on the effectiveness of spectral weighting are mixed. The experiments show very clearly that the impact of spectral weighting is small. They also seem to show, however, that small gains in performance are possible by using spectral weighting.

5.3.1.7 The Effects of Energy Weighting

As for the case of spectral weighting, energy weighting in the time domain appears to have very little effect on the performance of most spectral distance measures. The effect can be clearly observed from the data presented in Fig. 5.8, which shows correlation coefficients plotted as a function of the energy weighting parameter α for a number of linear and nonlinear spectral distance measures.

From Fig. 5.8, it is clear that energy weighting generally has very little effect on the performance of a spectral distance measure. This is true for the relatively high performance measures such as the $|\ |^{\delta}$ and log spectral distance measures as well as the poorer performing linear measures. It is noteworthy that in general no energy weighting (α=0) virtually always gives better performance than any form of en-

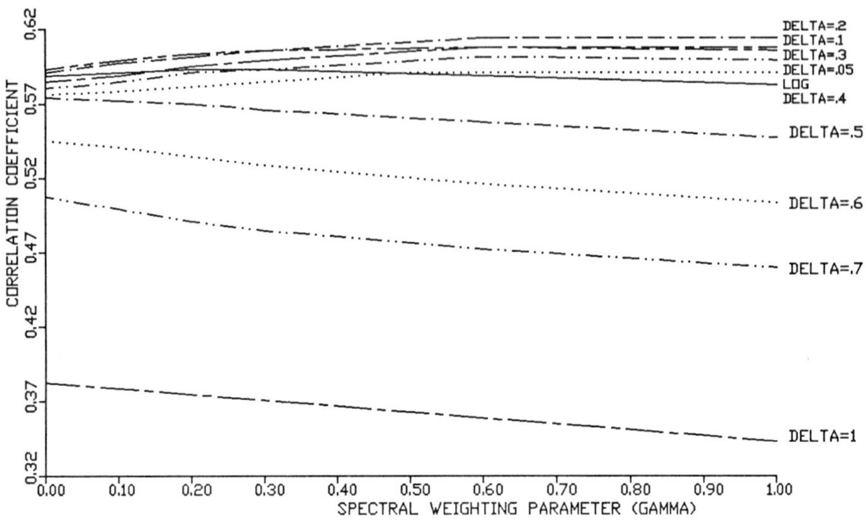

Figure 5.7: Correlation coefficients for CA across the ALL distortion subset plotted as a function of the spectral weighting parameter γ for $p = 1$, $\alpha = 0$ and several different $| \ |^{\delta}$, log, and linear spectral distance measures.

ergy weighting. Thus, energy weighting as a technique does not seem to improve spectral distance measures.

5.3.1.8 The Effects of Other Subjective Measures

Up to this section, most of the correlation results which have been reported have been between individual spectral distance measures and the composite acceptability (CA) isometric quality measure from the DAM. Composite acceptability is the most important DAM measure because it is intended to estimate the overall system performance. The DAM, however, provides a number of other measures of speech quality, and the ability of spectral distance measures to estimate these other subjective measures is also of interest.

Table 5.4 shows the maximum correlation magnitude value found for spectral distance measures over ALL and WBD for six different isometric subjective quality measures available from the DAM: Composite

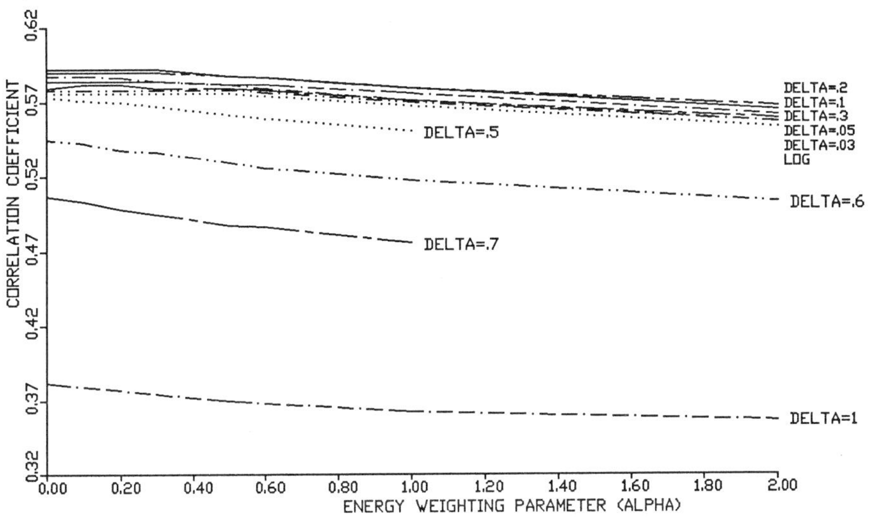

Figure 5.8: Correlation coefficients for CA across the ALL distortion subset plotted as a function of the energy weighting parameter α for $p = 1$, $\gamma = 0$ and several different $|\ |^{\delta}$, log, and linear spectral distance measures.

Acceptability (CA), Total System Quality (TSQ), Total Background Quality (TBQ), Pleasantness (P), Intelligibility (I), and Acceptability (A). Although only the maximum correlation values for all of the spectral distance measures tested are included in Table 5.4, these values are quite representative of the performance of the entire class of measures.

Two points are obvious from the data of Table 5.4. First, as before, TBQ is not tracked well by the objective measures. Since all spectral distance measures are essentially a measure of signal distortion, this is not unexpected. Second, the behavior is similar over all the measures, but the intelligibility measure (I) is estimated slightly better than the rest. The worst estimation of a signal quality is for pleasantness (P), with acceptability (A) showing intermediate behavior.

Figure 5.9 shows a plot of the maximum magnitude correlation coefficient for the same six isometric measures as a function of the individual type of spectral distance measure. The five types of spectral

Table 5.4: Maximum correlation over all spectral distance measures for different subjective measures.

Distortion Set		CA	TSQ	TBQ	P	I	A
ALL	$\hat{\rho}$	-.63	-.58	.17	-.53	-.66	.56
	$\hat{\sigma}_e$	7.6	8.7	7.1	8.5	8.1	8.1
WBD	$\hat{\rho}$	-.55	-.53	.24	-.47	-.71	-.64
	$\hat{\sigma}_e$	7.5	8.7	6.0	6.5	6.9	7.0

distance measure considered were $| \; |^{\delta}$, log, linear, inverted log, and inverted linear. The data in Fig. 5.9 provide no dramatic surprises. Once again, the inability of the spectral distance measures, in general, to predict perceived TBQ is clear and the performance of all of the different types of spectral distance measures on the other isometric measures is quite similar. All of the spectral distance measures have a slight advantage when predicting perceived intelligibility. In addition, the rank ordering of the different types of spectral distance measures is similar for all of the isometric measures. This not surprising, because CA, TSQ, I, P, and A are generally highly correlated with one another.

The results are quite different for parametric subjective measures. Figure 5.10 shows a plot of the maximum magnitude correlation coefficient for the parametric DAM measures for the same five types of spectral distance measures. The six signal parametric measures (SF, SH, SD, SL, SI, and SN) are shown on the left side of the graph, while the four background parametric measures (BN, BB, BF, and BR) are shown on the right. As with the isometric measures, there is generally much better performance on the signal distortions than on the background distortions. But unlike the isometric measures, the performance is not uniform.

In the parametric signal distortion, relatively good performance is obtained for SH and SL, relatively poor performance is obtained for SF and SI, and moderate performance is obtained for SD and SN. Upon reflection, these are not unreasonable results. The SH and SL measures are generally associated with lowpass filtering and highpass filtering, respectively. Since these are intrinsically spectral distortions, it is not

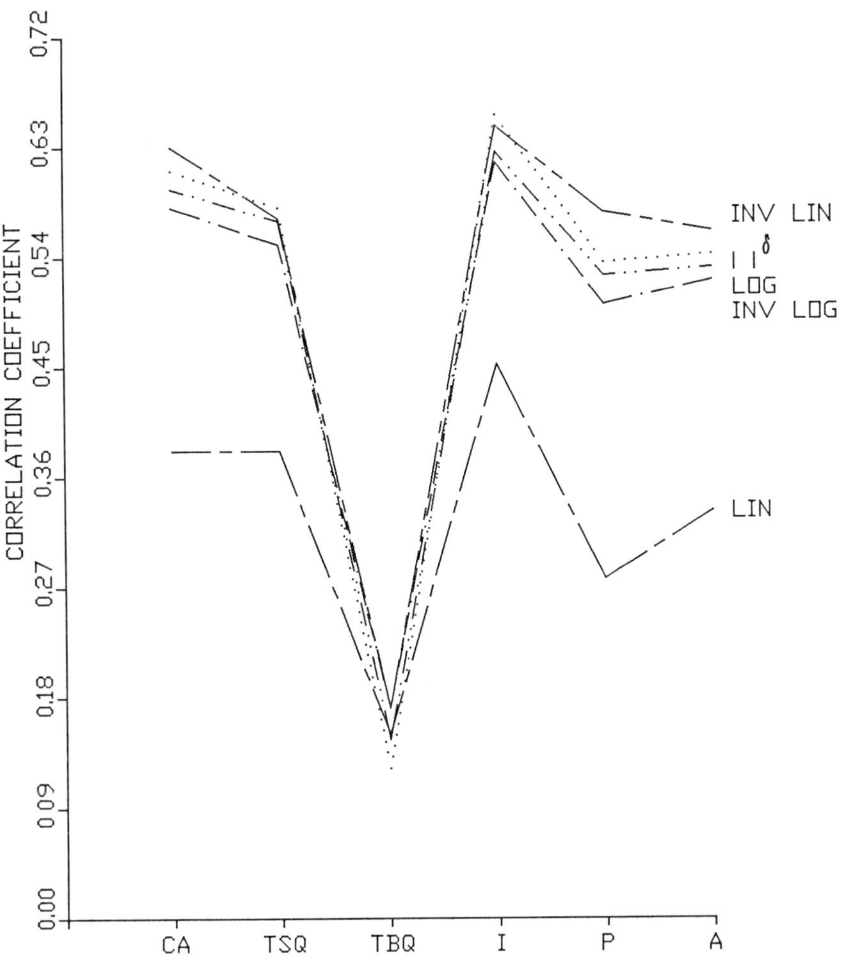

Figure 5.9: Maximum correlation coefficients for a variety of different isometric DAM subjective measures for the five types of spectral distance measures studied.

surprising that they can be predicted by spectral distance measures. SF and SI (fluttering and interrupted), on the other hand, are intrinsically time domain effects which cannot be well predicted by purely frequency domain measures.

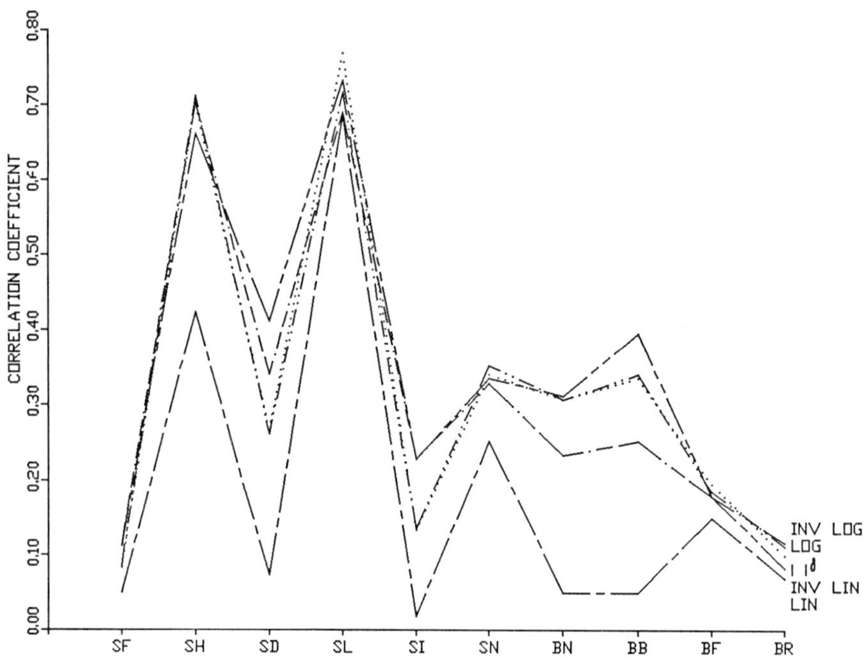

Figure 5.10: Maximum correlation coefficients for different parametric DAM subjective measures for the five types of spectral distance measures studied.

5.3.1.9 The Effects of Different Distorted Data Bases

Figure 5.11 illustrates the performance of the five types of spectral distance measures when the correlation analyses are performed over the various distortion subsets defined in Table 5.1. There are several important effects which are contained in this data. First, and probably most important, many of the subsets show essentially similar performance. The similar subsets include all distortions (ALL), wideband distortions (WBD), coders (CODE), controlled distortions (CON), banded distortions (BD), and pole distortions (PD). This suggests that spectral distance measures, as a class, are not particularly well suited to any particular type of distortions. In contrast, the performance is surprisingly good on two distortion subsets: waveform coders (WFC), and narrowband noise (NBN). Since waveform coders are such an important class of distortions, this may prove to be an important result. Finally,

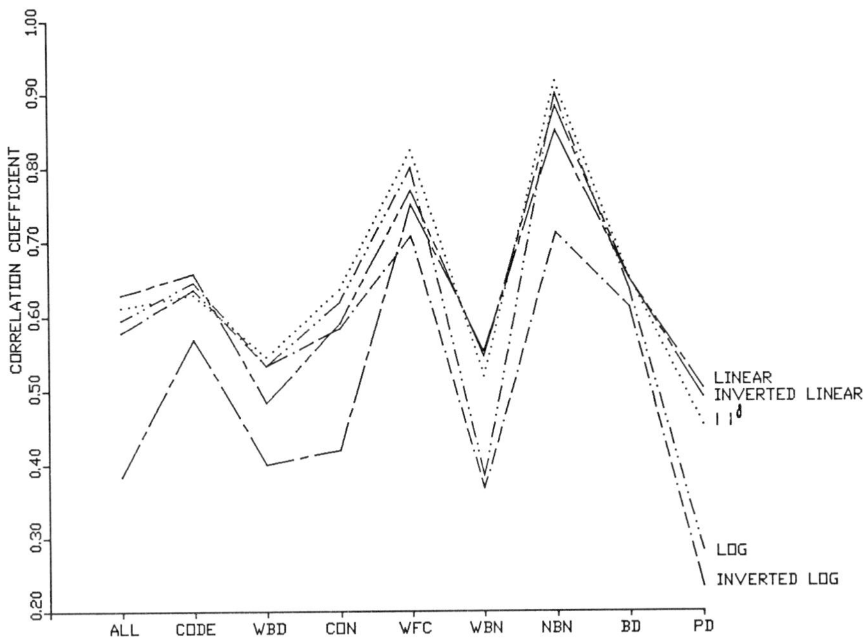

Figure 5.11: Maximum correlation coefficients between CA for the five types of spectral distance measures studied and for the distortion subsets defined in Table 5.1.

the performance is quite low on wideband additive noise (WBN). This effect is particularly surprising because of the very small number of distortions in this class and because of the good performance on NBN.

5.3.1.10 The Effects of Polynomial Regression Analysis

In order to study the effects of using higher order regression analysis, the CA, TSQ and TBQ subjective measures were tested using third and sixth order regression analyses across the ALL and WBD distorted data bases for a set of about 400 of the spectral distance measures under consideration (none of which used inverted pointwise nonlinearity). Table 5.5 gives a compilation of the best spectral distance measures for each distortion subset and each regression order in this study. In both the CA and TSQ cases, it appears that good improvements in performance are possible by using higher order regression. One qualification

Table 5.5: The effects of polynomial regression analysis on spectral distance measures.

CA

		1st Order	3rd Order	6th Order
ALL	$\hat{\rho}$	.60	.69	.80
	$\hat{\sigma}_e$	7.8	7.1	5.8
WBD	$\hat{\rho}$	.63	.73	–
	$\hat{\sigma}_e$	7.0	6.1	–

TSQ

		1st Order	3rd Order	6th Order
ALL	$\hat{\rho}$	.57	.64	.75
	$\hat{\sigma}_e$	8.8	8.2	7.0
WBD	$\hat{\rho}$	.64	.70	–
	$\hat{\sigma}_e$	7.7	7.1	–

TBQ

		1st Order	3rd Order	6th Order
ALL	$\hat{\rho}$	.14	.28	.44
	$\hat{\sigma}_e$	6.0	6.9	6.4
WBD	$\hat{\rho}$	.23	.42	–
	$\hat{\sigma}_e$	6.0	5.5	–

must be applied in the analysis of these results, however. While it is obvious that the higher the order of the polynomial model, the better the results, at the same time the results become more sensitive to artifacts. In this sense, the numbers presented here should be considered approximate upper limits on the performance of measures based on higher order regression models. In spite of the above warning, these results are very promising, and at least the third order effects should be attainable in a real system.

5.3.2 Simple LPC-Based Parametric Distances

As in the case of spectral distance measures, the LPC-based parametric distance measures assume that the distorted and undistorted speech signals have been divided into frames, given by $x_n(m,s,\phi)$ and $x_n(m,s,d)$ where n is the frame index, m is the sample index, s is the speaker index, d is the distortion index, and ϕ indicates no distortion. For each parametric distance measure, a set of L parameters, $\epsilon(n,s,d,l)$, $l=1,\ldots,L$, is derived from the corresponding speech frame $x_n(m,s,d)$. As in the case of spectral distance, a function F for use in Eq. (5.1) is derived for each case, given by

$$F = \left\{ \frac{1}{10} \sum_{l=1}^{10} |\epsilon(n,s,d,l) - \epsilon(n,s,\phi,l)|^p \right\}^{1/p} \tag{5.22}$$

where once again the L_p norm is taken, and as before, p is an object of study for each parametric distance measure.

All seven classes of the parametric distance measures studied were derivatives of LPC analysis. The first two were based on the feedback coefficient set, $a(1),\ldots,a(10)$, which is described in Eq. (5.5). The first form, the *linear feedback* measure is given by

$$F = \left\{ \frac{1}{10} \sum_{l=1}^{10} |a(n,s,d,l) - a(n,s,\phi,l)|^p \right\}^{1/p} \tag{5.23}$$

and second form, the *log feedback* measure is given by

$$F = \left\{ \frac{1}{10} \sum_{l=1}^{10} \left| 20 \log_{10} \left| \frac{a(n,s,d,l)}{a(n,s,\phi,l)} \right| \right|^p \right\}^{1/p} \tag{5.24}$$

Neither of these measures was expected to be of much interest, but they were included for completeness.

The third and fourth measures were based on the PARCOR coefficients, $K(m)$, as defined in Eq. (5.5). These two measures are given by

$$F = \left\{ \frac{1}{10} \sum_{l=1}^{10} |K(n,s,d,l) - K(n,s,\phi,l)|^p \right\}^{1/p} \tag{5.25}$$

and

$$F = \left\{ \frac{1}{10} \sum_{l=1}^{10} \left| 20 \log_{10} \left| \frac{K(n,s,d,l)}{K(n,s,\phi,l)} \right| \right|^p \right\}^{1/p} \tag{5.26}$$

where $K(n,s,d,l)$ and $K(n,s,\phi,l)$ are the lth PARCOR coefficients derived from the nth frame of the distorted and undistorted speech from speaker s, respectively.

The fifth and sixth measures were based on the area ratio functions

$$AR(n,s,d,l) = \frac{1 + K(n,s,d,l)}{1 - K(n,s,d,l)} \tag{5.27}$$

These measures are given by

$$F = \left\{ \frac{1}{10} \sum_{l=1}^{10} |AR(n,s,d,l) - AR(n,s,\phi,l)|^p \right\}^{1/p} \tag{5.28}$$

and

$$F = \left\{ \frac{1}{10} \sum_{l=1}^{10} \left| 20 \log_{10} \left[\frac{AR(n,s,d,l)}{AR(n,s,\phi,l)} \right] \right|^p \right\}^{1/p} \tag{5.29}$$

The final parametric measure of interest is called the *energy ratio* measure. This measure was first suggested by Itakura [5.11], and has been widely used as a quality measure [5.12], [5.13]. In computing this measure, a frame-by-frame LPC analysis is performed on both the undistorted and distorted speech, as shown in Fig. 5.12. Then the undistorted speech is passed through two *vocal tract inverse filters* given by

$$H_\phi(z) = 1 - \sum_{l=1}^{10} a_\phi(l) z^{-l} \tag{5.30}$$

$$H_d(z) = 1 - \sum_{l=1}^{10} a_d(l) z^{-l}$$

The output signal of each channel is then squared and summed, giving $< e^2(n,s,\phi) >$ and $< e^2(n,s,d) >$. The energy ratio is then given by either

$$F = \frac{< e^2(n,s,d) >}{< e^2(n,s,\phi) >} \tag{5.31}$$

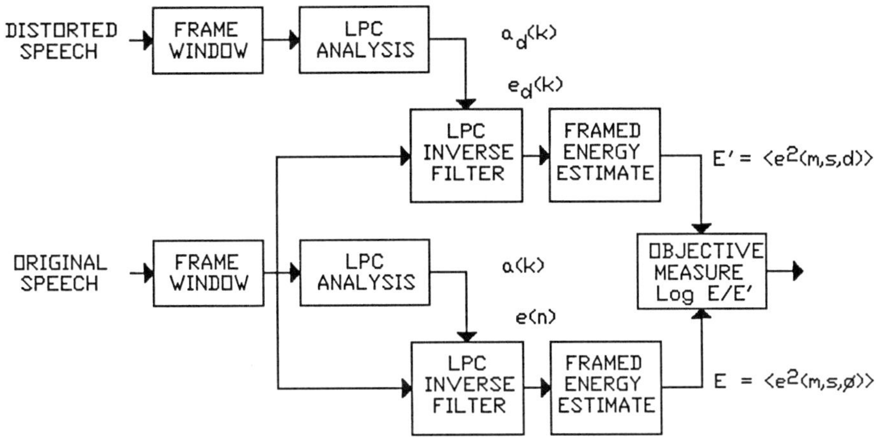

Figure 5.12: Computation of the residual energy measure.

or

$$F = 10\log_{10}\left[\frac{< e^2(n,s,d) >}{< e^2(n,s,\phi) >}\right] \tag{5.32}$$

In general, it is possible to compute these measures in a way which is much more efficient than the procedure suggested by Fig. 5.12. In particular, it can be shown that

$$\frac{< e^2(n,s,d) >}{< e^2(n,s,\phi) >} = \left[\frac{\vec{a}^{\,\mathrm{T}}(n,s,d)\mathbf{R}(n,s,\phi)\vec{a}(n,s,d)}{\vec{a}^{\,\mathrm{T}}(n,s,\phi)\mathbf{R}(n,s,\phi)\vec{a}(n,s,\phi)}\right] \tag{5.33}$$

where

$$\mathbf{R} = \begin{bmatrix} R(0) & R(1) & \cdots & R(10) \\ R(1) & R(0) & \cdots & R(9) \\ \vdots & \vdots & \ddots & \vdots \\ R(10) & R(9) & \cdots & R(0) \end{bmatrix} \tag{5.34}$$

and $R(k)$ is defined by Eq. (5.4) and

$$\vec{a} = \begin{bmatrix} 1 \\ -a(1) \\ -a(2) \\ \vdots \\ -a(10) \end{bmatrix} \tag{5.35}$$

where $a(k)$ is defined by Eq. (5.5). The three forms of this measure which were studied are given by Eqs. (5.31) and (5.32), plus

$$F = \left| \frac{< e^2(n,s,d) >}{< e^2(n,s,\phi) >} \right|^{\frac{\delta}{2}} \tag{5.36}$$

The LPC-based parametric distance measures can be divided into four classes: feedback coefficient distance measures, PARCOR coefficient distance measures, area ratio distance measures, and the energy ratio measure. In the experimental study, the first three classes measures were studied as a group, while the energy ratio measure, because of its wide use, was studied separately. In all, thirty-eight forms of the energy ratio measure and thirty-six forms of the other measures were studied. The overall experimental philosophy was the same for these measures as for the spectral distance measures, and a similar set of experiments was conducted. As before, the ALL and WBD distortion subsets are used to quantify the performance of the parametric objective measures.

Within each of the seven groups of parametric distance measures, the particular distance may be described by two conditions: the value of δ in the $| \ |^\delta$ for the energy ratio measure or the value of p in the L_p norm of the other measures, and the energy weighting parameter, α. In terms of these parameters, Table 5.6 describes the measures tested for each of the seven groups.

5.3.2.1 The Best LPC-Based Parametric Distance Measures

The best parametric distance tested was found to be the L_1 log area ratio measure without energy weighting. This measure has a correlation coefficient of -.62 for CA across ALL, and -.65 for CA across

Table 5.6: Parameters for parametric distance measures.

	p	Energy (α)	Number
Linear Feedback	1, 2	0, 1, 2	6
Log feedback	1, 2	0, 1, 2	6
Linear PARCOR	1, 2	0, 1, 2	6
Log PARCOR	1, 2	0, 1, 2	6
Linear area ratio	1, 2	0, 1, 2	6
Log area ratio	1, 2	0, 1, 2	6
	$\mid \mid^{\delta}$ (δ)		
Energy ratio	0.25, 0.5, 1, 2, 4, 8	0, 0.25, 0.5, 1, 2, 4, 8	38
log			

WBD. This is a very important result, for it shows that this parametric distance measure *performed essentially as well as any of the spectral distance measures* even though it is an order of magnitude less complex to compute.

Table 5.7 shows the highest magnitude correlation coefficient found for each of the seven classes of parametric measures for CA across ALL and WBD. The only two measures in this set which show any real promise are the log area ratio distance measure and the energy ratio distance measure. The results for these two measures will hence be presented in more detail.

5.3.2.2 The Log Area Ratio Measure

The results for the log area ratio measure tests are summarized in Table 5.8, which shows the maximum correlations obtained against the CA, TSQ, TBQ, I, P, and A subjective measures. Figure 5.14 shows a plot of all the correlation results of all the log area ratios measures tested for CA. The remarkable point concerning this data is its similarity to the results obtained for spectral distance measures. In each case, the log area ratio measure performs comparable to, or better than, the corresponding spectral distance measure. As with the spectral distance

Table 5.7: Maximum correlation over ALL and WBD for different parametric objective measures.

		Linear Feed-back	Linear PARCOR	Linear Area Ratio	Log Feed-back	Log PARCOR	Log Area Ratio	Energy Ratio
ALL	$\mid \hat{\rho} \mid$	-.06	-.46	.24	-.11	-.11	-.62	-.59
	$\hat{\sigma}_e$	9.8	9.3	9.6	9.8	9.8	7.7	7.9
WBD	$\mid \hat{\rho} \mid$	-.14	-.43	.32	-.33	-.32	-.65	-.65
	$\hat{\sigma}_e$	8.9	8.1	8.5	8.5	8.5	6.8	6.9

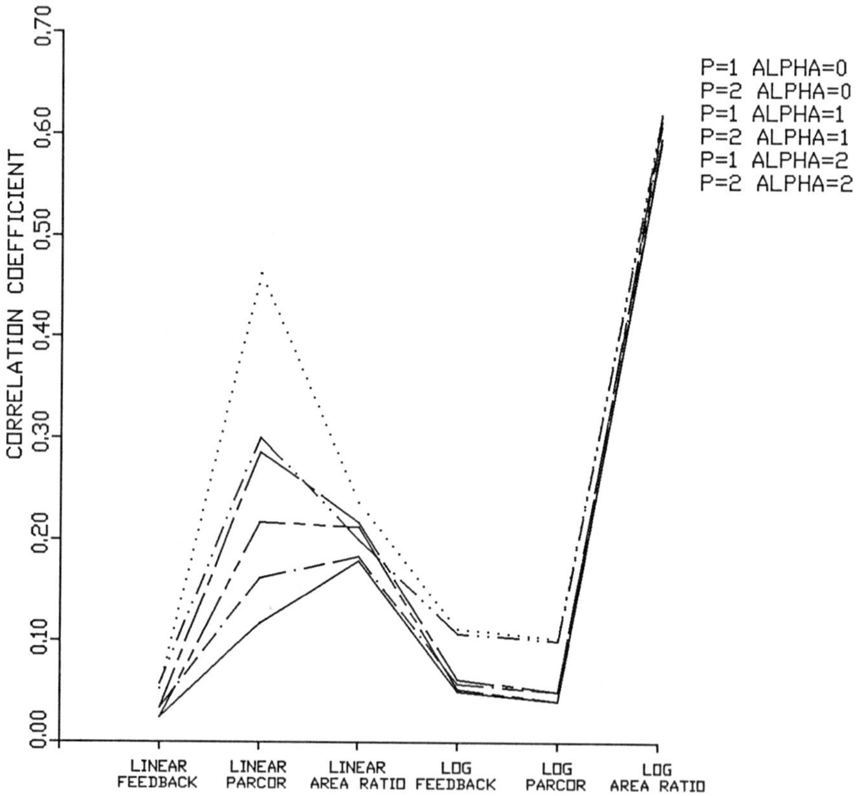

Figure 5.13: Maximum correlation coefficients between CA for the six types of parametric measures studied as a function of p and α.

Table 5.8: Maximum correlation over ALL and WBD for the log area ratio measure for different subjective measures.

Distortion Set		CA	TSQ	TBQ	P	I	A
ALL	$\hat{\rho}$	-.62	-.58	.11	-.55	-.64	-.63
	$\hat{\sigma}_e$	7.7	8.7	7.0	7.6	6.9	8.1
WBD	$\hat{\rho}$	-.65	-.52	.15	-.49	-.66	-.64
	$\hat{\sigma}_e$	6.8	8.7	6.1	6.7	7.3	7.0

Table 5.9: Performance of the log area ratio measure across different distortion subsets.

Distortion Subset	$\mid\hat{\rho}\mid$	$\hat{\sigma}_e$
ALL	.62	7.7
WBD	.65	6.8
WFC	.64	6.9
CODE	.62	6.2
CON	.65	8.2
WBN	.40	7.0
NBN	.91	3.8
BD	.58	6.0
PD	.53	6.9

measure, performance was relatively poor for TBQ.

Table 5.9 shows the maximum correlations for the log area ratios across the other distortion subsets for CA. Here, once again, the results are equivalent to those from the spectral distance measures. Table 5.10 shows the effects of third order and sixth order nonlinear regression. Improvements here are also comparable to those for spectral distance measures.

Examination of the data also shows other similarities to the spectral distance results. For the log area ratio, no energy weighting is best, followed by energy weighting to the first and second powers.

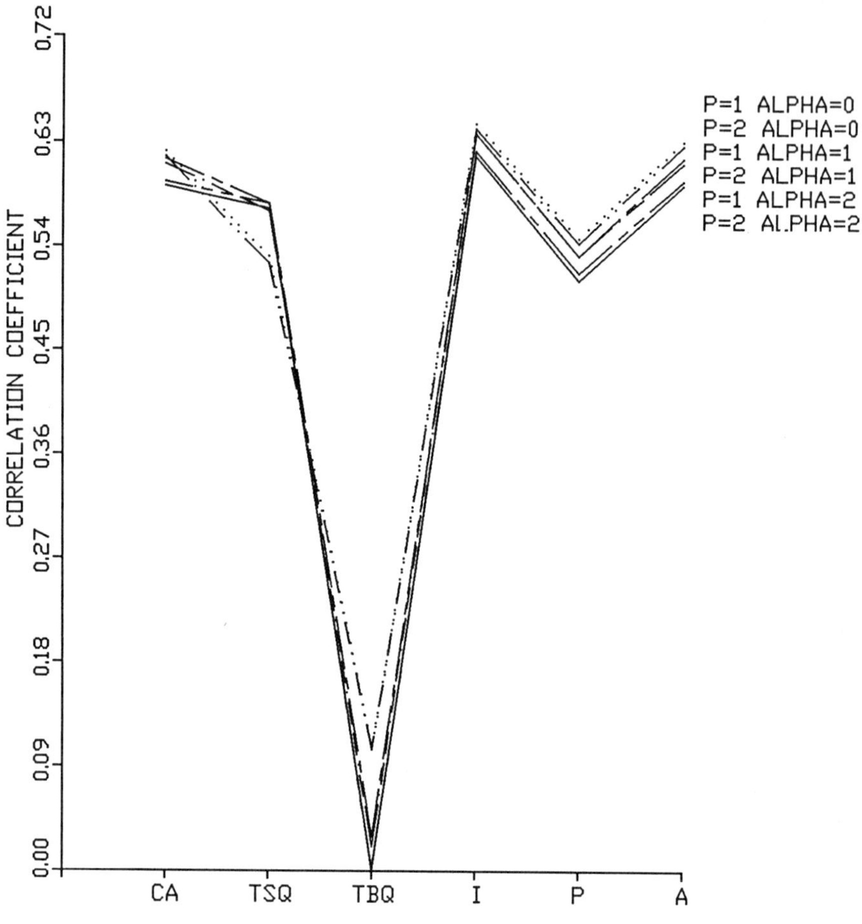

Figure 5.14: Correlation coefficients for a variety of different isometric DAM subjective measures for the six different log area ratio distance measures.

5.3.2.3 The Energy Ratio Distance Measure

The results for the energy ratio distance measure are summarized in Tables 5.11, 5.12, and 5.13. The first table gives maximum results for CA, TSQ, and TBQ over CA and WBD. The second table gives maximum results for CA over the other distortion subsets. The third table shows the results of nonlinear regression analysis.

Table 5.10: The effects of higher order regression analysis on the log area ratio distance measure.

	Analysis Order					
	1st Order		3rd Order		6th Order	
	$\lvert\hat{\rho}\rvert$	$\hat{\sigma}_e$	$\lvert\hat{\rho}\rvert$	$\hat{\sigma}_e$	$\lvert\hat{\rho}\rvert$	$\hat{\sigma}_e$
CA (ALL)	.62	7.7	.64	7.5	.69	7.1
CA (WBD)	.65	6.8	.66	6.7	.79	5.5
TSQ (ALL)	.58	8.7	.59	8.7	.72	7.4
TSQ (WBD)	.62	7.0	.63	7.0	.72	6.1
TBQ (ALL)	.11	7.2	.24	7.0	.42	6.6
TBQ (WBD)	.15	6.1	.35	8.4		

Table 5.11: Performance of the energy ratio measure.

	CA		WBD	
	$\lvert\hat{\rho}\rvert$	$\hat{\sigma}_e$	$\lvert\hat{\rho}\rvert$	$\hat{\sigma}_e$
CA	0.59	7.9	0.65	6.9
TSQ	0.54	9.0	0.62	7.0
TBQ	0.12	7.1	0.24	6.8

Table 5.12: Performance of the energy ratio measure across different distortion subsets.

Distortion Subset	$\lvert\hat{\rho}\rvert$	$\hat{\sigma}_e$
ALL	.59	7.9
WBD	.61	6.9
WFC	.58	6.7
CODE	.53	6.7
CON	.59	8.7
WBN	.47	6.7
NBN	.80	5.5
BD	.60	5.9
PD	.57	6.7

Table 5.13: The effects of higher order regression analysis on the energy ratio distance measure.

	Analysis Order					
	1st Order		3rd Order		6th Order	
	$\mid \hat{\rho} \mid$	$\hat{\sigma}_e$	$\mid \hat{\rho} \mid$	$\hat{\sigma}_e$	$\mid \hat{\rho} \mid$	$\hat{\sigma}_e$
CA (ALL)	.59	7.9	.64	7.6	.64	7.5
CA (WBD)	.65	6.9	.66	6.7	.68	6.6
TSQ (ALL)	.54	9.0	.38	8.8	.60	8.6
TSQ (WBD)	.62	7.0	.63	7.0	.65	6.8
TBQ (ALL)	.12	7.1	.24	7.0	–	–
TBQ (WBD)	.24	6.8	.30	6.7	.36	6.4

The energy ratio distance measure with the $| \, |^{.25}$ nonlinearity and $\alpha = 0$ performs the best in these tests, but it is not quite able to match the performance of either the log area ratio measure or the best spectral distance measure. The general performance of all three of these measures is very similar, with the energy ratio measures being the poorest of the three. This is probably because all three of these objective measures are actually responding to very similar features of the speech distortion. Note that Eq. (5.32) with α set to 0.0 defines the conventional log likelihood measure, for which the correlation coefficient was -0.49.

5.3.3 Simple Noise Measures

For many years, the signal-to-noise ratio (SNR) has been used as a quality measure for systems in which it is an applicable concept. In digital communications, the model of signal plus noise is meaningful in systems where the received signal is designed to be a sample-by-sample copy of the input signal. These systems include all forms of waveform coders, such as CVSD, ADM, DPCM, ADPCM, and APC, as well as the new techniques of subband and adaptive transform coding. These systems do not include the vocoder and "vocoder-like" systems such as LPC, VEV's of all types, channel vocoders, etc.

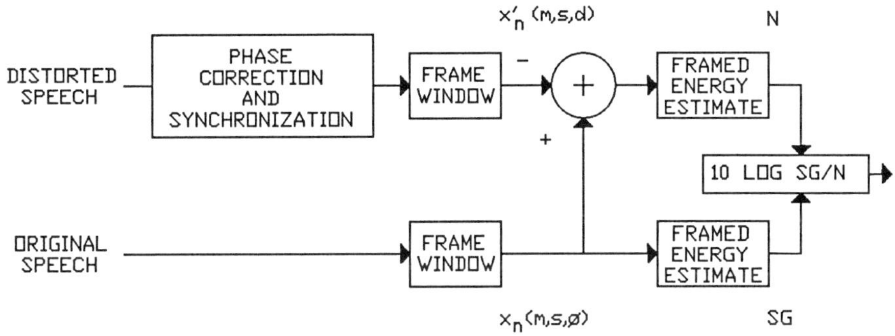

Figure 5.15: System for computing short-time SNR.

In this study, two types of broadband noise measurements were investigated. The first was the traditional SNR. In this system (see Fig. 5.15) any linear or nonlinear phase variations introduced in processing are first corrected. Since all of the distortions in the study were produced by computer simulation, this process was a straightforward procedure, although for actual digital communication systems, the synchronization and nonlinear phase correction problem could be severe. Once the phase corrected signals are available, the frame noise energy, $N(n, s, d)$ is computed as

$$N(n,s,d) = \left\{ \frac{1}{W} \sum_{m=-\infty}^{\infty} [x_n(m,s,d) - x_n(m,s,\phi)]^2 \right\} \tag{5.37}$$

where $x_n(m, s, d)$ and $x_n(m, s, \phi)$ are the windowed distorted signal and undistorted signal, respectively, as defined by Eq. (5.2), and W is the window length. Note that in practice the limits on m are finite because of the windowing process. In the same terms, the signal energy is defined as

$$SG(n,s) = \left\{ \frac{1}{W} \sum_{m=-\infty}^{\infty} [x_n(m,s,\phi)]^2 \right\} \tag{5.38}$$

and the traditional SNR can be defined as

$$\text{SNR} = D(d) = 10 \log_{10} \left\{ \frac{\sum_{s=1}^{4} \sum_{n=1}^{N} SG(n,s)}{\sum_{s=1}^{4} \sum_{n=1}^{N} N(n,s,d)} \right\} \tag{5.39}$$

where $D(d)$ indicates that this is an objective measure and the definition of terms is the same as in Sec. 5.2.

The second class of measures of interest were *short-time* or *framed* noise measurements, in which a frame-by-frame signal-to-noise ratio is computed, and a global average is computed from Eq. (5.1). In this measurement,

$$F = 20[\log_{10} G(n,s,d)]^{\delta} \tag{5.40}$$

where

$$log_{10}[1 + G(n,s,d)^2] = log_{10} \left[1 + \frac{SG(n,s,d)}{N(n,s,d)} \right] \tag{5.41}$$

and δ is a parameter for study. These short-time signal-to-noise ratios have been shown to be more highly correlated with subjective results than are traditional SNR measurements [5.14], [5.15].

Four simple noise measures were studied: the ordinary SNR and the short-time SNR with $\delta = 0.5$, 1, and 2. In all the short-time studies, the frame interval was taken to be 256 points. Previous researchers [5.15] have indicated that this measure is relatively insensitive to the frame interval. These measures, of course, are only meaningful over the waveform coders and those controlled distortions that can be thought of as being additive noise, and hence, were only tested across WFC (waveform coders) and ND (noise distortions). Table 5.14 shows the results of these experiments.

An obvious conclusion that can be made from the results is that the traditional SNR is a poor objective measure. By comparison, all forms of the short-time SNR consistently perform better. The performances of all the measures are comparable over the WFC and ND distortion sets, and should be a good estimate of their expected performance in actual coding tests. The best value of δ was found to be 1, although the differences among the three tested values were small.

Table 5.14: Results for SNR and short-time SNR for CA across WFC and ND.

	WFC		ND	
	$\lvert \hat{\rho} \rvert$	$\hat{\sigma}_e$	$\lvert \hat{\rho} \rvert$	$\hat{\sigma}_e$
SNR	.24	8.8	.31	8.8
Segmental SNR ($\delta = .5$)	.76	5.6	.77	5.9
Segmental SNR ($\delta = 1$)	.77	5.7	.78	6.0
Segmental SNR ($\delta = 2$)	.75	5.5	.77	5.9

5.4 Frequency Variant Objective Measures

A major hypothesis of the study was that since that the perception of sound in humans is a frequency variant process, frequency variant objective measures would perform better as predictors of subjective results than would their frequency invariant counterparts. One method of testing this hypothesis has already been discussed in the section on simple spectral distance measures, that of weighting the spectral distance measure by a function of the spectrum of the undistorted speech (see Eqs. (5.15) and (5.17)). This section offers a different approach to frequency weighting, in which the frequency weights are set so as to give maximum correlation between the objective and subjective measures.

Regression analysis, described in Sec. 4.4.1, is used to compute the frequency weights of the measure. The method can be described as follows. First, a frequency sampled objective measure is defined which uses k bands in its frequency analysis. In this study, two such measures, spectral distance and short-time frequency variant signal-to-noise, were used. These two measures will be described in the next two sections. Then for each distortion, k different objective measures, $D_j(d)$, where j is the band index and d is the distortion index, are computed. The subjective results for distortion d are estimated by a linear sum of the frequency variant objective measures as

$$\hat{y}(d) = b_0 + \sum_{j=1}^{k} b_j D_j(d) \tag{5.42}$$

where $\hat{y}(d)$ is the estimate of the subjective measure $y(d)$ and the b_j's are a set of unknown constants. This is simply a re-statement of Eq. (4.56). If b_j, $j = 0, \ldots, k$, are chosen to minimize the squared error, then the maximum correlation between $y(d)$ and $\hat{y}(d)$ is achieved. Solving for the b_j's in this manner is described in Sec. 4.5.1.

Several points should be discussed here. First, the coefficient of correlation between $y(d)$ and $\hat{y}(d)$ after the b_j's have been found must be considered a limit on the correlation obtainable by weighted frequency analysis. This is true because the data itself is being used to compute both the correlation and the weights. Second, since many of the distortions in the data base have inter-band distortions which are highly correlated with one another, the results of this analysis cannot be considered a direct estimate of the underlying optimal physical weights. This is the reason that a large subset of the distorted data base is made of narrowband frequency distortions. Estimates based on this subset would have more physical validity than those taken across the entire data base. Finally, since the optimization of Eq. (5.42) may be done against any of the different parametric subjective results (see Chap. 3), these measures may be "tuned" to predict specific parametric subjective results as well as isometric subjective results. With many of the parametric subjective results being frequency variant in nature, such tuning could be very effective.

There are two basic classes of frequency variant measures studied as part of this research: frequency variant spectral distance measures, and frequency variant noise measures. For both cases, the frequency range 200–3400 Hz is divided into six bands, as shown in Table 5.15. The individual measures for each of the bands are then computed, and the overall objective measure is formed as an optimally weighted sum of the subband results.

5.4.1 Frequency Variant Spectral Distance Measures

One of the two types of frequency variant distance measures considered is the frequency variant spectral distance measure. From Sec. 5.3, recall that in frequency invariant measures the frequency index is defined as $\theta_l = (\pi l)/(L)$ for $l = 0, \ldots, L-1$. For a k band analysis, the total frequency band is divided into k subbands where band j spans the range (θ_{j-1}, θ_j). In this study, k was equal to six.

The frequency variant spectral distance measures were computed

Table 5.15: Frequency bands used for the frequency variant objective measures.

Band Number	Range (Hz)
1	200–400
2	400–800
3	800–1300
4	1300–1900
5	1900–2600
6	2600–3400

by the same techniques as discussed in Sec. 5.3.1 but using the reduced ranges (θ_{j-1}, θ_j). In this analysis, two types of spectral normalization were computed. First, the spectra $V(n, s, d, \phi)$ were normalized to have an area of one in the set of six bands. Second, the spectra were normalized to have an area of one in each individual band. Since this second method reduces the effect of spectral tilt, it was expected to give better correlation results.

The parameters controlling the frequency variant spectral distance measures are the same as those controlling the spectral distance measures. These include four conditions. First, the distance may be between linear spectra, log spectra, or spectra taken to the δ power. Second, the distance may be frequency weighted by the energy spectrum $V(n, s, d, \theta)^\gamma$. Third the measure may be time weighted by the energy of the frame taken to the α power. And fourth, an L_p norm is applied. In all, a total of ninety-six variations of these measures were studied. These measures are summarized in Table 5.16.

Table 5.17 shows the results for the best log spectral distance measures. As can be seen, the use of frequency weighting improves the spectral distance results by roughly 0.1 points in the correlation measure. Also, it was found that the same log spectral distance measures which did well in the frequency invariant cases did well in the frequency variant cases as well.

Table 5.18 shows the results for the best linear spectral distance measures. Here, the improvement from the frequency invariant case is remarkable. Not only is the frequency variant linear spectral distance measure as good as the frequency invariant case, it is better that the

Table 5.16: Parameters for frequency variant spectral distance measures.

Linear Spectral Distance Measure	
Spectral weighting parameter (γ)	0, 1, 2
Energy weighing parameter (α)	0, .5, 1, 2
Lp Norm (p)	1, 2, 4, 8
TOTAL	48

Log Spectral Distance Measure	
Spectral weighting parameter (γ)	0, 1, 2
Energy weighting parameter (α)	0, .5, 1, 2
Lp Norm (p)	1, 2, 4, 8
TOTAL	48

Table 5.17: Best system for each category for log frequency variant spectral distance measures.

Condition	$\mid \hat{\rho} \mid$	$\hat{\sigma}_e$	Spectral Weighting (γ)	Energy Weighting (α)	Lp Norm (p)
CA (ALL)	.68	7.2	1.0	0.0	4
CA (WBD)	.72	6.2	1.0	0.0	2
TSQ (ALL)	.61	8.5	1.0	1.0	2
TSQ (WBD)	.64	7.7	2.0	2.0	8
TBQ (ALL)	.23	6.0	2.0	0.0	1
TBQ (WBD)	.35	5.8	2.0	0.0	1

log measure also.

Figure 5.16 shows a summary plot of the performance of all the linear frequency variant spectral distance measures tested. The plots are made as a function of the energy weighting parameter, α. The first obvious point in Fig. 5.16 is that performance of all the measures decreases as the value of α increases. Thus in frequency variant spectral distance measures, as for frequency invariant measures, no energy weighting is the best choice. Second, at $\alpha = 0$, the there is only a

Table 5.18: Best system for each category for linear frequency variant spectral distance measures.

Condition	$\|\hat{\rho}\|$	$\hat{\sigma}_e$	Spectral Weighting (γ)	Energy Weighting (α)	Lp Norm (p)
CA(ALL)	.68	7.2	0.0	2	1
CA(WBD)	.72	6.2	0.0	2	1
TSQ(ALL)	.61	8.5	0.5	2	1
TSQ(WBD)	.68	7.3	0.0	2	1
TBQ(ALL)	.24	7.0	0.0	2	1
TBQ(WBD)	.38	5.7	0.0	0	2

range of roughly 0.024 in correlation between the best measure and the worst measure. Hence, the results are very insensitive to both p and γ. Finally, there is a slight advantage for low values of p, and there is a slight advantage for $\gamma = 0$.

In order to take advantage of the "tunable" nature of the frequency variant spectral distance measures, correlation analyses were performed for the parametric subjective categories of SF, SH, SD, SL, SI, SN, BN, BB, BF, and BR across ALL and WBD. Table 5.19 shows some results from that study.

Qualitatively, these results are relatively easy to understand. The frequency variant spectral distance measures performed well on frequency variant parametric subjective measures (SF, SH, SL, SN, BN, and BB) and poorly on the non-spectrally-related subjective measures (SD, SI, BF, and BR). The performance on several of the measures (SF, SH, SL) can be said to be very good, while the performance on the others is moderate or poor.

5.4.2 Frequency Variant Noise Measures

The frequency variant noise measures are the second class of frequency variant measures considered in this study. Like all noise measures, these were only applied to the subset of the distorted data base for which noise measures are meaningful.

The computation of the frequency variant noise is illustrated in

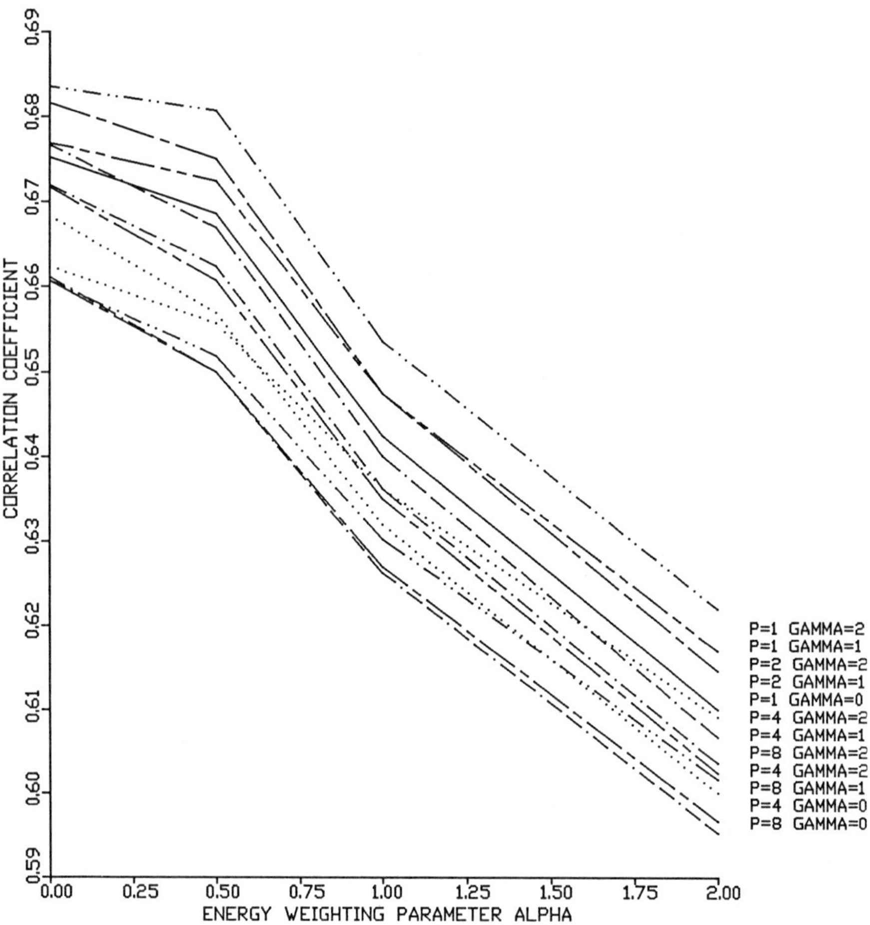

Figure 5.16: Correlation coefficients with CA for all of the linear frequency variant spectral distance measures studied as a function of the energy weighting parameter α.

Table 5.19: Sample of results for frequency variant spectral distance measures used for predicting parametric subjective results.

Parametric Subjective Measure	ALL $\lvert\hat{\rho}\rvert$	ALL $\hat{\sigma}_e$	WBD $\lvert\hat{\rho}\rvert$	WBD $\hat{\sigma}_e$
SF	.61	3.8	.74	4.2
SH	.63	3.9	.73	3.9
SD	.42	6.3	.26	6.7
SL	.72	3.6	.81	3.6
SI	.17	5.6	.19	7.9
SN	.45	3.8	.55	4.2
BN	.48	5.1	.23	4.0
BB	.43	4.0	.26	3.5
BF	.18	6.5	.38	5.4
BR	.27	2.8	.21	1.8

Fig. 5.17. As can be seen, the noise is computed in the usual way and then the results are filtered into six separate bands. If the frequency variant time signal is given by $x_j(m, s, d)$ and the windowed frequency variant time signal is given by

$$x_{n,j}(m, s, d) = x_j(m, s, d)w(m - nI) \tag{5.43}$$

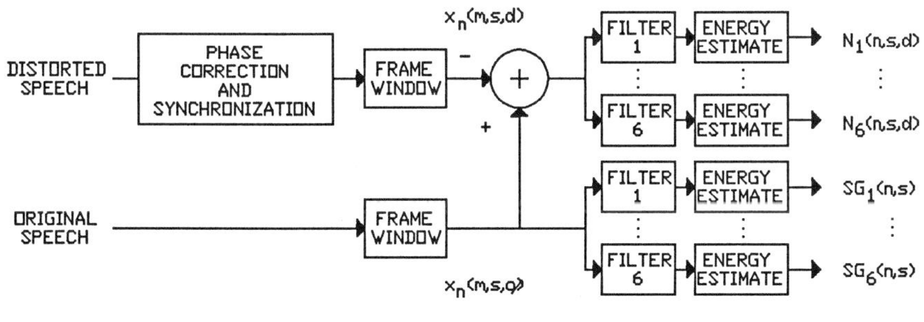

Figure 5.17: Computation of the frequency variant noise measure.

where $w(m)$ is the window function, j is the band index, and I is the frame interval as before, then the frequency variant noise energy for the nth frame of the sth speaker of the dth distortion is given by

$$N_j(n,s,d) = \left[\frac{1}{W}\sum_{m=-\infty}^{+\infty}[x_{n,j}(m,s,d) - x_{n,j}(m,s,\phi)]^2\right] \tag{5.44}$$

where, as before, the limit on m is in practice finite. The frequency variant signal energy, $SG_j(n,s)$, is given by

$$SG_j(n,s) = \left[\frac{1}{W}\sum_{m=-\infty}^{+\infty} x_{n,j}(m,s,\phi)^2\right] \tag{5.45}$$

In this context, the frequency variant frame measure is computed from

$$F_j = 20[\log_{10} G_j(n,s,d)]^{\delta} \tag{5.46}$$

where

$$log_{10}[1 + G_j(n,s,d)^2] = log_{10}\left[1 + \frac{SG_j(n,s,d)}{N_j(n,s,d)}\right] \tag{5.47}$$

as before. In these studies, δ is a parameter for study.

The frequency variant noise measures studied include both the frequency variant forms of the ordinary and segmental SNR. Only one form of the frequency variant SNR was tested. However, forty-nine versions of the short-time frequency variant SNR were tested. These different measures are characterized by two parameters: the energy weighting parameter α controls the time domain weighting by the energy of the original speech, and δ controls the power to which the log of the measure is taken (see 3.4.2). In terms of these parameters, the forty-nine cases studied are shown in Table 5.20.

Of course these, like all noise measures, cannot be used across the whole distorted data base. Hence, these tests were only run across WFC, WBN, NBN, BD, and PD. The most important is WFC (waveform coders), since it represents an estimate of the measures' performance in an actual coding environment.

Table 5.21 shows the results for WFC. The first noteworthy point is that these are outstanding results, with the best measure having a $|\hat{\rho}| = 0.93$ and a $\hat{\sigma}_e$ of only 3.28. Note that several forms of the measure come close to this performance. Finally, as before, the best energy weighting is no energy weighting.

Table 5.20: Parameters for short-time frequency variant signal-to-noise ratio.

Banded Short-Time SNR	
Energy Weighting (α)	0, .25, .5, 1, 2, 4, 8
Power of log (δ)	0, .25, .5, 1, 2, 4, 8
TOTAL	49

Table 5.21: Best results for frequency variant short-time SNR measure across WFC.

	$\mid\hat{\rho}\mid$	$\hat{\sigma}_e$	Energy Weighting (α)	Power Parameter (δ)
CA (WFC)	.93	3.3	0.0	.25
TSQ (WFC)	.81	3.6	0.0	.25
TBQ (WFC)	.93	2.9	0.0	.25

5.5 Time Variant Measures

It is well known that the performance of many speech coding techniques, particularly analysis/synthesis methods, varies with the segmental character of the speech signal [5.16]. For example, a speech coding system may faithfully reproduce most vowels, but introduce significant distortion in high front vowels or in stop consonants. Hence, a possible method of improving any objective measures is first to classify the speech waveform into the appropriate phonemic categories and then apply the measure separately to each of the classes.

In this study, the four time variant measures which were investigated are the log spectral distance, $|\ |^{\delta}$ spectral distance, log area ratio, and energy ratio [5.17]. In each of these measures, the time variance was achieved by segmenting the waveform into four phonemic classes: "vocalic," "fricative," "nasal," and "silence." Segmentation was fully objective, and was based on two parameters: frame energy s^2, and short-time prediction gain G_p. Since the linear prediction analyses used for the distance measures and segmentation were computed at a 16 ms frame rate, all segments were a multiple of 16 ms in length.

The four classes are determined by the use of two thresholds on G_p, denoted as G_u and G_l, and one on s^2. Segments with a frame energy s^2 of less than 11 dB are classified as silence, with all others being nasal, vocalic, or fricative. Fricative segments are those with G_p less than $G_l = 22$ dB. Vocalic segments are those with G_p greater than G_l but less than $G_u = 27$ dB. Nasal segments are those with G_p greater than G_u.

The frame-level distortion for each of the measures was formulated in exactly the same fashion as were the corresponding simple or LPC-based parametric measures described in previous sections of this chapter, specifically in Eq. (5.13)–(5.19), (5.29), and (5.36). Energy weighting was applied to each frame, as specified in Eq. (5.1), but with weighted sums of frame-level measures computed only within each segment class. Let these measures be denoted as $D(s,d,c)$, where c is the class index. The within-class measures are combined into a single global measure using

$$D(s,d) = \sum_{c=1}^{4} b_c D(s,d,c) \tag{5.48}$$

where the weights b_c are determined using regression analysis, so as to maximize the measure's correlation with the actual subjective scores.

As with other measures in this study, the time variant measures were parametrically varied to form a family of similar measures, and Table 5.22 lists the parameter values used. The performance of these measures and comparable time invariant measures are listed in Table 5.23.

The general result of this procedure is that phonemic segmentation of measures slightly improves the performance for all measures considered. Another important result is that, for all measures considered, the within-class measure $D(s,d,c)$ with the highest correlation to CA was consistently that associated with the nasal or vocalic sound class.

Although improvement in performance was relatively small in all cases, the consistent improvement suggests that segmental classification is a useful method in implementing objective speech quality measures. Furthermore, segmentation into finer intervals, classification into larger numbers of phonemic categories, and the use of class dependent distance measures may all increase time variant measure performance.

Table 5.22: Parameters for time variant measures.

Log Spectral Distance	
γ	1
p	1, 2, 4, 6, 8, 10
α	0

Nonlinear Spectral Distance	
γ	1
p	1, 2, 4, 6, 8, 10
δ	0.1, 0.2, 0.3, 1.0
α	0

Log Area Ratio	
p	1, 2
α	0, 1, 2

Energy Ratio	
δ	0.25, 0.5, 1.0, 2.5
α	0, 0.25, 0.5

Table 5.23: Best results for time variant measures as compared to time invariant measures, both across ALL.

Time Variant	Measure	Parameters	$\mid \hat{\rho} \mid$
yes	log spectral	$\gamma = 1$, $p = 10$, $\alpha = 0$	0.63
no	log spectral	$\gamma = 0.5$, $p = 2$, $\alpha = 1$	0.60
yes	nonlinear spectral	$\gamma = 1$, $p = 10$, $\delta = 0.1$, $\alpha = 0$	0.65
no	nonlinear spectral	$\gamma = 1$, $p = 1$, $\delta = 0.2$, $\alpha = 0$	0.61
yes	log area ratio	$p = 1$, $\alpha = 0$	0.63
no	log area ratio	$p = 1$, $\alpha = 1$	0.62
yes	energy ratio	$\delta = 0.5$, $\alpha = 0$	0.61
no	energy ratio	$\delta = 0.5$, $\alpha = 0$	0.59

5.6 Summary

In this chapter, a large number of variations of several basic quality measures were tested for correlation across an extensive data base. On the whole, performance of the measures studied was modest, with the best frequency invariant measure only accounting for roughly 36% of the variance in the subjective data. Such a level of performance, though of some value, is certainly less than what would be required for the practical use of a measure. A few interesting results emerged, however, which are important in understanding the underlying problems.

First of all, nearly the best performance of any frequency invariant measure was that of the log area ratio measure, with a correlation coefficient magnitude of 0.62. This is classified as a frequency invariant spectral measure since, in its derivation from linear prediction analysis, a polynomial model is fit to a spectrum with all frequencies weighted equally. The computation of the log area ratio is an order of magnitude more efficient than is a direct spectral distance and should be the measure of choice where fast computation is required. A more commonly used measure, the energy ratio measure, should also be displaced by the log area ratio measure if an LPC-based measure is required.

The short-time SNR turned out to be a measure with vastly superior performance to that demonstrated by the traditional SNR. This result is not too surprising since a wide variety of speech coding techniques already derive great advantage from the use of the segmental signal energy properties of speech.

When frequency variant measures are employed, roughly 0.1 improvement in correlation coefficient magnitude can be realized, with the best of the measures having $\hat{\rho} = -.68$. When segmental SNR's were computed within multiple frequency bands and combined, the correlation coefficient achieved was -.93. This an excellent result, but two qualifications need to be made. The first is that this measure is only applicable to waveform coders. The second is that the undistorted speech is required to have the same phase alignment as is found in the distorted speech, which is done by passing it through the waveform coders with no quantization. As such, it does not follow the form of other objective measures, which assume that only the unprocessed input speech is available.

A number of research avenues are suggested by the results of this chapter, and they will be the topics of Chaps. 6, 7, 8, and 9. The first question one might ask relates to why so many seemingly disparate measures, when used with the right parameters, gave similar correlation results. One answer might be related to the fact that most (excepting SNR measures) were parameterizations of functions derived from LPC analysis. The frequency variant measures showed that improvement is possible if the analysis is modified slightly to make it better match the behavior of the auditory system. Therefore, a model which used a parameterization that is not based on linear prediction analysis, but rather on psychoacoustic principles, may lead to substantially improved measures.

The success of the short-time frequency variant SNR on waveform coders suggests that if a distortion could be classified in advance as belonging to some particular distortion family, the most suitable measure could be invoked. The avenue of research has two problems areas: how distortions can be objectively classified, and which measures should be used on each family.

In the very design of the DAM, it was acknowledged that direct measurement of quality via a single scale was undesirable and prone to measurement errors. It is for this very reason that the DAM requires multiple scales for rating distorted speech. Of these several scales, the parametric quality scales not only contribute to the measurement of composite acceptability, but also give a very specific characterization of the distortion under test. In a similar manner, objective measures which estimate the parametric subjective qualities of the DAM could be used as a diagnostic tool for characterizing, understanding, and possibly correcting the specific distortions associated with a speech coder or communication system.

Lastly, just as the DAM arrives at composite acceptability through a merging of the parametric ratings, an objective measure of acceptability could be a combination of many simple measures, forming a composite measure. One approach to this end would be to combine different measures from this chapter, with the hope that each measures a different property of the distortion under test. Another method, which assures that each constituent measure tracks a different aspect of the distortion, is to combine objective measures which predict the parametric subjective measures of the DAM.

The remaining chapters in this book will deal with these four areas for possible improvement. It will be shown that all of these ideas work to some extent, with the greatest improvements exhibited in the area of composite measures.

References

[5.1] T. P. Barnwell, "Objective Measures for Speech Quality Testing," *J. Acous. Soc. Amer., vol. 66*, pp. 1655–63, Dec. 1979.

[5.2] B. S. Atal and M. R. Schroeder, "Optimizing Predictive Coders for Minimum Audible Noise," *Proc. 1979 IEEE ICASSP*, Washington D.C., pp. 453–6, Apr. 1979.

[5.3] J. L. Flanagan, *Speech Analysis, Synthesis and Perception*, Second Edition, Springer-Verlag, New York, 1972.

[5.4] F. Itakura and S. Saito, "On the Optimum Quantization of Feature Parameters in the PARCOR Speech Synthesizer," Proc. 1972 Conf. Speech Comm. Process., pp. 434–7, 1972.

[5.5] B. S. Atal and S. L. Hanauer, "Speech Analysis and Synthesis by Linear Prediction of the Speech Wave," *J. Acous. Soc. Amer., vol. 50*, 637–55, 1971.

[5.6] J. D. Markel, "Digital Inverse Filtering—A New Tool for Formant Trajectory Estimation," *IEEE Trans. Audio Electroacoustics, vol. AU-20*, pp. 129–37, 1972.

[5.7] A. V. Oppenheim and R. W. Schafer, "Homomorphic Analysis of Speech," *IEEE Trans. Audio Electroacoustics, vol. AU-16*, pp. 221–6, 1968.

[5.8] T. P. Barnwell, A. M. Bush, R. M. Mersereau, and R. W. Schafer, "Speech Quality Measurement," Final Report, DCA Contract No. RACD-TR-78-122, Jun. 1977.

[5.9] A. V. Oppenheim, "Speech Analysis-Synthesis System Based on Homomorphic Filtering," *J. Acous. Soc. Amer., vol. 45*, pp. 458–65, 1969.

[5.10] A. H. Gray, Jr.. and J. D. Markel, "Distance Measures for Speech Processing," *IEEE Trans. Acous. Speech, Sig. Proc., vol. 24*, pp. 380–90, 1976.

[5.11] F. Itakura, "Minimum Prediction Residual Principal Applied to Speech Recognition," *IEEE Trans. Acous. Speech, Sig. Proc., vol. 23*, pp. 67–71, 1975.

[5.12] M. R. Sambur and N. S. Jayant, "LPC Synthesis Starting from White Noise Corrupted or Differentially Quantized Speech," *Proc. 1976 IEEE ICASSP*, Philadelphia, PA, pp. 237–40, Apr. 1976.

[5.13] B. J. McDermott, C. Scagloila, and D. J. Goodman, "Perceptual and Objective Evaluation of Speech Processed by Adaptive Differential PCM," *Proc. 1978 IEEE ICASSP*, Tulsa, OK, pp. 581–4, Apr. 1978.

[5.14] P. W. Noll, "Adaptive Quantization in Speech Coding Systems," *IEEE Int. Zurich Sem. on Dig. Comm.*, Oct. 1976.

[5.15] P. Mermelstein, "Evaluation of a Segmental SNR Measure as an Indicator of the Quality of ADPCM Coded Speech," *J. Acous. Soc. Amer., vol. 66*, no. 6, pp. 1664–67, Dec. 1979.

[5.16] W.D Voiers and C.P. Smith, "Diagnostic Evaluation of Intelligibility of Present-Day Digital Vocoders," Record, *1972 Conf. on Speech Communications and Processing*, Newton, Mass., pp. 170–4, Apr. 1972.

[5.17] P. Breitkopf and T.P. Barnwell III, "Segmental Preclassification for Improved Objective Speech Quality Measures," *Proc. 1981 IEEE ICASSP*, Atlanta, Ga., pp. 1101-4, Mar. 1981.

Chapter 6

Aural Models for Objective Measures

6.1 Background and Theory

A distortion of speech resulting from coding can only be detected if the magnitude of the distortion is greater than the resolution of the human auditory system. Once a distortion is perceivable, subjective evaluation of the degree of distortion is affected by the scaling properties of the auditory system. (The auditory system includes both peripheral and central components.) The modeling approach to be discussed in this chapter will not deal specifically with speech perception, but rather with some of the basic psychophysics of hearing. A specific restriction imposed will be to look only at differences in coded and uncoded speech and to quantify these differences. This approach obviously cannot address all issues, but for speech coders it should be quite useful. Since only the basic properties of hearing will be modeled, it is expected that these quality assessments will most readily agree with subjective results for waveform coders rather than for more complex distortions. Some of the key issues which pertain to the modeling of hearing are frequency, temporal, and intensity resolutions as well as their perceptual scalings. Each of these will be discussed in turn.

Frequency differentiation appears to be comprised of at least two separate phenomena, one relating to stimuli composed of harmonically related components (pitch perception) and another relating to more general stimuli.

In some experiments, pitch perception can be accurate to within 0.3 percent, but this is true only for strictly periodic signals. When complex tones (stimuli composed of multiple sinusoids) have roughly seven or more inharmonic components, the individual tone frequencies cannot be perceived. For such signals, the pitch detection ability of human observers becomes too confused to function. Recent experiments suggest that pitch perception is a highly central neural process which must be modeled at a level much beyond the auditory periphery. It is for this reason that frequency resolution approaching that observed for pitch will not be included in our models.

Frequency resolution in more general signals is much poorer than that of pitch perception in periodic signals and is probably determined by other basic mechanisms. Many theories include the notion of critical bands, which corresponds to the presumed filtering action of the peripheral auditory system. Although there have been many attempts to explain the properties of critical bands based solely on peripheral auditory physiology (up through the auditory nerve), none have been satisfactory. It is probable that a portion of this filtering takes place in more central neural mechanisms. Such data as auditory nerve tuning curves obtained by probing auditory nerve fibers directly do not appear to provide a sufficiently complete model for auditory perception, and a preferable alternative is to base frequency analysis on psychoacoustical measurements. Table 6.1 lists the characteristics of one set of experimentally determined critical bands which span a large fraction of the audible spectrum. The bandwidths and center frequency spacings are non-uniform and roughly correspond to a 1/6-octave filter bank.

A well known property of linear filters is the inverse proportionality of temporal and frequency resolution (bandwidth versus risetime): as a filter's bandwidth increases, greater temporal resolution is possible. Nerve latency data suggests a lower limit for auditory resolution of approximately 2 ms [6.1]. Low frequency stimuli are more poorly resolved temporally than this value due to the fact that the low frequency channels have narrower bandwidths. Here, the temporal resolution is worse than 10 ms. Since a lower limit of 2 ms is achieved only for such artificial stimuli as high bandwidth clicks, and since such acuity would have the undesirable effect of allowing pitch periods to be resolved temporally, an analysis resolution of 10 ms was chosen.

Table 6.1: Critical band center frequencies and bandwidths.

Filter Number	Center Freq.	Bandwidth	Filter Number	Center Freq.	Bandwidth
1	50	70	13	1020	127
2	120	70	14	1148	140
3	190	70	15	1288	153
4	260	70	16	1442	168
5	330	70	17	1610	183
6	400	70	18	1794	199
7	470	70	19	1993	217
8	540	70	20	2221	235
9	617	86	21	2446	255
10	703	95	22	2701	276
11	798	105	23	2978	298
12	904	116	24	3276	321
			25	3597	346

Masking, a phenomenon closely related to frequency resolution, is a mechanism undoubtedly arising from both peripheral and central processing. Critical bands are often measured by using steady-state signals to mask other signals, and critical band decompositions serve to model this simultaneous masking. Another form of masking occurs between signals separated in time. Many of the theories which describe this non-simultaneous masking involve exponential decay of masking functions with time, some with frequency-dependent time constants. A fuller discussion of this will be given when describing the analysis.

In many models, intensity is perceived as the sum of intensities in the various critical bands, which are themselves nonlinear functions of the short-time energies within the bands. A short-time energy can be approximated by application of a nonlinearity to the filter output followed by a lowpass filter. This operation results in a slowly varying signal for each channel. The nonlinear function of the energies can be any one of a variety of relations such as logarithm, raising to a non-integer power, or in terms of JND's (just noticeable differences).

6.2 Analysis Procedures

For better assessment of the quality of coded and distorted speech using aural models, an attempt has been made to take into account the perception of differences in the signals. Since the assumption has been made that all of the distortions in the study are perceivable, the task becomes one of quantifying these differences.

The spectral analysis performed by the ear strongly suggests that the reference (original) speech and the distorted speech should be compared in the frequency domain. In this study, analysis paralleling the frequency decomposition of the ear, or critical band filtering, was performed. Of the many alternatives for the computation of critical band spectra, such as using linear prediction, DFT's of windowed speech (Time Dependent Discrete Fourier Transforms), or filter bank analysis, the last was chosen. Since the ear shows little sensitivity to phase as long as components are not within critical bands, and appears to respond to energy as a function of frequency, the analysis involved short-time spectral densities. Denote the positive square root of this quantity as $V(n, s, d, m)$ where n is the time index, s the speaker, d the distortion ($d = \phi$ indicates no distortion) and m is a discrete variable representing the critical band over which the energy is summed. In the LPC method, a high density DFT of the LPC spectrum would be computed, with the energy within each critical band summed. The windows for summation in the frequency domain are of the form shown in Fig. 6.1, for auditory modeling. This method of analysis results in a pre-emphasis of roughly 3 dB/octave due to the increasing bandwidths, which would have to be compensated. A potential problem with this method is that although bandwidths increase with frequency, time resolution is not proportionally enhanced due to constant analysis window size. To allow added flexibility, the analysis consisted of digital filtering and envelope detection instead, where critical band energies were sampled fast enough to represent even the wider bandwidth channels.

Once critical band spectra were computed for original and undistorted waveforms, comparisons were made. Sensation levels, masking functions, and auditory nerve firing rates can be approximated only if a nonlinear scaling of the energy envelopes is applied. This would suggest that for an isolated filter's energy at an isolated time (one frame), the distance between the reference and distorted speech frames for channel

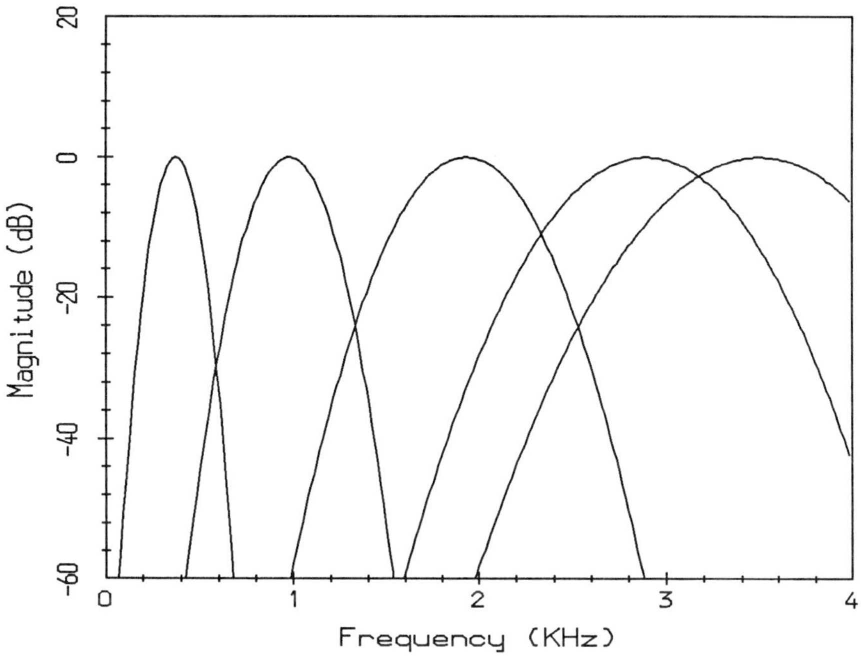

Figure 6.1: Critical band filters.

m should be a monotonic function of D_m, the magnitude difference of the nonlinearly scaled energies in the two spectra. Here, the distance would be of the form:

$$D_m = (|f_1(V(n, s, \phi, m)) - f_1(V(n, s, d, m))|) \tag{6.1}$$

where $f_1(\ )$ is a non-linearity such as a logarithm or power function. Combining the contributions of the frequency bands into an overall frame distance requires the inclusion of the monotonic function applied to D_m, as well as a weighting which may depend on the band's energy.

$$F_n = \sum_{m=0}^{L-1} \{f_2(V(n, s, d, m) f_3(D_m)\} \tag{6.2}$$

where the index of summation, m, covers all critical bands. Previous work suggests that $f_3(\ )$ should be $|\ |^p$ where p is a positive integer, and

$f_2(\ )$ a monotonic nonlinearity such as a logarithm or a power function. Combining frames to arrive at an overall measure is accomplished in a similar manner

$$\text{Overall distance} = \frac{\sum_n W_n f_4(F_n)}{\sum_n W_n} \tag{6.3}$$

where W_n is a weighting function denoting frame n's overall importance, and $f_4(\ )$ is generally the inverse function of $f_3(\ )$. In this study, only $f_3 = |\ |^p$ and $f_4 = |\ |^{1/p}$ were used. Note that these choices amount to computing weighted L_p norms for the L dimensional vectors comprised of the nonlinearly transformed magnitude spectral differences.

This established framework allows for a large number of hypotheses to be tested. As examples, the $f_1(\ )$ nonlinearity can be modeled by the JND structure for bands, or alternatively by the form that perceived loudness assumes as a function of intensity. In the first case, a logarithm is approximately correct, and in the second, a non-integer power function is appropriate. By the same logic, $f_2(\ )$ should have a similar form, although the two nonlinearities need not be the same. The functions can also involve maximum and minimum values. As mentioned, if $f_3(\ )$ and $f_4(\ )$ are of the form $|\ |^p$ and $|\ |^{1/p}$, frequency based combinations follow L_p norm measures. Another more complex set of measures called *weighted slope spectral distance measures* (WSSD) were employed, and will be described in more detail later in this chapter.

6.2.1 Log Spectral Distance Measures

According to the notion that the perceived intensity difference between two stimuli is proportional to the ratio of the two intensities (Fechner's law) or that intensity resolvability is proportional to intensity (Weber's law) [6.2], the f_1 nonlinearity should be logarithmic. With the notation that $F_n = F(n, s, d)$ is the frame distance for speaker s, frame n, and distortion d,

$$F_n = \left| \frac{\sum_{m=0}^{L-1} |V(n,s,\phi,m)|^\gamma |\log\{V(n,s,\phi,m)/V(n,s,d,m)\}|^p}{\sum_{m=0}^{L-1} |V(n,s,\phi,m)|^\gamma} \right|^{1/p} \tag{6.4}$$

was used.

6.2.2 Power Function Spectral Measures

Psychophysical measurements point to significant modeling errors obtained from application of Fechner's or Weber's laws. A more accurate model states that the perceptual intensity doubles for every N dB increase (N is usually set to 10). Therefore, if we let i_{1p} and i_{2p} be the perceived intensities, and i_1, and i_2 be the actual intensities, the relation is:

$$\log_2(i_{1p}/i_{2p}) = (N/10)\log_{10}(i_1/i_2) \tag{6.5}$$

or

$$\begin{aligned}(i_{1p}/i_{2p}) &= 2^{(N/10)\log_{10}(i_1/i_2)} \\ &= 2^{(N/10)(\log_2 10)(\log_2(i_1/i_2))} \\ &= 2^{\log_2\{(i_1/i_2)(.3N/10)\}} \\ &= (i_1/i_2)^{.03N}\end{aligned} \tag{6.6}$$

Therefore perceived intensity grows as magnitude to the .06N power. If $N = 10$, this exponent becomes 0.6. A general form in which the exponent δ is left a free parameter would result in:

$$f_1(x) = x^\delta \tag{6.7}$$

Therefore:

$$F_n = \left| \frac{\sum_{m=0}^{L-1}\{V(n,s,\phi,m)\}^\gamma |V(n,s,\phi,m)^\delta - V(n,s,d,m)^\delta|^p}{\sum_{m=0}^{L-1} V(n,s,\phi,m)^\gamma} \right|^{1/p} \tag{6.8}$$

6.2.3 Articulation Index Approximation

Although the goal is to characterize the quality of speech rather than its intelligibility, there should be some similarities in estimation methods for the two. One set of procedures useful for predicting intelligibility from a description of signal-to-noise ratio as a function of frequency falls under the category of articulation theory. The computed articulation index value can be calculated in a variety of ways. Kryter's method [6.3] divides the frequency scale into 1/3-octave bands. Signal-to-noise ratios (SNR's in dB) are computed for each band, with a maximum of 30 dB, and a minimum of 0 dB allowed in each band. Specific weights

Table 6.2: Articulation index weights.

Band Number	Weight	Band Number	Weight
1	.003	14	.032
2	.003	15	.034
3	.003	16	.035
4	.007	17	.037
5	.010	18	.036
6	.016	19	.036
7	.016	20	.033
8	.017	21	.030
9	.017	22	.029
10	.022	23	.027
11	.027	24	.026
12	.028	25	.026
13	.030		

for each band, are applied to each SNR, and these weighted values are summed.

There are a number of differences between Kryter's method and the current approach. First of all, rather than 1/3-octave bands, the current filters are critical bands. If anything, this analysis should be an improvement over that of Kryter, whose original intent was to approximate critical bands. The weights which were used for the 1/3-octave filter bank were interpolated from Kryter's weights to produce the appropriate weights for the current procedure and are listed in Table 6.2. Second, in this study, only approximate SNR's were computed. This was accomplished by observing the differences in the original and the distorted filter bank signal energies. And, third, long term SNR's were not directly computed, but rather approximated as the average SNR over many frames. With these differences in mind, the new version of the articulation index gives a frame measure of:

$$\begin{aligned} F_n &= \textstyle\sum_{m=0}^{L-1} W_m \max\{0, \min[20\log_{10} V(\phi,m) - \\ &\qquad 10\log_{10}|V^2(\phi,m) - V^2(d,m)|, 30]\} \qquad (6.9) \\ &= \textstyle\sum_{m=0}^{L-1} W_m N_m \end{aligned}$$

where N_m is the coefficient of W_m in Eq. (6.9). So that additional degrees of freedom could be incorporated into the model, an energy dependent frequency weighting was allowed as well as an L_p norm for frequency band combinations. The resulting frame distances:

$$F_n = \left| \frac{\sum_{m=0}^{L-1} |V(n,s,\phi,m)|^{\gamma} (N_m)^p}{\sum_{m=0}^{L-1} |V(n,s,\phi,m)|^{\gamma}} \right|^{1/p} \tag{6.10}$$

appear similar to the log spectral distances.

6.2.4 Forward Masking Models

Simultaneous masking of signals is modeled by critical band analysis, which describes masking as a function of frequency separation. Another masking phenomenon is temporal masking, or masking of one stimulus by another which is separated in time. Because the effect of temporal masking is more dramatic when the masker precedes the target (forward masking) than in the reverse case (backward masking), only forward masking was considered. Various experiments indicate that masking level decays exponentially in dB with linear time separation [6.4], with the time constant for a 1000 Hz stimulus being roughly 75 ms. In other words, if the masking level of a stimulus is 80 dB at $t = 0$, at $t = 75$ it will be $80/e$ dB $= 30$ dB. Denote τ_1 the time constant for frequency m. If the masking level at time t for frequency m and a stimulus which is no longer present is $M(t,m)$, it would be $M(t,m)/e$ at $t = t + \tau_1$ or $M(t+t_1,m) = M(t,m)/e$. This amounts to a frequency dependent smoothing for each filter's envelope which can be accomplished by:

$$M(n+1,s,\phi,m) = r(m)M(n,s,\phi,m) + 20\log_{10} V(n+1,s,\phi,m) \tag{6.11}$$

The constant $r(m)$ specifies the amount of smoothing and is frequency dependent. The new values, $M(n+1,s,\phi,m)$, can be placed into the same framework as V( ) in the log spectral distance measures.

6.2.5 Weighted Slope Spectral Distance Measures

One interesting frame distance measure which was originally formulated for speech perception modeling has been presented by Klatt [6.5].

This measure, the weighted slope spectral distance (WSSD), was based on the observation that certain distortions (e.g., addition of a spectral tilt) may result in large psychoacoustic differences, but change the perceived phonetic units very little. Four basic guidelines were proposed by Klatt:

1. Frequency decomposition should be based on critical-bands.
2. Intensities within the frequency bands should be measured in dB SPL.
3. The slopes of the log critical-band spectra should be the quantities compared rather than the spectra themselves.
4. Differences in slopes of the log critical-band spectra should be weighted in a manner which weights peaks more than valleys.

Klatt's basic distance was of the form:

$$D_{12} = \sum_i W(i)\{S1(i) - S2(i)\}^2 \tag{6.12}$$

where $S1$ and $S2$ were spectral slopes and $W(i)$ was the weighting for each band.

By suitable adjustment of free parameters, correlation between experimentally obtained phonetic distance judgments using isolated, synthetic, steady-state vowels and objective measures based on the above measure, achieved a value of .93. Although this task is quite different from objective quality assessment, some of the same factors involved in subjective phonetic distance judgments may be useful in subjective quality evaluation.

6.3 Objective Measures

In this section, the implementation of the objective measures introduced earlier will be described.

6.3.1 Filter-Bank Analysis

The critical-band filters were designed in accordance with measurements and theory presented by Patterson [6.6]. Filter shapes were

Gaussian, with the center frequencies and bandwidths listed in Table 6.1. Twenty-five filters were used to cover the spectrum 0-4000 Hz. All filters were linear phase and were designed using a 97-point Hamming window. Finite impulse response filtering was performed on the original and distorted waveforms, and RMS values were computed every 10 ms using a 20 ms Hamming window.

6.3.2 Frame Combinations

The main concentration of objective measures work involved exploration of how the 10 ms frames from the distorted and original speech signals should be compared. Barnwell and Voiers had found that weighting frames by some function of their energies did not improve the performance of the objective measures tested [6.7]. Therefore, for a given set of frame distances, $F(n, s, d)$, objective quality was computed by simply averaging $F(n, s, d)$ over n.

6.3.3 Frequency Weighted Objective Measures

In the log spectral measures, frame distances were of the form shown in Eq. (6.4). Here $L = 25$, and the m index denotes the different critical band channels. The free parameters were γ and p. The values used were $\gamma = 0$, 0.2, 0.4, 0.6, 1.0, and $p = 0.2$, 0.5, 1, 2, 3.

The power function spectral measures were as in Eq. (6.8), with free parameters γ, p, and δ. The values used were $\gamma = 0$, 1; $p = 1$, 2, 3; and $\delta = 0.2$, 0.3, 0.6, 1.0, 1.5, and 2.0.

The articulation index approximation as in Eq. (6.10) left the free parameters $\gamma = 0$, 0.2, 0.4, 0.6, 1.0, and $p = 0.2$, 0.5, 1, 2, and 3. In order to investigate the effect of the value of the weighting vector W listed in Table 6.2, all experiments were repeated with no weighting, i.e., a weighting vector with all elements of W equal to 1.

Forward masking models in accordance with Duifuis [6.4] allowed exponential decay of the log intensities. The frame measure was generated as shown in Eq. (6.4) but with M from Eq. (6.10) substituted for V. Because of earlier results, γ was fixed at 0, and p and $r(m)$ (specifying rate of decay – see Eq. (6.11)) were allowed to vary. The range for p was 0.2, 0.5, 1, 2, and 3, and $r(m)$ varied over the range 0, 0.2, 0.5, 0.9, 0.95. Note that the value 0 is the extreme case of no masking, or a time constant of 0, and the other values lead to time constants of 6, 14, 95, and 195 ms respectively.

For the WSSD measures, Klatt's basic form as listed in Eq. (6.11), was used with slight modification. The slope of the spectrum was defined as:

$$S(n,s,d,m) = 20\log_{10}[V(n,s,d,m+1] - 20\log_{10}[V(n,s,d,m)] \quad (6.13)$$

where $V(\ )$ is as before. Due to the fact that 25 spectral values were available, the index varied between 1 and 24. Eq. (6.11) was modified to allow the parameter, p, in the L_p norm to vary. The frame distance was:

$$F(n,s,d) = \left| \sum_{m=1}^{24} W(m)|S(n,s,\phi,m) - S(n,s,d,m)|^p \right|^{1/p} \quad (6.14)$$

where $W(\ )$ depended on both the distorted and original frames, and was specified by

$$W(m) = [W(\phi,m) + W(d,m)]/2 \quad (6.15)$$

where $W(d,m)$ depended solely on the spectrum $V(n,s,d,k)$, for $k = 1$ to 24.

$$\begin{aligned} W(d,m) &= C_1/\{C_1 + {}^{\max}_{k} V(n,s,d,k) - 20log_{10}[V(n,s,d,m)]\} \\ &+ C_2/\{C_2 + {}^{local}_{max\ m} V(n,s,d,k) - 20log_{10}[V(n,s,d,m)]\} \end{aligned} \quad (6.16)$$

The ${}^{\max}_{k}$ $V(n,s,d,k)$ term indicated the maximum value $20log_{10}(V(n,s,d,k))$ achieves as k is varied, and ${}^{local}_{max\ m}$ $V(n,s,d,m)$ indicated the value $20log_{10}(V(n,s,d,k))$ takes on at the closest peak to frequency band m. The free parameters were C_1, C_2, and p. Values chosen were $C_1 = 10$, 20, 30, 40, 50, 60, 100, and 1000, $C_2 = 0.5$, 1, 2, 10, 100, and 1000, $p = 0.5$, 1, 2. Please note that for the cases C_1 and C_2 large, the weighting approaches 1 for all frequencies.

6.3.4 Trained Measures

Although little auditory interaction takes place outside of critical bands, significant correlation exists across these bands in speech. It is therefore reasonable to expect individual frequency band distances to show some correlation with each other. A way of accounting for this would be to find the best linear combination of frequency based distances for predicting subjective quality. This procedure would amount to choosing a weighting vector, $W(m)$, $m = 1, 2, ..., 25$, to maximize objective and subjective quality correlation. In this study, optimum vectors were computed for four contexts. The first two contexts weighted different frequency bands for the log spectral measure as in Eq. (6.4), but with the constraints that $\gamma = 0$ and $p = 2$, giving the form:

$$F_n = \left| \frac{\sum_m W_m |\log\{V(n,s,\phi,m)/V(n,s,d,m)|^2}{\sum_m W_m} \right|^{1/2} \tag{6.17}$$

In the first, all twenty-five bands were employed, whereas in the second, the five bands used were determined by summing filter energies in groups of five at a time. A similar procedure was performed for the power-law spectral distance, where γ, δ, and p for Eq. (6.8) were set to 0, 0.2, and 2 respectively, giving frame distances of:

$$F_n = \left| \frac{\sum_m W_m |\{V(n,s,\phi,m)^{.2} - V(n,s,d,m)^{.2}|^2}{\sum_m W_m} \right|^{1/2} \tag{6.18}$$

As before, both five and twenty-five band analyses were performed.

The two results of interest in each analysis are the actual weighting vectors and the correlations achieved.

6.4 Results

The computed objective measures estimated the DAM composite acceptability scale, described in Chap. 3. The figure-of-merit used in this portion of the research was the magnitude of the estimated correlation coefficient, $\hat{\rho}$.

6.4.1 Log Spectral Distance Measures

When log spectral distance measures of the form given in Eq. (6.4) were tested using the free parameters given in Sec. 6.3.3, the following observations were made:

1. For p held fixed, and γ varied, the best correlation resulted from $\gamma = 0.0$, for all values of p. Furthermore, the degree of correlation invariably decreased as γ moved further away from 0.0 in value.
2. For γ held fixed, and p varied, the best correlation resulted from $p = 2$ or $p = 3$ with $p = 1$ giving reasonably close performance. Values of p less than 1 were inferior in performance to the larger values in all cases.
3. Of the twenty-five combinations of parameters, the top five were:

Rank	p	γ	$\|\hat{\rho}\|$
1.	2.0	0.0	.715
2.	2.0	0.2	.707
3.	3.0	0.0	.705
4.	1.0	0.0	.703
5.	3.0	0.2	.702

Subsets of the distortions which fit into particular categories were observed also. ADPCM and CVSD coding distortions led to coefficients higher than 0.99, as one might expect since the set is highly restrictive. Larger sets which included pole distortions, coding distortions, wide-band distortions, controlled distortions, added colored noise, added white noise, and banded distortions, were also tested. Each of these included a minimum of six sets of distortions (most contained more) giving at least 144 data points per distortion type for correlation analysis. Listed below are the best set of parameters for each set of distortions.

Distortion	γ	p	$\|\hat{\rho}\|$
Waveform Coders (WFC)	.4	2	.71
Pole Distortions (PD)	1.0	.2	.16
Coding (CODE)	0	3	.51
Wide-band (WBD)	.2	1	.58
Controlled (CON)	0	2	.72
Colored Noise (FN)	0	2	.93
Banded (BD)	0	2	.72

Most of these fit the pattern of small γ and p larger than 1. The fact that pole distortions were not matched well at all by any set of parameters can be attributed to the small spread of the corresponding subjective composite acceptability results. In general, however, results are fairly consistent across distortions. The high correlation of objective quality with composite acceptability of additive noise distortions no doubt reflects the fact that audibility of noise and perceived quality are closely related.

6.4.2 Power Function Spectral Distance Measures

Power function distance measures with frame distances of the form given in Eq. (6.8) were computed with parameters listed in Sec. 6.3.3. After running correlation analyses, the following observations were made:

1. For γ and p held fixed, correlation was always best for $\delta = 0.2$, with $\delta = 0.3$ yielding comparable but slightly worse results. In addition, as γ increased in value, performance monotonically decreased.
2. For p and δ held fixed, performance was generally best for $\delta = 0$. Only when p and δ were far from their best values did $\gamma = 1.0$, give better correlation than $\gamma = 0.0$, and then only slightly better.
3. For δ and γ held fixed, performance was generally superior for $p = 2.0$, although $p = 1.0$ and $p = 3.0$ were not much worse.
4. The best five combinations of parameters were:

Rank	γ	δ	p	$\lvert\hat{\rho}\rvert$
1.	0.0	0.2	2.0	.721
2.	0.0	0.2	1.0	.719
3.	0.0	0.3	1.0	.714
4.	0.0	0.2	3.0	.712
5.	0.0	0.3	2.0	.695

When subsets of distortions were observed as described in the previous section, the best set of parameters in terms of correlation were:

Distortion	γ	δ	p	$\|\hat{\rho}\|$
WFC	0.0	0.6	1.0	.77
PD	0.0	0.6	3.0	.60
CODE	0.0	0.2	2.0	.52
WBD	0.0	0.2	1.0 or 2.0	.58
CON	0.0	0.2	2.0	.74
FN	0.0	0.2	2.0	.92
BD	0.0	0.3	1.0	.71

Again, a consistent picture emerged in that γ should be 0.0 and p could be 1.0, 2.0, or 3.0 with little difference. Only waveform coders and pole distortions led to a δ different from 0.2 or 0.3. As with log spectral measures, good prediction of colored noise distortion acceptability was possible.

6.4.3 Articulation Index

Measures of the form in Eq. (6.10) were tested with the parameters as described in Sec. 6.3.2. When weighted by the vector in Table 6.2, the following results were noted:

1. Very little variation in performance existed for the entire set of parameters, with 0.67 being the best correlation coefficient and and 0.58 the worst.
2. For γ held fixed, the best value for p was either 0.2 or 0.5.
3. For p fixed, the best values of γ tended to be small, although, not always zero.
4. The top five systems were:

Rank	γ	p	$\|\hat{\rho}\|$
1.	0.0	0.5	.67
2.	0.2	0.2	.67
3.	0.4	0.2	.67
4.	0.0	0.2	.67
5.	0.2	0.5	.67

The unweighted measures were also tested in an identical manner with the same values for the parameters. Results which were very similar to the previous tests were achieved.

1. The top five systems were:

Rank	γ	p	$\|\hat{\rho}\|$
1.	0.2	0.2	.67
2.	0.4	0.2	.67
3.	0.0	0.2	.67
4.	0.0	0.5	.67
5.	0.6	0.2	.67

2. For p held fixed, better results where generally achieved with γ small.
3. For γ held fixed, in all cases, correlation was a monotonically decreasing function of p.
4. The spread was much larger than in the weighted case.

If the original articulation index characterization were simulated as closely as possible, the parameters $\gamma = 0$ and $p = 1$ would have been used. These values led to scores of 0.65 and 0.64 for the weighted and unweighted cases, respectively, which were not far from the maxima. In the regular log spectral distance measure, $\gamma = 0$ and $p = 1$ led to a correlation of 0.70, which is also similar.

Distortion subsets were also tested on the unweighted measure with the following results:

Distortion	γ	p	$\|\hat{\rho}\|$
WFC	0	0.2	.70
PD	1	0.2	.30
CODE	0	0.5	.63
WBD	*	*	.40
CON	0	0.2	.54
FN	0	0.2	.90
BD	0	0.5	.68

where the * indicates that all values tested worked equally well. For all but the pole distortions (which as mentioned earlier, gave little spread in subjective quality) small values of γ were best. The prevalence of values of p less than 1 appears throughout. For the additive colored noise distortion, as expected, good correlation was achieved.

6.4.4 Forward Masking Models

Log spectral distance measures were also formulated to use frequency dependent levels, where the levels were computed as in Eq. (6.11) with decay rates described in Sec. 6.3.3. In all cases, for p held fixed, maximum correlation was achieved for a time constant of 0 for all channels, or no additional forward masking. The same result was observed for all the distortion subsets. The best results for the various time constants are listed below:

Time Constant (ms)	$\lvert\hat{\rho}\rvert$
0	.717
6	.708
14	.694
95	.675
195	.627

6.4.5 Weighted Slope Spectral Distance Measures

Correlation tests were run on the WSSD measures as described in Sec. 6.3.2. The following points were noted:

1. For all combinations of parameters C_1 and C_2, using $p = 1$ gave superior correlation to using $p = 2$. In most cases $p = 0.5$ outperformed $p = 2$, and in a few instances outperformed $p = 1$.
2. For p fixed at 0.5, 1 and 2 rankings were as follows:

Rank	C_1	C_2	p	$\|\hat{\rho}\|$
1.	10.	0.5	2	.694
2.	10.	1.0	2	.693
3.	20.	0.5	2	.691
4.	10.	2.0	2	.691
5.	30.	0.5	2	.690
6.	20.	1.0	2	.689

Rank	C_1	C_2	p	$\|\hat{\rho}\|$
1.	40.	100.	1.	.736
2.	40.	1000.	1.	.736
3.	40.	10.	1.	.735
4.	50.	10.	1.	.735
5.	50.	100.	1.	.735

Rank	C_1	C_2	p	$\|\hat{\rho}\|$
1.	1000.	1000.	0.5	.735
2.	100.	1000.	0.5	.734
3.	60.	1000.	0.5	.733
4.	50.	1000.	0.5	.733
5.	40.	1000.	0.5	.733

For $p = 0.5$ or 1.0, many other combinations resulted in correlations of roughly 0.73.

The interpretation for the meaning of C_1 is that as it increases, the difference between the largest frequency band intensity and the intensity of the frequency band examined becomes less important. Similarly, as C_2 increases, the difference between the intensity of the examined band and that of the closest local maximum becomes less important. Note from Eq. (6.16) that since all intensities are in decibels, and differences are actually ratios, the measure is normalized for overall gain. Therefore, no terms similar to the energy weighting terms which were used in the previously described measures were used in this measure. The difference terms in Eq. (6.16) vary between 0 and 60, with the bulk confined to the 0 to 40 range.

The different values of p led to different choices for C_1 and C_2. In his initial experiments for phonetic distance, Klatt essentially used only $p = 2$. He found optimum values of C_1 and C_2 to be 20 and 1 respectively. As is evident from the table above, near maximum correlation for $p = 2$ was achieved with just such a combination. For $p = 1$, and C_1 fixed, C_2 tended to be larger, although a wide range was spanned. For $p = 1$ and C_2 fixed, C_1 tended to give best results when it was roughly equal to 40. When p was 0.5, maximum correlation was achieved for $C_2 = 1000$, and C_1 large. We find it interesting that when p was 0.5, the best weighting was none at all, while for $p = 1$, the weighting was moderate, and for $p = 2$, the best weighting was substantial. The most pleasing of these is the $p = 0.5$, $C_1 = 1000$, $C_2 = 1000$ case, which was one of the best combinations tested. Clearly, distance can be viewed as a combination of square roots of differences between spectral slopes with no weighting. Differences in slopes are the same as differences between the tangents of the corresponding angles. Since the inverse tangent function has much the same shape as the square root function, it may be that an important factor is something similar to the angle.

As with the other measures, various subsets of distortions were explored. The parameters giving the best correlation for some of them are listed below:

Distortion	C_1	C_2	p	$\lvert\hat{\rho}\rvert$
WFC	1000.	100.	2	.79
CODE	1000.	1000.	1	.53
WBD	1000.	0.5	0.5	.61
CON	100.	1000.	0.5	.73
FN	1000.	1000.	2	.90
BD	40.	1000.	1	.77

Good correlation was observed for additive noise and waveform coder distortions. Other types of distortions were not modeled as well with a notable deficiency in coding distortions.

6.4.6 Trained Measures

Measures as described in Sec. 6.3.4 were analyzed for optimum values for W_m. Table 6.3 lists the values computed for the twenty-five and

Table 6.3: Weights for trained measures.

Log Spectral Distance Measure			
25 Band Case			
Band	Weight	Band	Weight
1	-80.8	14	0.0
2	106.2	15	-27.6
3	0.0	16	0.0
4	0.0	17	-48.4
5	103.1	18	0.0
6	-140.5	19	0.0
7	0.0	20	15.5
8	0.0	21	0.0
9	0.0	22	76.4
10	-32.9	23	0.0
11	0.0	24	0.0
12	0.0	25	25.3
13	0.0		
5 Band Case			
Band	Weight	Band	Weight
1	9.7	4	-10.7
2	-16.3	5	-9.7
3	-4.8		

five band cases for log spectral distance. Given optimum weightings, substantially better performance was observed for the twenty-five band case. Also, comparing optimum weighted performance with unweighted for the twenty-five band case, improvement can be seen in log spectral measures from $|\hat{\rho}| = 0.72$ to 0.78. The five band weighted log spectral measure gives results close to the twenty-five band optimum.

No clear interpretation for the meaning of the weights in Table 6.3 can be seen. Weights which are not statistically different from zero are shown as zeros in the table, and indicate the high degree of correlation among channels for the distortion set in the data base.

A few experiments were conducted in an attempt to see if the optimum weights were robust. First, various subsets of distortions were evaluated for correlation of objective and subjective data. The results are listed below:

Distortion	$\lvert\hat{\rho}\rvert$
WFC	.81
CODE	.60
WBD	.69
CON	.83
FN	.94
BD	.70

In almost all cases, correlations were superior to those reported in Sec. 6.4.1. This shows that the weights give improvements across the board, giving some hope of robustness.

Another simple experiment consisted of extending the duration over which the measure was computed by roughly 40 percent. Objective and subjective quality were then recorrelated with a resulting coefficient of 0.717. This number is almost identical to that achieved with the unweighted log spectral distance over the same interval. When weighted measures were calculated over the interval not used in training, the correlation coefficient was only 0.56. Also unweighted log spectral distances computed over the same interval as was used to train the weighted measures resulted in a correlation coefficient of 0.75. One conclusion which can be drawn from these data is that the training of the weights gives only a minor improvement (e.g., 0.75 to 0.78) when testing occurs over the same intervals used in training. When additional speech outside of these intervals is included, the trained measures lose their advantage. It therefore is likely that the weighting coefficients computed in training have little or no meaning in themselves.

6.5 Conclusions

The measures tested in the current study were in many cases similar to those used in Chap. 5. However the unique aspect of the auditory-based measures was the critical band based spectral analysis.

Tests similar to the log spectral and power law measures but using uniformly spaced samples of LPC spectra were made on the same data base in Chap. 5. In both cases, optimum parameters closely matched those observed here. For example, $\gamma = 0$ in both sets of measures was best. Both studies also found the best exponent for power law spectral distances to be 0.2. With these values the same, however, critical band spectral analyses led to correlations of 0.72 whereas, LPC spectral distances led to correlations of 0.63. Clearly the nonuniform spacing of bands was preferable. In the earlier study, nonuniformly spaced LPC spectral samples were also computed by lumping thirty-two uniformly spaced samples into six. Both log spectral and power law measures achieved maximum correlations of 0.68, which are comparable to critical band performance. Another factor involves the fact that the LPC spectral analysis had poorer time resolution than the critical band analysis, which will be discussed below.

The articulation index approximation sought to measure short-time signal-to-noise ratios using critical band spectra. A wider class of distortions could be tested than with a time domain short-time SNR, but at the expense of precision. This is evident from a result reported in Chap. 5 in which time domain short-time SNR's had correlations of 0.78 with subjective acceptability of waveform coders. The articulation index measure achieved correlation of only 0.70 with the same subset. However, a correlation of 0.67 was possible for the set of all distortions, where the time domain system could only be used on a few of them. The weighting function applied to the traditional articulation index was shown here to give no more than slight improvement over unity weighting, which demonstrates a possible discrepancy between quality and intelligibility requirements.

The forward masking models tended to diminish the time resolution of the spectral analysis. A time constant of zero amounted to the 10 ms time resolution of the critical band analysis. Considering the degradation that occurred when this was extended to 16 ms ($\hat{\rho} = 0.717$ went to $\hat{\rho} = 0.708$), it may be possible that the 10 ms frames were too wide. The frequency variant measures of Chap. 5 had a resolution of 15 ms. The critical band analysis smoothed to 16 ms resolution gave a result of 0.708 compared to the uniform band results of 0.68. In view of this fact, one may question the importance of the precision with which the spectral analysis was formulated, and argue that any reasonable frequency variant spectral analysis choice would lead to similar results. The filter bank approach appears to have been worth a few percentage points in correlation, perhaps because of the increased time resolution, which could possibly be compensated by a smaller LPC analysis window. It seems clear, however, that some benefit was derived from time resolution better than 16 ms.

The trained measures give an upper limit on what is possible for the particular measures tested. Although the results are hard to interpret, they allude to the fact that not all 25 filter bands are necessary. This result is highly dependent on the distortion set used, and enough degrees of freedom existed with the weighting vector to encourage artifactual results. Again, however, this procedure tends to indicate that precise critical band analysis may be unnecessary for good results.

The weighted slope spectral distance measures performed best of all. Two factors may account for their superior results over the log-spectral measures: 1) use of spectral slopes rather than spectral magnitude, and 2) the particular weighting function used. Consider the log spectral distance with $\gamma = 0$, and the WSSD measure with C_1 and C_2 large. These measures are essentially identical except for spectral slopes being used in the latter case as opposed to log spectra in the former. For $p = 1$, the log spectral measure gave a correlation of 0.70 and the WSSD measure gave 0.73. However, for $p = 1$, the former gave $\hat{\rho} = 0.72$ and for the latter $\hat{\rho} = 0.67$. Therefore, simply converting from log spectra to slopes does not always lead to improvement. It should be noted, however, that given the same number of free parameters, the best WSSD measures outperformed the best critical band spectral distance measures. One of the best performing of the WSSD measures used unity weighting, (however with $p = 0.5$), which supports the idea that the slopes, rather than the weights, are important. One conclusion is that there is significant potential in this types of measure, and that it is the combination of slopes and weights which makes it unique.

Several statements can be made in summary:

1. Simple psychophysical models do not model subjective quality extremely well. For example, the psychoacoustical growth of loudness exponent of 0.6, when put into the critical band model, performed much worse than an exponent of 0.2. It is likely that degradations not modelable by simple distortions go much beyond the auditory periphery in their perception, and are inextricably linked to more central neural processes. The emergence of an exponent of 0.2 in several instances is quite puzzling, and possible

explanations need further consideration.

2. The precise Gaussian shaped critical band filter bank characteristics may be of little importance as long as a fair number of roughly logarithmically spaced channels are used.
3. Time resolution better than 10 ms may be desirable. One suggestion is that short windows allow differences in transient phenomena (e.g., bursts) to be measured.
4. Simple speech perception models, such as the WSSD measures, may be of great value in the task of predicting subjective quality. Therefore further expansion of this work to other models should have great potential.

References

[6.1] N. Kiang, *Discharge Patterns of Single Fibers in the Cat's Auditory Nerve*, Research Monograph No. 35. Cambridge, MA: The MIT Press, 1965.

[6.2] E. Galanter, "Contemporary Psychophysics," in *New Directions in Psychology I.* New York: Holt Reinhart and Winston, 1966.

[6.3] K. D. Kryter, "Methods for Calculation and Use of the Articulation Index," *J. Acoust. Soc. Amer., vol. 34.*, no. 11, pp. 1689–97, Nov. 1962.

[6.4] H. Duifuis, "Consequences of Peripheral Frequency Selectivity for Nonsimultaneous Masking," *J. Acoust. Soc. Amer., vol. 54*, pp. 1471–88, Dec. 1973.

[6.5] D. H. Klatt, "Prediction of Perceived Phonetic Distance from Critical-Band Spectra: a First Step," *Proc. 1982 IEEE ICASSP*, Paris, France, pp. 1278–81, May 1982.

[6.6] R. D. Patterson, "Auditory Filter Shaped Derived with Noise Stimuli," *J. Acoust. Soc. Amer., vol. 60*, pp. 640 51, Mar. 1976.

[6.7] T. P. Barnwell III and W. D. Voiers, "An Analysis of Objective Measures for User Acceptance of Voice Communication Systems," Final Report, DCA100-78-C-0003, Sep. 1979.

Chapter 7

Classified Objective Measures

7.1 Introduction

In previous chapters, the goal was to develop broadly based objective measures which were valid for a wide variety of speech distortions. In contrast, this chapter considers measures which are specifically designed for only a narrow class of distortions. This chapter demonstrates that when the set of distortions on which an objective measure is designed to operate is constrained to be of only one or a few types, the performance of the objective measure is often greatly enhanced. This effect can be used as the basis for an alternate approach to improving the performance of broadly based objective measures, resulting in objective quality measures called "classified" measures.

The procedure used in designing and implementing classified objective measures has three steps. In the first step, each speech distortion to be measured is assigned to a specific class of distortions. This classification can be done either subjectively or objectively, although objective classification is preferable for obvious reasons (but generally less accurate). Once all distortions are classified, separate objective measures are are designed for each of the distortion classes. Finally, the results of the classified objective measures are combined to form a single, composite measure which is valid across a broad range of distortions.

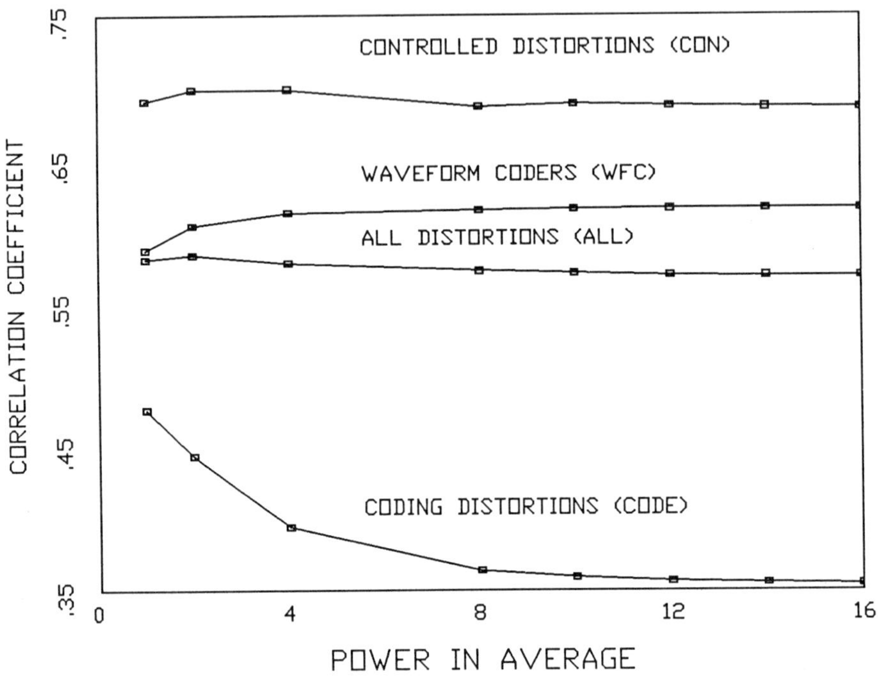

Figure 7.1: Results of classification.

7.2 Measures for Narrow Distortion Classes

It is easy to demonstrate that the classification of distortions produces large variations in the performance of simple objective measures. Figure 7.1 shows a plot of the correlation coefficient for a log spectral distance measure as a function of the value of p in the L_p norm (see Eq. (5.16)). The results are shown separately for four distortion classes: all distortions together, controlled distortions, waveform coder distortions and all coder distortions. As the graph shows, the log spectral distance measure produces much more accurate quality estimates for the class of *controlled distortions* than for the class of *all coder distortions.* The point is that if the distortions can be correctly divided into appropriate classes, then separate objective measures can be applied to the individual classes, with potentially great improvements in performance.

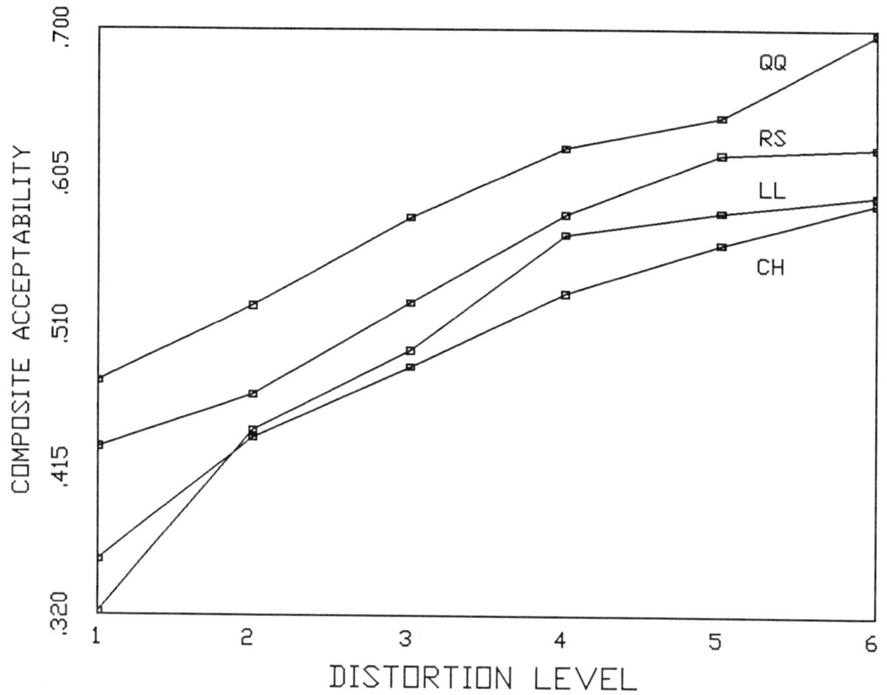

Figure 7.2: Subjective CA for CVSD.

It is also easy to demonstrate that the talker impacts the perceived distortion. Figure 7.2 shows the subjective composite acceptability scores for the six distortion levels of CVSD (see Sec. 4.2.1.1), in which the scores are plotted for each speaker separately rather than for all speakers averaged together. Note that there is a considerable and consistent spread between the quality scores of individual talkers in the graph. This spread is as much as ten points—much larger than the three point standard error typical of the DAM's composite acceptability estimate. Because of this disparity, it might well be appropriate to classify both talker and distortion. Clearly, either the objective measure must track the differences between talkers, or a separate objective measure should be designed and used for each individual talker.

An example of the improvement provided by classified objective measures is shown in Fig. 7.3, which illustrates the performance of narrowly classified objective measures. In this case only a single type of distortion, CVSD, is considered. Curve A is subjective composite acceptability (across all four speakers) as a function of distortion level. Curves B, C, and D are composite acceptability as estimated from: a short-time frequency variant SNR measure, a log spectral distance measure using simple regression, and a log spectral distance measure using polynomial regression. The performance of these measures is considerably improved as a result of considering only a narrow class of distortions. The magnitude of the coefficient of correlation between the objective and the subjective scores is greater than 0.94 for each of the three measures.

The performance of classified objective measures can be increased even more if more complex measures are considered. Figure 7.4 illustrates the performance of classified objective measures composed of multiple simple measures that are combined using regression analysis. Measure C is a combination of several simple measures, while measure D is a combination of powers of several simple measures (so that the regression analysis is in a sense a polynomial regression). Subjective composite acceptability A and the short-time frequency variant SNR measure B are the same as in the previous figure. One can see from the graphs that these combined measures, which are in themselves composite measures, have very good performance, and are better than the simple classified measures. In each case the correlation of objective to subjective measures is greater than 0.95, illustrating that composite measures generally result in improved estimates of subjective composite acceptability.

Taking this example one step further, Fig. 7.5 shows the composite acceptability estimates for a composite measure which is designed for a distortion class containing a single distortion (CVSD) and a single speaker. The graph shows both subjective composite acceptability and objectively estimated acceptability as a function of distortion level. The subjective and objective results form pairs of curves, one for each speaker. Very good performance results if the distorted speech is classified to this level, with a coefficient of correlation greater than 0.98 for all pairs of curves.

Two observations can be made based on the previous discussion. First, if the class of distortions of interest is narrow enough, it is pos-

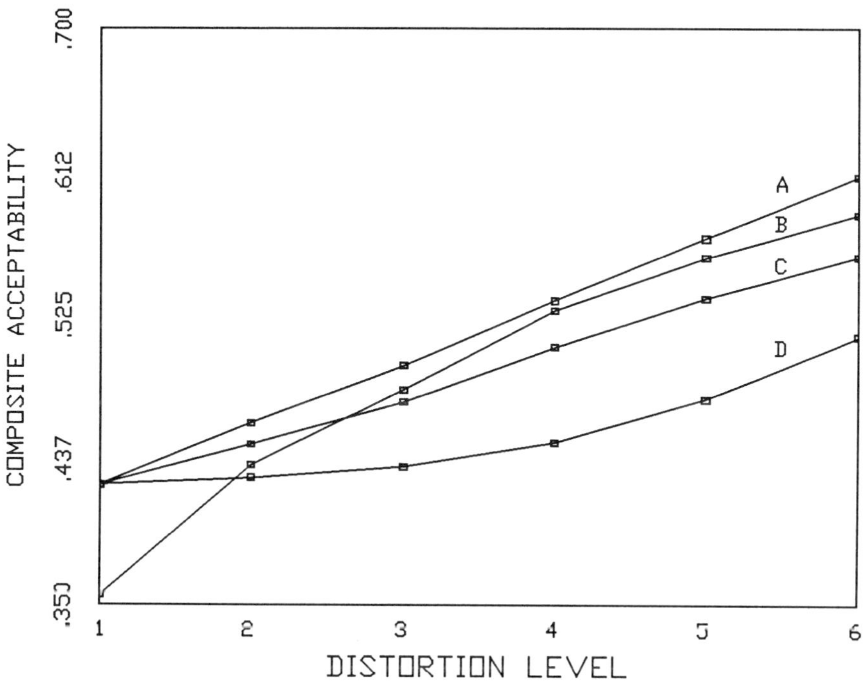

Figure 7.3: Objective estimates for CVSD from simple measures.

sible to design composite objective measures which predict subjective quality assessments with remarkable accuracy. This fact can be used to advantage if the goal is to determine the performance of a known type of speech coding system. Second, if the class of distortions of interest is broad, it is best to classify the distortions into more narrow classes and to then apply the appropriate objective measures to each class individually. In this case, the fundamental issue is that of finding procedures to objectively classify the distortions.

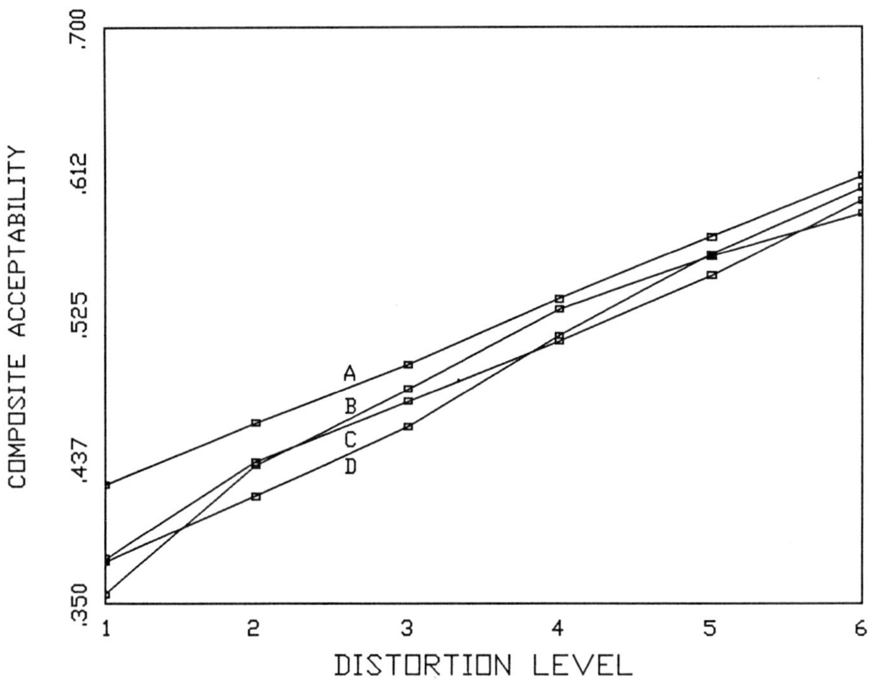

Figure 7.4: Objective estimates for CVSD from composite measures.

7.3 Identification of Homogeneous Subsets

7.3.1 Introduction

In the broad view, there are actually two possible approaches to searching for improved objective speech quality measures. The first, and most common, is to search for an objective quality measure given a set of speech distortions. A second approach, however, is to reverse the procedure and to search for a set of speech distortions given an objective quality measure. In both cases the criterion to be satisfied by the

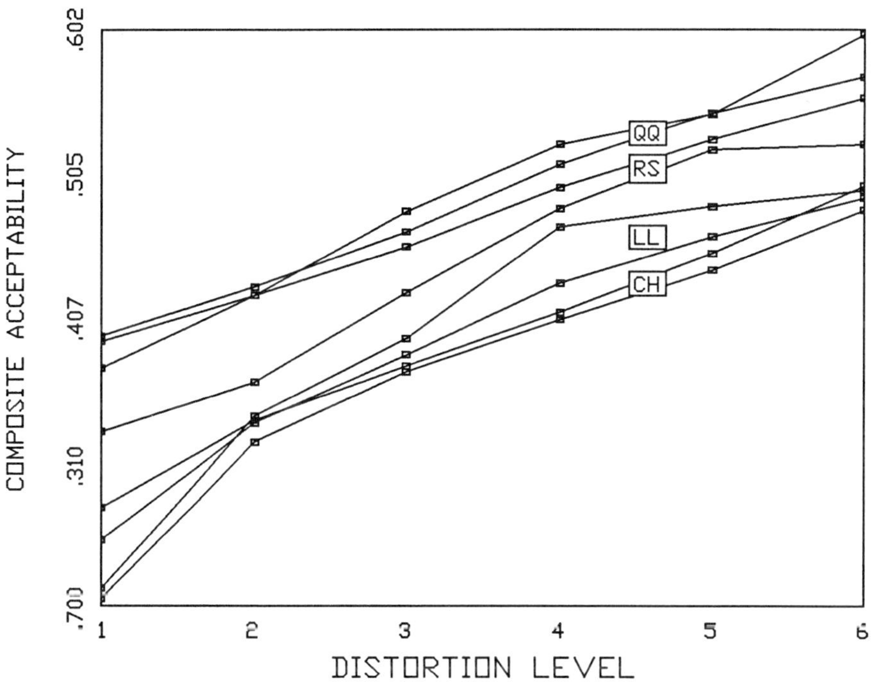

Figure 7.5: CVSD objective and subjective scores by speaker.

search is to maximize the correlation between the objective estimates of speech quality and the subjective estimates of speech quality over the speech distortions considered. The previous section discussed the first approach. The purpose of this section is to investigate the second approach.

The second approach begins with an objective quality measure, and then searches for a group of speech distortions for which the objective measure accurately predicts subjective results. This approach can be thought of as an objective classification procedure in which the available speech distortions are objectively categorized into two classes: one class contains the distortions used to assess the objective measure's performance and the other class contains the distortions to be ignored. The approach is similar to that of restricting objective measures to operate only on certain classes of distortions, such as waveform coders; but here the classes of distortions are objectively rather than heuristically specified. The intent is to select objectively a set of distortions which is homogeneous in the relationship between the objectively and subjectively measured qualities. The hope is that these homogeneous sets of distortions will provide two insights into the objective measure being studied: first, that they will show what kinds of specific distortions are best matched to an objective measure and, second, that they will indicate, by means of common features of the set's members, what overall physical characteristics of the distortions are being measured by the objective quality measure. The next step in this process would be, of course, to use these insights to adjust or reformulate an objective measure to give improved performance over a given class of speech distortions.

7.3.2 Dissimilarities Between Distortions

In order to further motivate the approach of searching for homogeneous subsets as a means of improving objective measures, consider the following experiment using the log area ratio objective measure. The experiment consists of three regression analyses. In the first analysis a sixth order polynomial regression model was used

$$CA_i = \beta_0 + \sum_{j=1}^{6} \beta_j (O_i)^j + \epsilon_i \tag{7.1}$$

in which the objective measure variable, O_i is the log area ratio measure (see Eq. (5.29)), and the dependent variable, CA_i, is composite acceptability. The regression coefficients, β_j, were estimated using the older set of forty-four speech distortions. Subscript j is an index over the order of the model and subscript i is an index over the speaker/distortion system combinations in the distorted speech data base. The resulting correlation of subjective composite acceptability

Table 7.1: Four distortions used in polynomial regression.

Adaptive differential pulse code modulation (ADPCM)
Adaptive pulse code modulation (APCM)
Continuously variable slope delta modulation (CVSD)
Adaptive predictive coder (APC)

to the regression model's estimated composite acceptability was 0.67, so that the log area ratio objective measure was able to account for 44.4 percent of the variance of composite acceptability. This result is comparable to the performance of several other simple objective measures, though this performance is not sufficient for providing reliable estimates of subjective speech quality.

The second regression analysis used the same form as Eq. (7.1), except that the data set was restricted: just four waveform coder distortions were included in the analysis, as specified in Table 7.1. The analysis resulted in an estimated correlation coefficient of 0.71, so that the log area ratio objective measure was able to account for 49.9 percent of the variance of composite acceptability. This is a surprisingly small improvement as compared to its performance over the entire set of speech distortions.

The central issue in this experiment is to find out why the log area ratio objective measure performed so poorly over the apparently homogeneous set of waveform coder distortions. One method of investigating this issue is to hypothesize that each distortion conforms to a distinctly different regression model as opposed to the single model of Eq. (7.1). A means to explore this hypothesis is to use an indicator variable regression model, stated as follows:

$$CA_i = (\beta_0+\beta_1 Z_1+\beta_2 Z_2+\beta_3 Z_3)+(\beta_4+\beta_5 Z_1+\beta_6 Z_2+\beta_7 Z_3)D_i+\epsilon_i \quad (7.2)$$

Note that this is a linear regression model as opposed to the polynomial regression model used in the previous analysis. The variables Z_j are indicator variables, and have the value either zero or one. They are called indicator variables because they indicate to which distortion data O_i belongs, as follows:

Z1	Z2	Z3	Waveform Coder Distortion
0	0	0	ADPCM
1	0	0	APCM
0	1	0	CVSD
0	0	1	APC

The indicator variables permit each distortion to have a unique slope and intercept in the regression model. The results of the analysis are shown in Table 7.2. The model has improved dramatically, in that it now accounts for 84 percent of the variance of composite acceptability. Hence the hypothesis that each distortion has a unique model is strongly supported. In particular, coefficients β_5 through β_7 are not statistically different from zero, so that the major difference between models for each distortion is that they each have a different intercept value. This is dramatically illustrated in Fig. 7.6. The solid lines are the regression curves for each of the four speech distortions. It is evident that the four curves have a similar slope but distinctly different intercepts. The dashed curve is the regression curve obtained from the previous sixth order polynomial regression analysis of this data set. The polynomial curve did not represent the underlying model of any of the distortions very well, and hence had poor performance.

What this experiment clearly illustrates is that a heuristically chosen group of speech distortions, such as a group of waveform coders, does not guarantee a set of distortions which are homogeneous with respect to their regression models. It therefore seems reasonable to use a purely statistical approach to select speech distortions which do have similar regression models and can therefore be grouped together and operated on by a single objective measure to estimate subjective composite acceptability.

7.3.3 The Objective Classification Procedure

This section describes a set of experiments whose goal was to identify the set of distortions whose subjective composite acceptability was best predicted by a given objective measure. These experiments were performed in the hope that, first, the resulting distortion partitions

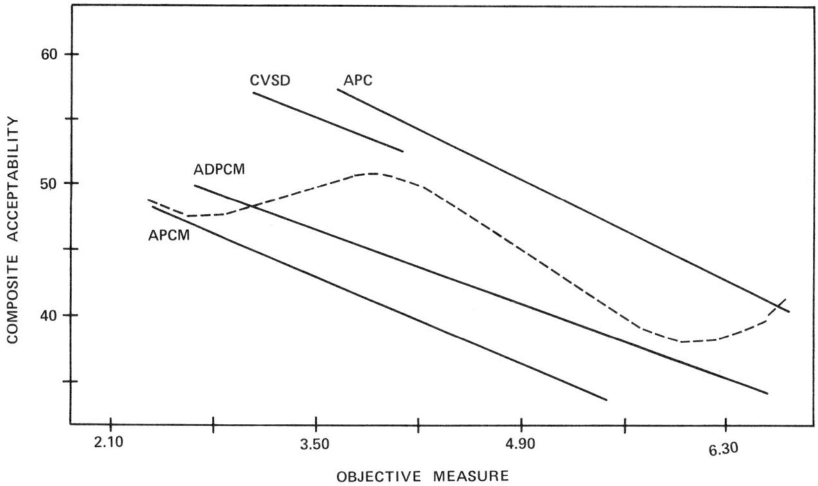

Figure 7.6: Indicator variable regression analysis.

Table 7.2: Results of indicator variable analysis.

Variable	Regression Coefficient
0	60.10
1	-4.75
2	-0.23
3	16.10
4	11.29
5	0.94
6	-0.45
7	0.26
R	.9126
R^2	.8329

could be used to realize new composite objective measures with significantly improved performance and, second, that the resulting distortion subsets could later be characterized objectively. As will be discussed below, the first of these hopes was realized, but the second remains an area for future research.

The distortion classification procedure assumes that the objective measure is given, and that it is not a composite measure. The objective measures that were considered were chosen from the best of the simple objective measures proposed by Barnwell [7.1]. Given the objective measure, the classification procedure finds forty-four separate distortion subsets, with from one to forty-four members respectively. Each subset contains the set of distortions which provide the best correlation between the objective measure and subjective composite acceptability, given the number of members in the subset. The procedure for determining the subsets can be divided into two parts. The first part of the procedure searches through all possible distortion subsets for the subset of size N which provides the greatest correlation between the selected objective measure and composite acceptability. The correlation is computed only over the members of the subset. Let this subset of size N be called S_N. This would be the only section of the procedure were it not for the very large number of computations involved as the number of members in the subset grows larger. In the investigation of all subsets of size N the number of correlations that must be computed is equal to the number of combinations of 44 items taken N at a time, or

$$\binom{44}{N} = \frac{44!}{(44-N)!N!} \tag{7.3}$$

An exhaustive search of all subsets of all sizes would then require a number of correlation calculations equal to the sum of 44 items taken N at a time for N equals one to 44, a number which exceeds 10^{12}. Because of this excessive number of calculations, the first part of the procedure was only done for subsets of size one through five.

The second part of the procedure circumvents the problem of burdensome calculations at the expense of being suboptimal (suboptimal in the sense that it will not necessarily find the subset of N members which provides the highest correlation between the objective measure and composite acceptability). This second part searches for a distortion not already a member of set S_{N-1} which, when added to S_{N-1}, produces a new set S_N which provides the largest possible correlation between the objective measure and composite acceptability. Again, the correlation is computed over the set S_N. This step is repeated for N equals 6 through 44. The entire algorithm is summarized in Fig. 7.7.

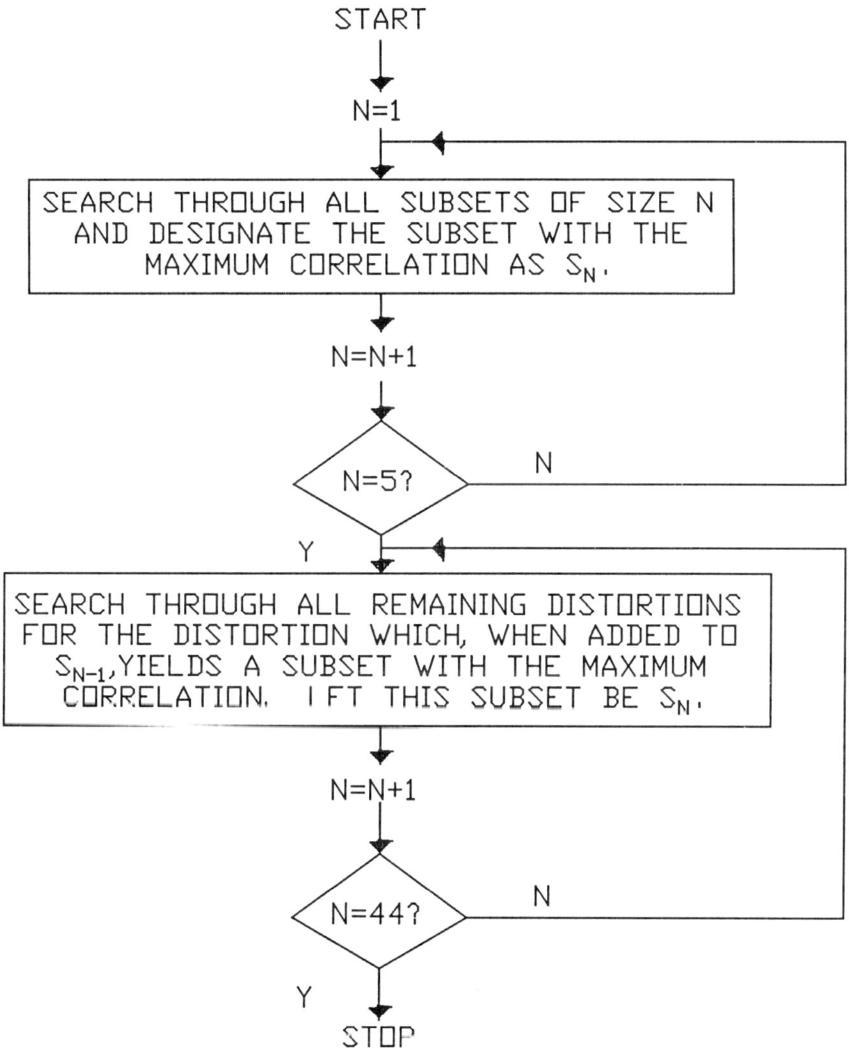

Figure 7.7: Algorithm for selecting distortion subsets.

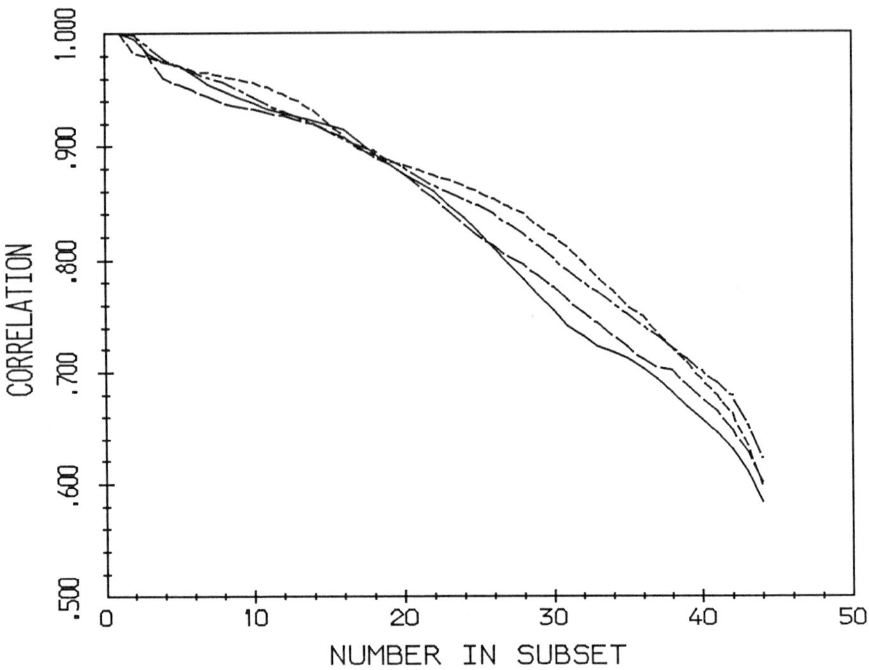

Figure 7.8: Results of homogeneous subset analysis.

7.3.4 Results of Objective Classification

The results of the subset classification experiment are, in general, inconclusive. The graph in Fig. 7.8 shows how the correlation coefficient for the best subset varies with the number of members in each subset for each of the objective measures studied. These results look quite promising. For each of the four measures, a subset of fifteen distortions, or one third of the total number of distortions, had a correlation of better than 0.90. Therefore all of these objective measures are producing very good estimates of subjective composite acceptability for each of the distortions in the subsets. The results are less encouraging, however, when the types of distortions contained in the subsets is examined. Table 7.3 lists the distortions contained in the subset of fifteen distortions for each of the objective measures presented in Fig. 7.8.

The most notable aspect of these distortion subsets is that each

Table 7.3: Homogeneous subsets.

Log Spectral Distance Measure:	Nonlinear Spectral Distance Measure:
center clipping	800 - 1300 Hz noise
400 - 800 Hz noise	PD 2600 - 3400, angular
PD 1900 - 2600, angular	PD 200 - 400, angular
PD 2600 - 3400, radial	400 - 800 Hz noise
ADPCM	VEV 13
PD 200 - 400, angular	VEV 7
BD 1300 - 1900	APC
APCM	BD 1300 - 1900
VEV 7	APCM
VEV 13	PD 2600 - 3400, radial
800 - 1300 Hz noise	0 - 400 Hz noise
peak clipping	BD 800 - 1300
PD 1300 - 1900, angular	center clipping
0 - 400 Hz noise	PD 1300 - 1900, angular
APC	quantization

Log Area Ratio Distance Measure:	Log Likelihood Measure:
bandpass filtering	800 - 1300 Hz noise
2600 - 3400 Hz noise	BD 1300 - 1900
PD 2600 - 3400, angular	PD 200 - 400, angular
PD 200 - 400, angular	ADPCM
BD 1900 - 2600	center clipping
1900 - 2600 Hz noise	PD 2600 - 3400, radial
BD 100 - 400	APCM
1300 - 1900 Hz noise	PD 1900 - 2600, angular
PD 2600 - 3400, radial	0 - 400 Hz noise
highpass filtering	peak clipping
BD 800 - 1300	PD 1300 - 1900, angular
lowpass filtering	BD 100 - 3500
BD 1300 - 1900	400 - 800 Hz noise
APC	PD 800 - 1300, radial
PD 1900 - 2600, angular	PD 400 - 800, angular

contains a very diverse group of distortions, which is contrary to the desired results. A close examination of each subset reveals that there are one or two groups of the same distortion type within each subset. For example, the subset associated with the log spectral distance measure contains three contiguous bands of additive narrowband noise distortions and two contiguous bands of angular pole distortions. Similarly, the subset associated with the log likelihood measure contains three contiguous bands of additive narrowband noise and four bands of angular pole distortions. The subset associated with the log area ratio distance measure contains three bandlimiting filtering distortions, three contiguous additive narrowband noise distortions and three contiguous narrowband frequency distortions. Though there are these limited similarities between distortions in the subsets, in general there is not enough commonality between distortions to make any firm conclusions regarding the type of distortions which are best suited for the objective measures. Since it is not clear what general qualities these distortions have in common, it is even less clear what physical qualities of those distortions are being measured to yield the undeniably good estimates of subjective composite acceptability. Hence it was not possible to advance any clear hypotheses about the underlying mechanisms which, in a statistical sense, make the subsets homogeneous.

7.3.5 Discussion

Intuitively, the statistical method for choosing homogeneous distortion subsets, as presented in this section, has merit in that it identifies, by the very nature of the algorithm, near-optimal distortion subsets. For all objective measures investigated the performance over subsets containing one third of the total number of distortions was very good, with correlation with composite acceptability exceeding 0.90 in all cases. These facts promote the statistical approach as opposed to a heuristic approach to choosing distortion subsets. Unfortunately, whereas a heuristic approach based on grouping common distortion types, by its very nature, yields physically homogeneous subsets, the statistical approach yields subsets which are fragmented, containing small groups of diverse distortion types. This is largely unsatisfying, in that no broad conclusions can be drawn as to the physical or perceptual nature of distortions which are best matched to the objective measure being investigated.

This is not to say that the statistical approach for grouping distortions is entirely rejected, but merely that it is inconclusive based on an initial set of experiments. The conclusion at this stage, however, is that insight into the underlying mechanisms which cause an objective measure to be a good match to a certain set of distortions, and hence permit the objective measure to make good estimates of subjective quality, are best found through other experimental approaches. In particular, it is felt that investigation of objective measures for estimating parametric subjective qualities would yield more insight into these issues.

References

[7.1] T.P. Barnwell III and W. D. Voiers, "An Analysis of Objective Measures for User Acceptance of Voice Communication Systems," Final Report, DCA100-78-C-0003, Sep. 1979.

Chapter 8

Parametric Objective Measures

8.1 Introduction

A fundamental problem in the objective estimation of speech quality is that broad ratings of quality, such as "acceptability," give no indication of the precise perceptual nature of the speech distortion. This means that two coders might have equal acceptability ratings, but produce speech which sounds very different. This ambiguity complicates the design of objective speech quality measures when the measures are to operate across a wide range of speech distortions, as in this research. This is because the objective measure must perform a many-to-one mapping, in which speech distortions which are vastly different both perceptually and physically must receive the same objective estimate of speech acceptability.

A clear example of this effect is shown in Fig. 7.6. It illustrates that the log area ratio measure, which is a fairly good objective measure, does not accurately perform this many-to-one mapping of objective distortion to subjective quality. As explained in Chap. 7, the solid lines in this figure show the linear regression model when four different waveform coders are allowed to have individual model parameters. The correlation between estimated and actual subjective quality for this model is 0.91, good enough to consider it a "correct" model. The problem is that when all four coders are considered simultaneously, a given objectively measured value does not map into a single subjective quality. In fact, the four regression lines for the four waveform coders are sufficiently far apart that there is no hope of producing accurate quality estimates when the four are grouped together. This is graphically illustrated by the dashed line in Fig. 7.6, which shows the polynomial regression model produced when all coders are considered as a group. Clearly, the log area ratio is not measuring all of the important factors which determine speech acceptability. Since the four distortions considered here are all waveform coders, more severe problems could be expected, in general, over a broader set of distorting systems.

It is doubtful that the measurement of any one physical parameter can ever account for all perceptually significant aspects of speech degradation. Hence, a reasonable approach is to extract a number of different physical parameters objectively and combine them into a single objective measure. However, even if this were done, the broad quality characterization "acceptability" does not give many clues as to the appropriate functional form for a corresponding objective measure. There are, however, other scales in the Diagnostic Acceptability Measure which are considerably more specific in scope than composite acceptability. Table 3.1 lists the entire set of quality assessments included in the DAM. Although the term "composite acceptability" does not immediately suggest a corresponding objective measure, many of the parametric subjective qualities scales do. It is therefore reasonable to expect that objective measures can be designed which will track these more specific parametric subjective qualities successfully. Once such specific objective measures are designed, they can be combined to form an estimate of composite acceptability, which, for most applications, is the quality of greatest interest.

Another reason for using parametric objective measures can be seen in the multidimensional scaling of Fig. 8.1. In fact, this figure provided the initial motivation for investigating the parametric quality scales of the DAM. In this scaling, each of the points are quality measures, with the subjective quality measures of the DAM represented by white dots, and several of the best objective measures of speech quality represented by black dots. In this analysis, the similarity between two quality measures is equal to the magnitude of the coefficient of correlation between them. Highly correlated measures are represented by points which are close together while relatively uncorrelated measures are represented by points which are far apart. The DAM quality measures are identified by their associated mnemonic, as listed in Table 3.1. The signal, background and total quality scales of the DAM have been circled, as have the points representing various parameterizations of the log likelihood ratio and spectral distance measures. The black dots which are not circled represent objective measures based on LPC parameters such as the log area ratio, PARCOR, and feedback coefficients. The stress for this configuration is 19 percent, which is rather high, but the configuration still has a very enlightening interpretation.

What is significant about the multidimensional scaling of Fig. 8.1 is that all of the various objective measures are grouped together in one sector of the perceptual space defined by the DAM's parametric quality scales. This means that even though the objective measure are arithmetically quite dissimilar, they all measure approximately the same perceptual aspects of speech quality, and, hence, little improvement can be expected when they are combined into composite objective measures. If, on the other hand, it was possible to design objective measures whose results were scattered throughout this multidimensional scaling space, then these new measures would clearly be measuring different perceptual aspects of speech quality. Hence, when optimally combined, such measures would be likely to produce improved estimates of composite acceptability. A good approach to achieving this result is to design such a new set of objective measures in which the individual members predict the individual parametric subjective quality scales of the DAM.

8.2 Analysis of DAM Quality Scales

8.2.1 Regression Analysis

Before proceeding to design parametric objective measures, it is important first to check the viability of using parametric subjective measures as a means of estimating composite acceptability. Clearly, if composite acceptability cannot be well estimated from the parametric subjective scales, it is unlikely that it can be well estimated from objective estimates of the parametric subjective scales. In order to determine the answer to this issue, a regression analysis was performed on the sub-

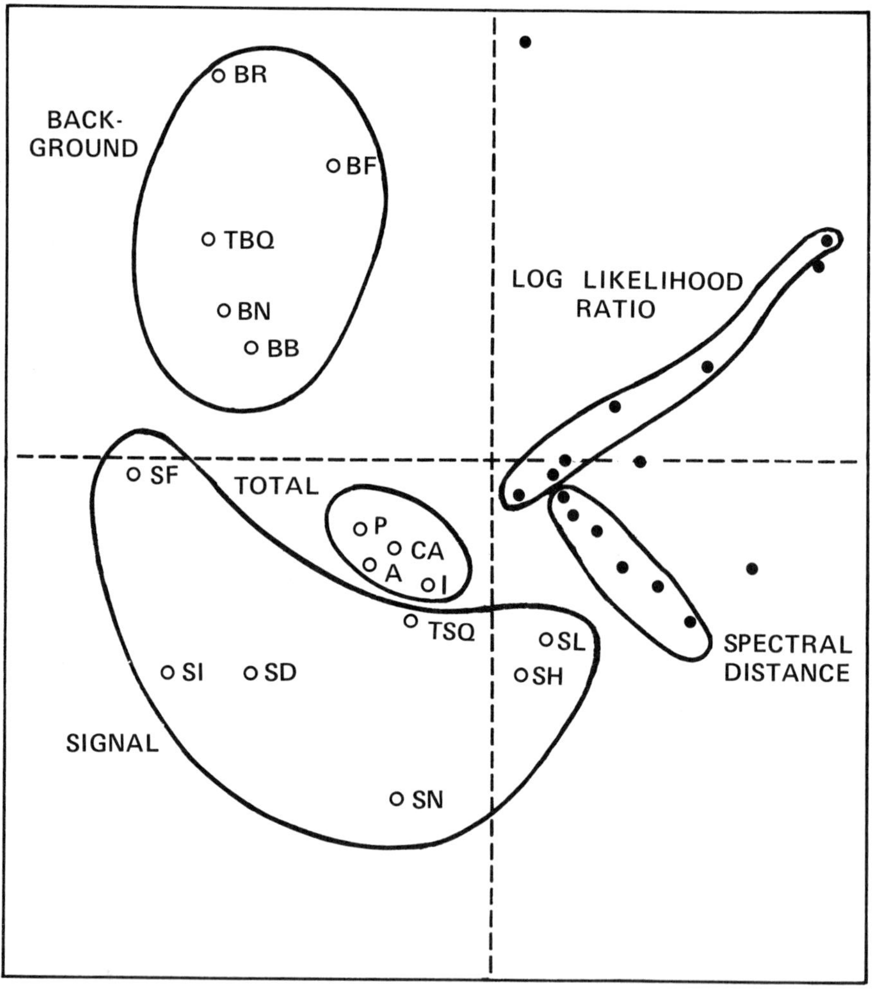

Figure 8.1: Multidimensional scaling of the DAM subjective measures and a sampling of objective speech quality measures.

jective quality data base. There were basically two questions to be answered. The first question involves how well composite acceptability can be estimated from the total set of parametric subjective qualities. If composite acceptability can be estimated well from the set of parametric subjective qualities, then it is reasonable to predict that it can also be estimated well from a set of parametric objective quality measures. The second question involves the independence of the individual parametric subjective qualities, or, more precisely, the question of which and how many parametric subjective qualities need to be used to estimate composite acceptability adequately. Fewer terms in the model of composite acceptability mean fewer objective measures to design and implement, with a corresponding reduction in the computational cost of implementing the composite acceptability objective measure. It is important to note, however, that if the regression analysis indicates that one or more of the parametric qualities are not needed, it may be a reflection of the data in the distorted speech data base and it may not be valid for all speech distortions. The distorted speech data base is very large, but of course does not represent all possible aspects of speech distortion. In this respect, parametric measures which are found to be insignificant in determining composite acceptability for this data should not necessarily be excluded when considering a very different mix of distorted speech data.

The model for estimating composite acceptability from the parametric subjective qualities is

$$\hat{y}_i = b_0 + \sum_{j=1}^{k} b_j x_{ij} \tag{8.1}$$

where $\hat{y}_i$ is the estimated composite acceptability score for distortion system i, the b's are the regression coefficients to be estimated, and x_{ij} is a parametric or isometric subjective quality score for distortion system i. In all cases the regression analysis was done over the set of 1056 distortion systems.

It should be noted that a regression model which included some parametric subjective qualities was used by Dynastat in generating the estimate for composite acceptability. However, the Dynastat model was multiplicative rather than additive, and gave large weights to the non-parametric qualities and small weights to the parametric qualities

Table 8.1: Independent variables.

MNEMONIC	DESCRIPTORS
SIGNAL QUALITY	
SF	fluttering, bubbling
SH	distant, thin
SD	rasping, crackling
SL	muffled, smothered
SI	irregular, interrupted
SN	nasal, whining
BACKGROUND QUALITY	
BN	hissing, rushing
BB	buzzing, humming
BF	chirping, bubbling
BR	rumbling, thumping
TOTAL QUALITY	
I	intelligibility
P	pleasantness
A	acceptability

when computing composite acceptability from the parametric subjective qualities.

In all, three regression studies were performed on the subjective data base. The first represents an upper limit on how well composite acceptability can be estimated based on all of the available information and using only linear regression models. Table 8.1 lists the parametric and isometric qualities used in this analysis. The results of the analysis, listed in Table 8.2, are that 99.9 percent of the variability of composite acceptability was explained by the included variables (R-squared= 0.9990). This is nearly perfect, indicating that the subjective qualities included in the model together contain all the information present in the composite acceptability quality.

The second analysis was limited to using only the parametric signal and background qualities as independent variables in the regression model for estimating composite acceptability. The regression analysis was slightly different in that forward stepwise regression was used as a means of identifying the most important of the parametric qualities. As the name implies, stepwise regression is an iterative technique used for independent variable selection. In the first step the variable which ex-

Table 8.2: Results of multiple linear regression analysis.

Multiple R	0.9995	Std. Error of Estimate	0.3153
Multiple R-Squared	0.9990		

Analysis of Variance

Source of Variation	Sum of Squares	Degrees of Freedom	Mean Square	F Ratio	Area in Tail
Regression	102252.63	13	7865.586	79135.60	0.0000
Residual	103.57	1042	0.099		

Objective Variable	Regression Coefficient	t-statistic	Two-tail Significance
Intercept	-60.6702		
SF	0.1159	53.82	0.000
SH	0.1021	31.54	0.000
SD	0.1243	57.88	0.000
SL	0.1720	42.25	0.000
SI	0.0571	24.07	0.000
SN	0.0159	5.08	0.000
BN	0.0993	38.84	0.000
BB	0.0307	9.89	0.000
BF	0.0710	29.60	0.000
BR	0.0471	12.22	0.000
I	0.3064	74.81	0.000
P	0.2754	49.23	0.000
A	0.2560	34.04	0.000

plains the most variability in the dependent variable is included in the model and all model statistics are computed. In all subsequent steps the variable which, when added to the current model, helps explain the most variability in the dependent variable is included in the model and all model statistics are computed. In this way a useful, though suboptimal, ranking of the independent variables is obtained by the degree to which the variables contribute to the model. In addition, at every step a regression model for the included independent variables is obtained.

Table 8.3 shows the results of this analysis. Listed are the parametric qualities in the order in which they entered the model, the multiple R, or estimated correlation coefficient, the multiple R-squared, or fraction of variance explained, and the increase in multiple R-squared.

Table 8.3: Results of stepwise regression.

Step No.	Entered	Multiple R	R^2	Increase in R^2
1	SD, rasping, crackling	.6541	.4278	.4278
2	SL, muffled, smothered	.8120	.6594	.2316
3	SF, fluttering, bubbling	.8648	.7478	.0885
4	BN, hissing, rushing	.9039	.8171	.0692
5	BF, chirping, bubbling	.9175	.8418	.0248
6	SI, irregular, interrupted	.9380	.8798	.0380
7	SH, distant, thin	.9494	.9014	.0216
8	BB, buzzing, humming	.9518	.9059	.0045
9	BR, rumbling, thumping	.9524	.9070	.0011

Table 8.4: Results of all possible subsets regression.

	Parametric Quality	Number in Subset									
		1	2	3	4	5	6	7	8	9	10
1	SD, rasping, crackling	X	X	X	X	X	X	X	X	X	X
2	SL, muffled, smothered		X	X	X	X	X	X	X	X	X
4	BN, hissing, rushing				X	X	X	X	X	X	X
6	SI, irreg., interrup.					X	X	X	X	X	X
5	BF, chirping, bubbling					X	X	X	X	X	X
7	SH, distant, thin						X	X	X	X	X
3	SF, fluttering, bubbling			X	X			X	X	X	X
8	BB, buzzing, humming								X	X	X
9	BR, rumbling, thumping									X	X
	SN, nasal, whining										X

Number in Subset	Multiple R^2	R
1	0.427	0.653
2	0.659	0.812
3	0.747	0.864
4	0.816	0.903
5	0.866	0.931
6	0.885	0.941
7	0.901	0.949
8	0.905	0.951
9	0.906	0.952
10	0.906	0.952

The results show that two qualities dominate the rest in terms of contribution to the model. These are SD, which by itself accounts for 43 percent of the variability of CA, and SL which, along with SD, accounts for 66 percent of the variability of CA. Histograms of frequency of occurrence of a given subjective quality rating, as plotted over the range of all possible subjective quality ratings (the upper graphs in Figs. 8.5 and 8.6), give an indication of the variance of SD and SL ratings for the speech quality data. Both SD and SL have a large variance relative to the other parametric subjective qualities, so they help to explain a larger portion of the variance in composite acceptability. Another encouraging result is that only seven parametric qualities are needed, SD through SH, to raise the multiple R-squared, or coefficient of multiple determination, for the model to above 0.90. Therefore only seven of the ten parametric subjective qualities are needed to explain 90 percent of the variability of composite acceptability, and the remaining three subjective qualities, when added to the model, explain less than one additional percent of the variability of composite acceptability.

The performance of this model is quite good, with the model error being comparable to the error in the DAM itself. The index of reliability, or test-retest correlation, of the DAM is 0.95, which means that $(1 - 0.95^2)$, or 10 percent of the variability of composite acceptability is due to random error. Any increase in the performance of the regression model would be meaningless, as it would be tracking random errors in the data. This analysis suggests that objective measures for only seven of the parametric subjective qualities need to be built, since the remaining subjective qualities contribute little to the estimation of composite acceptability in this data base.

The third regression analysis was an all possible subsets analysis, done to better support the conclusions reached by the stepwise regression analysis. Stepwise regression is, in general, a suboptimal method for independent variable selection. In a given step only those variables not yet included are examined, without regard for the appropriateness of the variables already included. In contrast, all possible subsets is an optimal method of variable selection since it examines all the independent variables at each step and chooses that subset of n variables (n being the step number) which best explains the variability in the dependent variable. Therefore this analysis method will find the set of parametric subjective qualities that will yield the best estimate of composite acceptability, under the restriction that the set contain only n members.

The results of this analysis are listed in Table 8.4. For each subset of size n, the table lists the corresponding multiple R-squared, multiple R and also indicates the parametric qualities included in that subset. In this method of analysis, a specific ordering of importance of parametric qualities is more difficult than with stepwise regression. Since the regression model is totally re-evaluated for each subset size, there is no one order of variable entry. The table lists parametric qualities in approximate order of entry under all possible subsets regression, and also indicates, by the numbers in the leftmost column, the order in which the qualities entered under stepwise regression. The most notable difference between the two types of analyses concerns the quality SF. Under stepwise regression this variable entered in step three, where, under all possible subsets, SF entered in subset three, dropped out in subset five, and re-entered in subset seven. Therefore stepwise analysis overemphasizes the importance of SF. However, for the remaining parametric qualities the two analysis methods yield quite similar results.

Two conclusions can be drawn from the results of regression analysis on the subjective data base. First, all subjective qualities can be used to construct a model which provides excellent estimates of subjective composite acceptability. And second, a subset of the parametric qualities can be used to construct a model which provides estimates of composite acceptability which are nearly as good as the estimates made by the full model. Given these conclusions, it is clearly desirable to construct objective measures which provide good estimates of the parametric subjective qualities, since such objective measures, if combined into one large model, would be expected to provide improved estimates of subjective composite acceptability.

8.2.2 Multidimensional Scaling Analysis of Subjective Measures

Multidimensional scaling was used on the subjective data base to explore the perceptual relationship between the parametric subjective qualities and the isometric subjective qualities, with particular interest in composite acceptability. Figure 8.2 shows the results of a non-metric multidimensional scaling analysis done on the DAM subjective quality data. Each white dot is a subjective measure, with the measures identified by their associated mnemonic (see Table 3.1). For this analysis, the similarity between subjective measures was equal to the magnitude of the correlation coefficient between the quality ratings of the two scales for all the distortion systems in the data base. A descending monotonic regression (see Sec. 8.3) was done on the similarities, so that a similarity nearly equal to 1.0 mapped into a distance nearly equal to zero. Because the transformation from similarity to distance was monotonic, the scaling was non-metric.

The scaling shown in Fig. 8.2 is for a realization space of two dimensions, with a stress of 16.9 percent. Although this value is not extremely low, indicating a less than perfect correspondence between the plot and the actual correlations between subjective qualities, the plot is easy to comprehend and the axes of the plot are amenable to perceptual interpretation. These two facts argue for using a two rather than a higher dimensional realization space.

The plot shows composite acceptability near the center of the space. The other high level qualities, intelligibility, pleasantness, and acceptability, are centered closely around composite acceptability suggesting that qualities in the center of the realization space are general in nature. The left side of the realization space contains most of the signal qualities while the right side contains the background qualities, suggesting that the horizontal axis measures a signal versus background quality degradation dichotomy. Similarly, the top of the plot contains qualities whose exemplars are mostly fluttering or interrupted, while the bottom of the plot has qualities which exemplify primarily noisy distortions. Therefore the vertical axis seems to measure a noisy versus fluttering quality degradation. Further, total signal quality and total background quality are both nearly centered within their respective signal or background parametric qualities.

A conclusion which can be drawn from this multidimensional scaling is that the parametric qualities in the subjective data base do, in fact, measure different subjective qualities since all the parametric qualities are widely spaced in the realization space. Parametric qualities closely spaced in the realization space would indicate a large degree of redundant information. Another point is that, in two dimensions, perceptual qualities can be associated with the axes of the realization

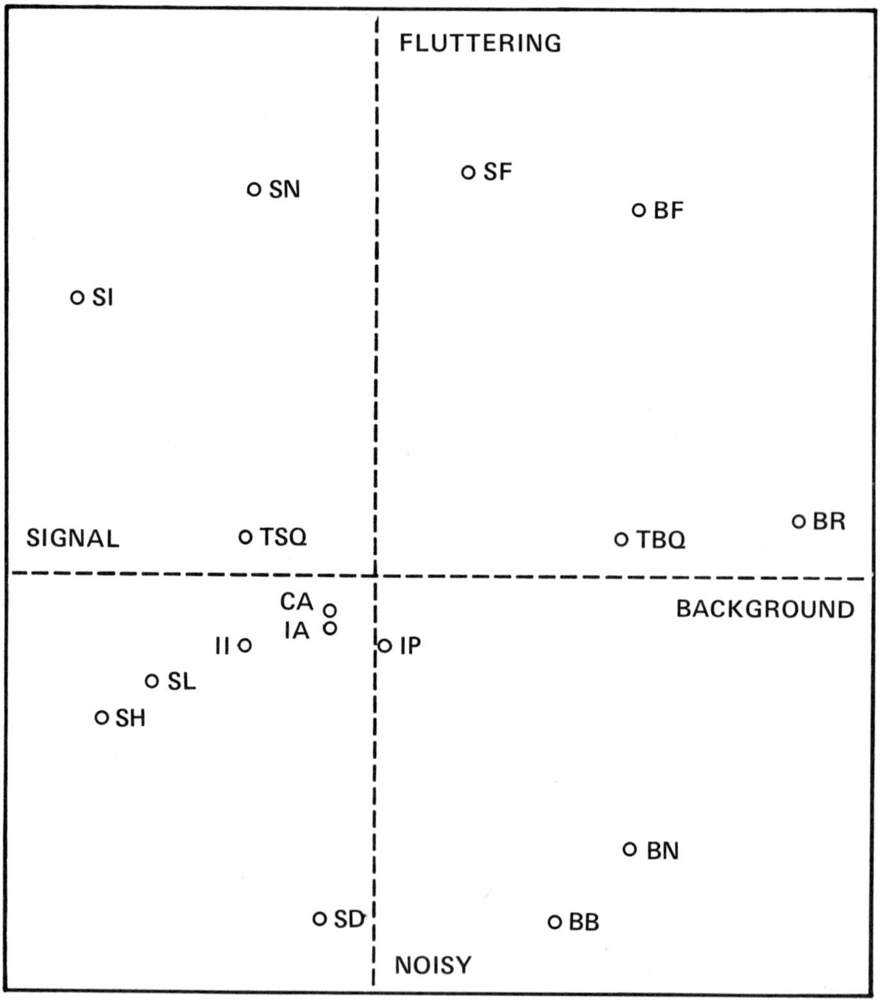

Figure 8.2: Multidimensional scaling of the subjective measures of the DAM.

space. And a final point is that composite acceptability falls nearly in the center of the realization space, which agrees with its role as an overall quality measure.

8.3 Monotonic Regression Analysis

Two of the objective measures (SL and SH) to be discussed in the next section make use of *monotonic regression analysis*. Because this method of data analysis is not widely used, a brief tutorial will be presented here.

Monotonic regression is similar to simple linear regression in that the goal is to pass a curve through a set of points such that a cost function is minimized. In the case of monotonic regression, however, the curve need not have a parameterized functional form, such as $y = ax + b$, but rather must simply be a monotonically non-increasing or non-decreasing curve. This is a special case of regression under order restrictions, and is thoroughly covered by R. E. Barlow [8.1]. In both types of regression there are three principal variables: the independent variable x_i, the dependent variable y_i, and the estimated dependent variable $\hat{y}_i$. The subscript i is the observation index, or index over all data points. As in regression, the objective function to minimize is the sum of the squared error over all observations, where the error is $e_i = (y_i - \hat{y}_i)$. However, in monotonic regression the only restriction on $\hat{y}_i$, besides minimization of squared error, is monotonicity, such that $\hat{y}_i < \hat{y}_{i+1}$ if $x_i < x_{i+1}$. The inequality relating the $\hat{y}_i$'s is "less than" for monotonically increasing regression and "greater than" for monotonically decreasing regression. Figure 8.3 part shows a monotonically decreasing curve (solid line) fit to a set of data.

The "Up-and-Down Blocks" algorithm, developed by J. B. Kruskal (in [8.1]) is an efficient method of computing a monotonic regression. In the following discussion assume that the dependent variable, x_i, is arranged in ascending order such that $x_i < x_{i+1}$ for all i from 1 to $N - 1$ and that there are N elements in the dependent variable data set. Understanding the algorithm first requires defining a few terms.

BLOCK - a set of consecutive elements y_j through y_k, $j < k$. The value of a block is equal to the average of the elements in that block.

UP-SATISFIED and DOWN-SATISFIED - consider three consec-

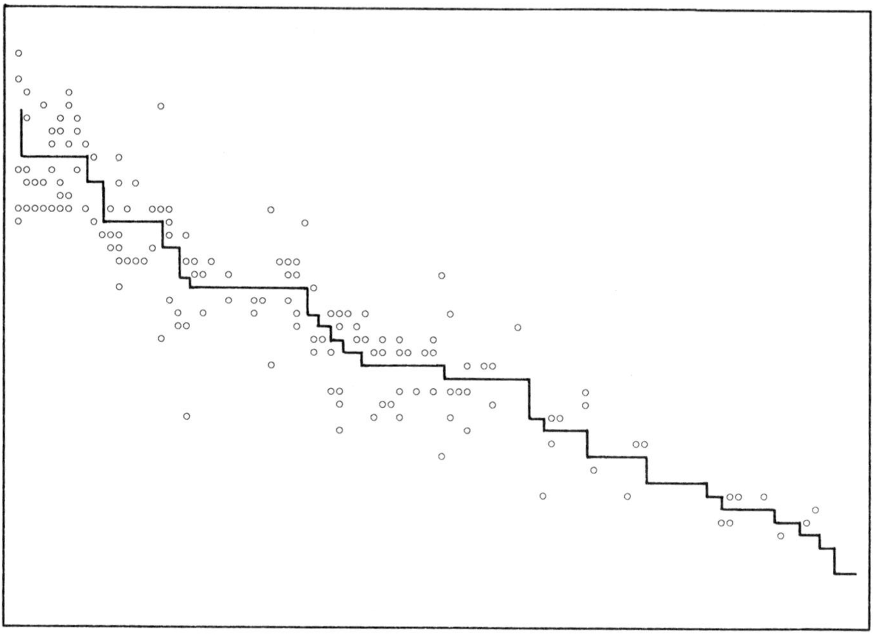

Figure 8.3: Monotonic regression curves.

utive blocks, $B-$, B, and $B+$. For a monotonically increasing regression, block B is said to be up-satisfied if the average of the elements of B is less than the average of the elements in $B+$. For a monotonically increasing regression, block B is said to be down-satisfied if the average of the elements of B is greater than the average of the elements in $B-$. For a monotonically decreasing regression the previous two inequalities are reversed. Additionally, any block containing y_N is automatically up-satisfied and any block containing y_1 is similarly down-satisfied.

A flowchart of this algorithm is shown in Fig. 8.4. The algorithm begins with the independent variable data set partitioned into N blocks of one element per block. At each stage in the algorithm one block is designated as "active." Three choices are available for an active block. If the active block is not up-satisfied then it is combined with the next higher block. If the active block is not down-satisfied then it is combined with the next lower block. If the active block is up-satisfied and down-satisfied then the next higher block becomes active. At the start the first block is active and the algorithm is terminated when the highest active block is up-satisfied. The values of the blocks at termination are the desired $\hat{y}_i$ and are the best monotonically increasing fit to the data y_i, subject to minimizing the sum of the squared error. If, at termination, a block contains more that one element, for example y_j through y_k, then each corresponding estimate of the dependent variable, $\hat{y}_j$ through $\hat{y}_k$, is equal to the value of the block containing y_j through y_k.

The most significant result of monotonic regression is the "stress" statistic, which is the error variance divided by the dependent variable variance. This can be expressed as

$$Stress = \left[\frac{\sum_{i=1}^{N} (e_i)^2}{\sum_{i=1}^{N} (y_i - \bar{y})^2} \right]^{1/2} \tag{8.2}$$

where $\bar{y}$ is the average of the y_i's over all i. The stress of a monotonically increasing regression provides a measure of how closely a set of y_i's conform to a monotonically increasing function. If the set is perfectly monotonically increasing, then the resultant stress is zero, and if the set is perfectly monotonically decreasing then the resultant stress is one.

An straightforward extension of monotonic regression is unimodal regression, which fits a unimodal curve to the data set under the constraint that the sum of the squared error is minimized. A discussion of the details of this technique is in [8.1]. As before, stress is computed from Eq. (8.2).

8.4 Parametric Objective Measures

This section discusses specific objective measures for estimating parametric subjective quality. The parametric qualities which will be objectively estimated are the first seven qualities in Table 8.3, which were found to be most significant in determining composite acceptability for the data used in this research.

The ultimate goal in the research is to develop improved objective measures of speech acceptability, with an intermediate step being the design of parametric objective measures. Table 8.19 summarizes the performance of the parametric objective measures discussed in this

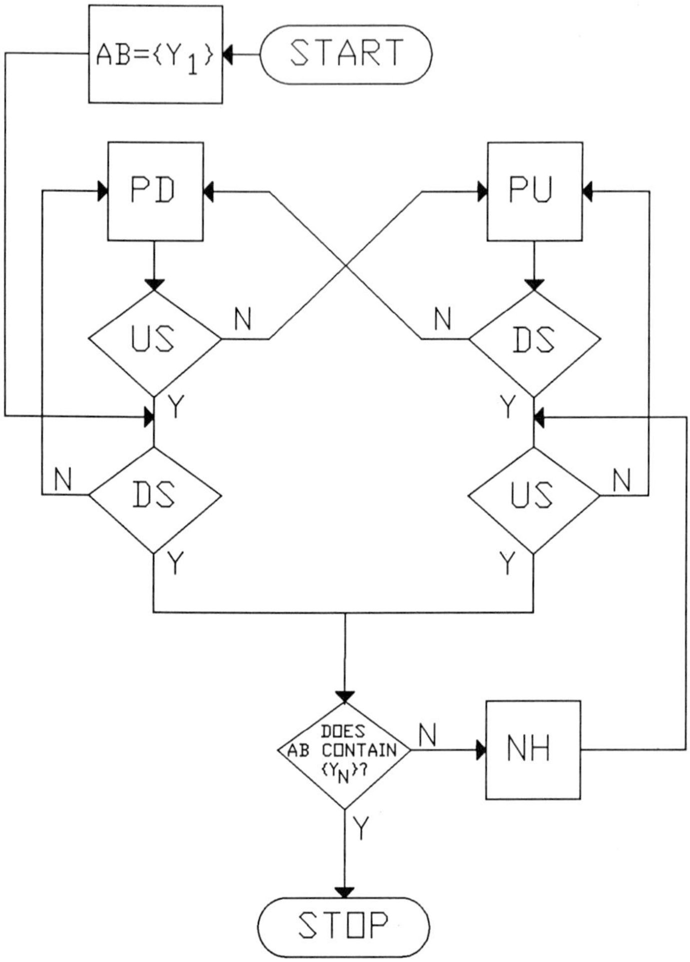

Figure 8.4: Up-and-down blocks algorithm. Symbols in the diagram are explained below. (From [8.1]).

AB Active block

US Up Satisfied

DS Down Satisfied

PD Pool down. The current block is merged with the next lower block.

PU Pool up. The current block is merged with the next higher block.

NH The active block is now the next higher block.

chapter. Their performance ranges from a minimum correlation of 0.44 to a maximum correlation of 0.91. Clearly, several of these measures are producing inaccurate estimates of parametric subjective quality. However, the performance of the majority of these objective measures is quite good when compared to objective measures commonly used to evaluate speech communication systems. For example, a common measure is the log spectral distance, listed in Table 2.4, which has a correlation to composite acceptability of only 0.60 over the speech quality data base.

It is difficult to specify what level of performance in the parametric objective measures is adequate for developing a good composite measure of acceptability. The criterion that was adopted is not to insist on a specific correlation level for each parametric measure, but rather to judge the parametric measures based on their ability, when combined into a single regression model, to predict subjective composite acceptability. If accurate estimates of composite acceptability result, then the parametric objective measures must be measuring all of the perceptually significant aspects of speech quality for the data set considered. The performance of the parametric measures presented in this chapter definitely satisfies this requirement, since the composite measure formed from the parametric measures, to be discussed in Chap. 9, produces estimates with a correlation of 0.82 to subjective composite acceptability.

The approach used in designing an objective measure was first to understand the parametric subjective quality it was to estimate. The subjective scores provide a key to this understanding. Distortions which register a subjective quality score widely deviating from the global average are exemplary of that quality, and hence provide insight into the physical or objective nature of that subjective measure. The application of this approach to understanding the meaning of subjective quality scales will be discussed in detail for each of the parametric qualities identified as important by the regression analysis of the DAM quality scales.

Because of computer mass storage constraints, only the first of each of the twelve sentences per speaker in the distorted data base was stored on disk. Therefore, all objective measures were computed only on the first of the set of twelve sentences from each speaker. This is to say, only the original and distorted utterances for the number one sentences in Table 4.1 entered into the calculation of objective variables. These sentences had a duration of approximately three seconds.

Many of the objective measures to be discussed use the bank of twenty-five filters described in Sec. 6.2 to estimate the short-time speech spectrum. In order to model the perceptual importance of each frequency interval, these filters follow the critical band model of the ear's frequency acuity, as indicated by the center frequency and bandwidths listed in Table 6.1. The root mean square values of the filter outputs were computed every 10 ms using a 20 ms Hamming window, and were used as an estimate of the short-time speech spectrum. Since only the first of the set of twelve sentences was processed, the filtering yielded approximately 300 frames of speech per sentence .

In the remainder of this chapter, the six levels of each distortion for all four speakers will be collectively referred to by the term *distortion.* This is a departure from the terminology used elsewhere in this book.

8.4.1 SD: Signal Rasping, Crackling

The subjective quality SD (signal distortion) describes the degree to which speech is rasping or crackling. Table 8.5 lists the distortions which excite the signal distorted scale. For each distortion the minimum, maximum, and range of quality scores associated with that distortion are listed. The degree to which a distortion exemplifies a parametric quality is related to either the range, or spread, of the distortion on the parametric quality scale or to the maximum quality score on that scale. The latter case is most important when a distortion does not have a large range, but instead scores uniformly low on the subjective quality scale. This indicates that the entire distortion exemplifies that quality. The list in Table 8.5 is ordered according to the range of the distortion quality scores so that, in general, the distortions most exhibiting the subjective quality fall near the top of the list.

The dominant physical or objective characteristics the distortions in Table 8.5 have in common are that they involve nonlinearities which distort the waveform and therefore smear energy across the spectrum. This smearing is particularly noticeable at higher frequencies where the speech level is naturally lower and more easily dominated by noise from nonlinearities. Also present are additive noise distortions, which support the hypothesis that noise, either correlated to the speech power

Table 8.5: Distortions which most prominently excite subjective quality SD, listed in order of decreasing significance.

SD rasping, crackling

DISTORTION	MAX	MIN	RANGE
center clipping	83.90	50.70	33.20
CVSD	85.40	53.40	32.00
ADM	85.40	57.70	27.70
peak clipping	81.50	55.70	25.80
quantization	71.90	47.80	24.10
400 - 800 Hz noise	83.40	61.80	21.60
1900 - 2600 Hz noise	85.10	65.00	20.10
1300 - 1900 Hz noise	86.80	68.60	18.20
BD 400 - 800	79.70	61.70	18.00
800 - 1300 Hz noise	85.80	68.90	16.90
APCM	77.70	60.90	16.80
BD 2600 - 3400	83.30	68.10	15.20
2600 - 3400 Hz noise	84.40	69.40	15.00
LPC	83.00	69.50	13.50
broadband additive noise	85.10	73.90	11.20
ECHO	87.60	76.40	11.20
0 - 400 Hz noise	86.20	75.10	11.10
lowpass filtering	85.10	74.10	11.00
BD 100 - 400	91.60	80.80	10.80
VEV 7	76.90	66.20	10.70
ADPCM	78.50	67.90	10.60
PD 1900 - 2600, radial	87.20	76.80	10.40
BD 100 - 3500	73.40	63.50	9.90

and arising from nonlinearities or uncorrelated and arising from an additive process, is the objective character of this subjective quality.

The SD, or signal distorted, scale accounts for a very large fraction of the variance of composite acceptability, some 60 percent. This is principally because of the large number of distortions which excite the signal distorted scale. Figure 8.5 provides another perspective on this issue. The top panel of this figure is a histogram of the frequency of occurrence of SD scores, and the bottom, larger, panel is a scatterplot of SD scores versus CA scores. In both plots the horizontal axis is the SD subjective quality score. Recall that a subjective quality of 85 is very good, or nearly completely free of the quality SD, while a 20 is very poor, or highly distorted. In the histogram, the vertical axis is the

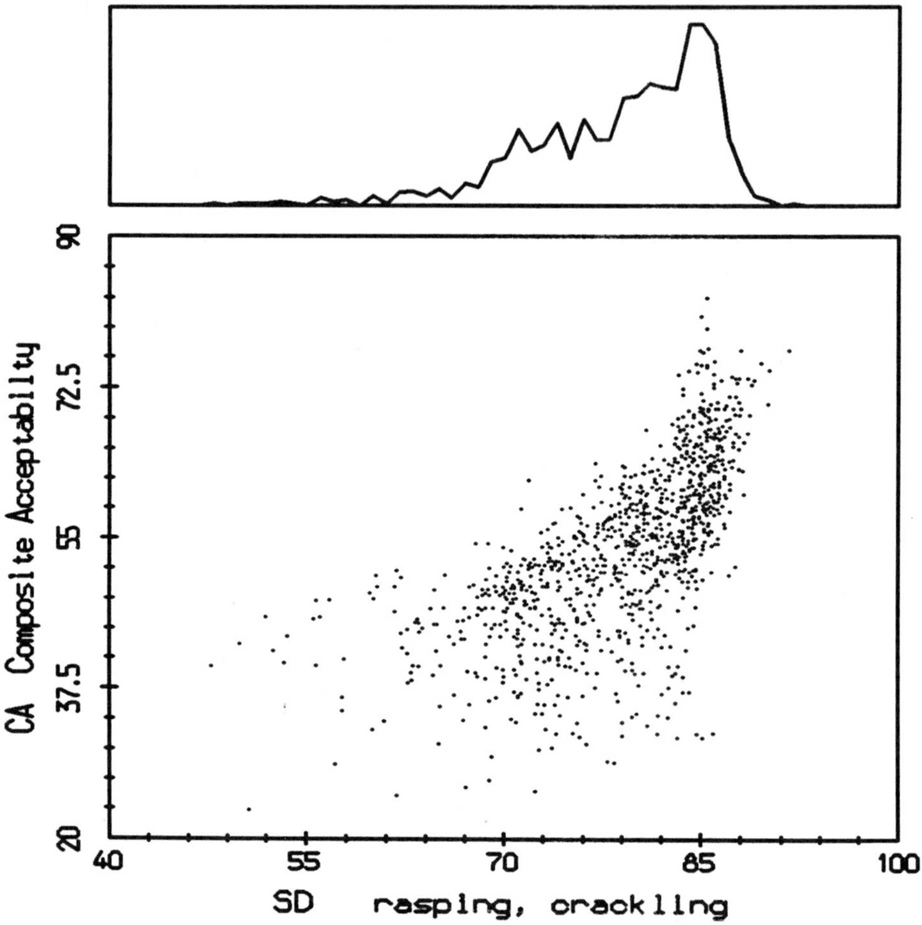

Figure 8.5: Scatterplot and histogram for subjective quality SD.

frequency of occurrence of a given value of the SD quality score when taken over all speakers and all distortion systems in the data base. A case by case examination of the data in this histogram would show that points which fall in the left tail of the distribution are among the distortions listed in Table 8.5. The scatterplot shows how SD scores vary with CA scores. There is a definite trend observable in the data and there is a large range of values on the SD scale. This indicates that SD is important in determining CA.

Since the majority of distortions which excited this scale are nonlinear distortions, a combination of objective classifier and spectral distance measure is used to measure the spectral error in the high end of the speech spectrum. The spectral distance for a 10 ms frame of speech is computed as

$$d(p,m,s,d,k) = (1/M)\left[\sum_{m=1}^{M}\left|\log_{10}\frac{V(m,s,d,k)}{V(m,s,\phi,k)}\right|^{p}\right]^{1/p} \tag{8.3}$$

where m is the frame index and p is the power in the norm. The short-time spectrum estimate is $V(m,s,d,k)$, which is the critical band filter output for the original and distorted speech. Index k is the critical band filter, and s and d are the indices for speaker and distortion.

The spectral distance per frame is combined to form an overall measure of spectral distance in a given frequency band as follows:

$$D(p,s,d,k) = \sum_{m=1}^{M} d(p,m,s,d,k) \tag{8.4}$$

where M is the number of frames in the utterance. M was conservatively chosen to be 250 so as to exclude the silence interval at the end of each sentence.

The function of the objective classifier is to separate the distortions into two groups. The first group contains systems which produce spectral distortions but do not excite the SD scale, while the second group contains all other systems. Of the distortions in this data base, the bandlimiting filter distortions (lowpass, bandpass and highpass), the additive noise distortions (both broadband and narrowband), and the interruption distortions all produce severe spectral distortions, but are not perceived as signal distorted. Therefore these distortions are put in the first group.

The objective classifier identifies the distortions in the first group by means of any of three objective criteria. The first is severe spectral distortion, as would result from filtering. Since the filtering distortions are time invariant, the time averaged critical band spectrum is used in this measurement. If, for any band, the distorted speech spectrum is less than the reference speech spectrum by a specified threshold value, filtering distortion is assumed and the distortion is classified into the first group. A value A is computed as

$$A(s,d) = \text{MAX}\left(-10\log_{10}\left\{\frac{V(\cdot,s,d,k)}{V(\cdot,s,\phi,k)}\right\}\right) \text{ for } k = 1,...,25 \quad (8.5)$$

where MAX is the maximum operator and $V(\cdot,s,d,k)$ and $V(\cdot,s,\phi,k)$ are the long-time average critical band spectrum estimates for the distorted and original speech. If A is greater than 20, the distortion is assigned to the first class.

The second criterion is the presence of additive noise. If the energy in the distorted speech is greater than a specified threshold during known silence intervals in the reference speech, then additive noise distortion is assumed. A value, B, is computed as

$$B(s,d) = 10\log_{10}(\sigma(s,d)) \quad (8.6)$$

where $\sigma(s,d)$ is the root mean square value of the distorted speech waveform across all silent frames in the undistorted speech waveform. If B was greater than 3, additive noise distortion was present and the distortion was classified into the first group.

The third criterion is periodic interruption. The distorted speech signal is examined for intervals of zero energy, as would occur if the waveform were interrupted. The distorted waveform is considered interrupted if the RMS value of two consecutive samples is less than 0.1 percent of full scale. The number of samples for which this condition continues is the duration of the interruption. The standard deviation of the lengths of the interruption intervals is computed, and if this value is less than a threshold value of 5, the distortion is assumed to be periodic interruption.

The objective measure of SD is a combination of objective classifier and spectral distance measure. The objective estimate of SD for distortions in the first group is arbitrarily set equal to zero, since these systems produce spectral errors which are not perceived as SD in nature. For distortions in the second group, the objective variables used in estimating SD are the spectral distances computed in Eq. 8.4 for bands 24 and 25. This can be summarized as follows:

$$D_{SD} = \begin{cases} b_0 + \sum_{j=1}^{2} b_j D(p,s,d,j+23) & \text{if } d \text{ in group one} \\ 0 & \text{if } d \text{ in group two} \end{cases} \quad (8.7)$$

Table 8.6: Summary statistics for SD objective measure.

Multiple R	0.6480	Std. Error of Estimate	5.2798
Multiple R-Squared	0.4199		

Prediction Statistics

Multiple R	0.6449	Std. Error of Prediction	5.2900
Multiple R-Squared	0.4159		

Analysis of Variance

Source of Variation	Sum of Squares	Degrees of Freedom	Mean Square	F Ratio	Area in Tail
Regression	21244.22	2	10622.11	381.05	0.0000
Residual	29353.32	1053	27.87		

Objective Variable	Regression Coefficient	t-statistic	Two-tail Significance	Contribution to R-Square
Intercept	82.4463	393.25	0.000	
D(1)	-23.9800	-1.78	0.076	0.0017
D(2)	-84.8302	-6.12	0.000	0.0206

where the b_i's are estimates of the regression coefficients.

Table 8.6 shows the results of the multiple linear regression analysis used to formulate D_{SD}. This is for p equal to 2, which yielded the best measure. The performance of this measure is fair, with a correlation to SD of only 0.65. This means that D_{SD} is able to account for 42 percent of the variability in composite acceptability. This is a significant shortfall since SD is the most important factor in determining composite acceptability for this data base. Nevertheless, no better measure of SD was discovered.

The prediction statistics listed in Table 8.6 show the model's performance for the jackknife validation method. For this measure, and all the parametric objective measures, the performance in predicting subjective quality scores not used in computing the regression model was always comparable to the performance in fitting the entire data set. This is clear from the fact that the multiple R-squared is nearly equal for both analysis and prediction. The R-squared for prediction is the squared correlation between actual and predicted subjective quality. In the jackknife method a single data point is predicted based on all remaining data (see Chap. 4). Perhaps a better measure of performance is the standard error of prediction, or prediction mean square error, which is only slightly larger than the standard errors of estimate for all data together.

The analysis of variance table shows that the regression coefficients are all different from zero at a .001 significance level. The two-tail t-test on each regression coefficient shows that all coefficients are non-zero at better than a 0.10 level of significance. The contribution to R-squared listed for each variable is the amount by which R-squared would be reduced if that variable were removed from the regression equation. This gives an indication of the importance of the variable in the regression model.

Though this objective measure for SD has only fair performance, many measures were investigated. In particular, since SD is primarily a time-domain distortion, signal-to-noise measures were tried. Other studies have been successful in predicting the performance of waveform coders using segmental signal-to-noise ratio objective measures. Since waveform coders do excite the SD scale, segmental SNR measures are good candidates for estimating SD. However the data base used in this research makes it extremely difficult to implement SNR measures. The calculation of noise power requires precise alignment of distorted and reference waveforms. This is difficult to achieve even for some of the waveform coder distortions because of the phase delays produced by the coders. In the very successful frequency variant segmental SNR measure listed in Table 1.1, the "undistorted" speech was actually the original speech passed through the same waveform coders which were used to produce the distorted speech, but with no quantization of the coded parameters. In this way both the undistorted and the distorted speech experienced the same phase delay from the coders, enabling accurate calculation of the noise power. SNR measures also assume that the distorted waveform is a facsimile of the reference waveform. This assumption is certainly not valid for many of the coders in the speech quality data base. For these reasons the segmental SNR measure is inappropriate and could not be used as an objective measure of SD for this data base. Clearly, this is caused by the broad definition of the problem considered in this research, and for applications which only involve waveform coders, an appropriately parameterized frequency variant segmental SNR should perform very well as a predictor of SD.

Table 8.7: Distortions which most prominently excite subjective quality SL, listed in order of decreasing significance.

SL muffled, smothered

DISTORTION	MAX	MIN	RANGE
lowpass filtering	83.20	46.30	36.90
CVSD	87.50	62.40	25.10
bandpass filtering	77.60	53.40	24.20
ADM	87.10	68.30	18.80
center clipping	84.10	66.70	17.40
highpass filtering	79.20	62.40	16.80
400 - 800 Hz noise	85.50	69.20	16.30
800 - 1300 Hz noise	86.20	73.00	13.20

8.4.2 SL: Signal Muffled, Smothered

This subjective quality describes the extent to which speech is perceived as muffled or smothered. Table 8.7 lists the distortions which excite this subjective scale. Most prominent of these is the lowpass distortion which eliminates high frequencies and therefore fits well with the subjective quality of muffled. The center frequency bandpass distortions also produce a similar muffled quality. Other distortions appear to fit better with the subjective quality of smothered. The highpass and the high bands of the bandpass distortions with high center frequencies eliminate or diminish speech energy in the middle of the 0 to 3600 Hz speech band, which produces the perceptual effect of smothered. The two waveform coders, CVSD and ADM also diminish the mid-band energy of the coded speech with respect to the original speech and hence produce the same smothered effect. The remaining two distortions listed in Table 8.7 are narrowband additive noise, both injecting noise in the low to middle part of the speech spectrum. These distortions can be thought of as smothered in that they produce a noise masking of the speech.

Like the subjective quality SD, the SL quality is generated by a relatively diverse mix of distortions. On the SL scale there are, however, far fewer types of distortions which produce severe SL quality degradations. This can be seen from the relatively small number of entries in Table 8.7 and from Fig. 8.6. The top panel of the figure shows the frequency of occurrence of a specific level of the quality SL across the ensemble of all distortions. It is strikingly different from the corresponding figure for SD in that the main lobe for the quality SL is narrower and its left tail is longer and lower. This indicates that the same range of quality degradation is provided by fewer types of distortions. This is also illustrated in the scatterplot, in which there is a single mass of points centered at 80 on the SL scale, but spreading across the CA scale. However, there are also points, having low CA scores, which spread across the low range of the SL scale. These few points account for most of the spread in the SL scores.

There are primarily two types of distortions which excite the subjective scale SL. These are bandlimiting distortions and narrowband noise distortions. This suggests that a composite objective measure would be appropriate for tracking subjective quality SL. The objective measure designed for this purpose has, for its first component, a frequency variant spectral distance measure and, for its second component, a frequency variant noise measure. An important point as yet unmentioned concerning SL is that the bandlimiting and additive noise distortions which exemplify SL are all time invariant systems. Therefore their distortion characteristics should be recoverable from the time averaged spectrum of the reference and distorted speech waveforms. The method used to estimate the average spectrum is to square the RMS critical band filter outputs, sum over all frames in the utterance and divide by the number of frames in the sum. The square root of the sum is the long-time RMS value in a given band, and the sum itself is the energy in a given band. In this way an estimate of the power in frequency bands for an entire utterance is obtained. The power in bands can be combined, as appropriate, to provide coarser estimates of the reference and distortion spectrum. Broader bands were found to produce objective measures which were more easily interpreted.

The spectral distance objective measure has the following form:

$$D_1(s,d,k) = \log_{10}\left[\min\left\{\max\left\{\frac{V(\cdot,s,d,k)}{V(\cdot,s,\phi,k)}, TH_{\min}\right\}, TH_{\max}\right\}\right] \tag{8.8}$$

where $V(\cdot,s,d,k)$ and $V(\cdot,s,\phi,k)$ are the mean square values in the band k for the distorted and reference waveforms, respectively. Again, this average is taken over the entire utterance. $TH_{\min}$ and $TH_{\max}$ are parameters of the measure. Setting $TH_{\min}$ equal to 0.001 and

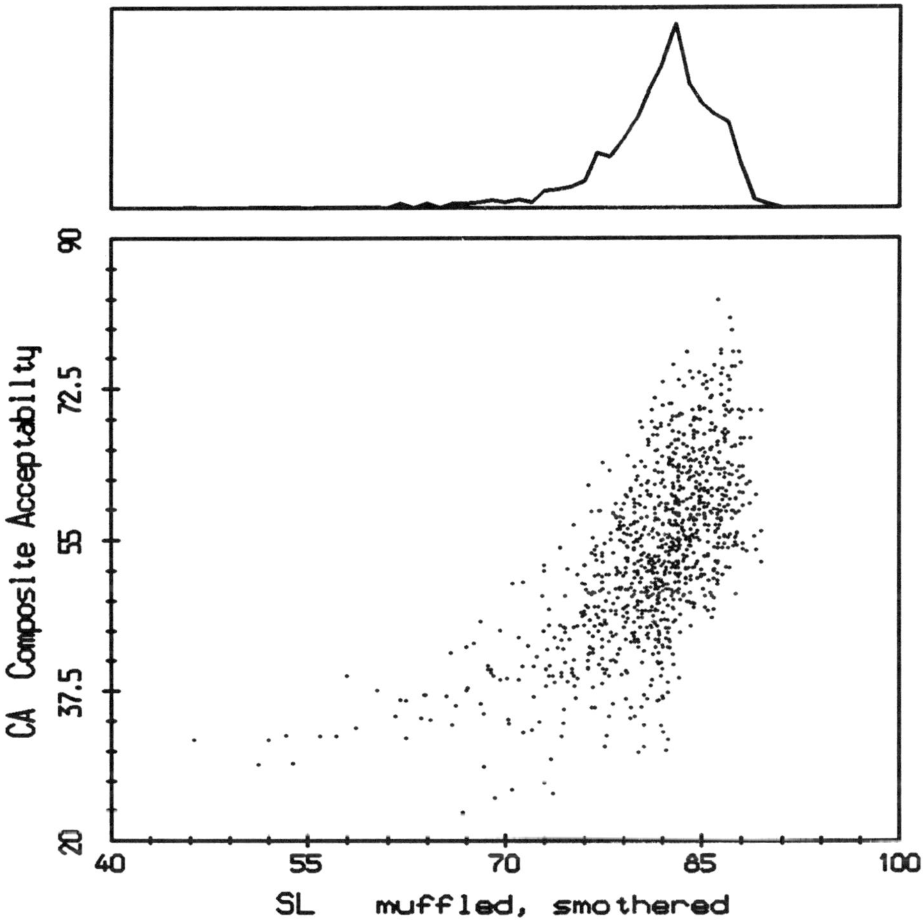

Figure 8.6: Scatterplot and histogram for subjective quality SL.

TH_{max} equal to 1.0 produced the best measure. The objective variables $D_1(s,d,k)$ were transformed into new distance variables, $D_1'(s,b,k)$, which have coarser frequency resolution. Instead of having 25 bands D_1' had only four bands, and is obtained by summing $D_1(s,d,k)$ as follows:

$D'_1(s,d,j)$ Band No.	$D_1(s,d,k)$ Band Nos.
1	1 – 10
2	11 – 15
3	16 – 20
4	21 – 25

In addition, monotonic and unimodal regressions were done on the distance function $D_1(s,d,k)$ and stresses for the functional forms lowpass, highpass, bandpass and band reject were computed. Computing stress for the functional form of lowpass requires computing a monotonically increasing regression, while highpass requires a decreasing regression. Bandpass requires computing a global maximum unimodal regression, and band reject requires a global minimum unimodal regression. The motivation for computing these stresses was to measure the extent to which the distortion applied to the speech had one of these bandlimiting functional forms. The total number of independent variables used in this objective measure was six: four spectral distance variables for four frequency bands and two stress variables, one for the functional form highpass, represented as $D'_1(s,d,5)$, and one for band reject, represented as $D'_1(s,d,6)$. The remaining stress variables did not significantly contribute to the regression model.

The second part of the composite measure is an additive noise measure. The functional form of this measure is as follows:

$$D_2(s,d,k) = \log_{10}\left[1 + \frac{1}{NF}\sum_{f_S} V(f,s,d,k)\right] \tag{8.9}$$

where f_S are all silent frames in the reference utterance and NF is the number of silent frames. The 25 bands in $D_2(s,d,k)$ are combined to form two bands in a new additive noise measure, $D'_2(s,d,k)$, as follows:

$D'_2(s,d,j)$ Band No.	$D_2(s,d,k)$ Band Nos.
1	6 – 15
2	16 – 25

The two variables in this measure are the noise power in the extended bands as measured during intervals of known speech inactivity in the distorted signal.

The two measures were combined in a linear function with weights determined by regression analysis. The resultant measure was formulated as

$$D_{SL}(s,d) = b_0 + \sum_{j=1}^{6} b_{1j} D_1'(s,d,j) + \sum_{j=1}^{2} b_{2j} D_2'(s,d,j) \tag{8.10}$$

where $D_{SL}(s,d)$ represents the objective estimate of subjective quality SL and the b_j's are the estimated regression coefficients.

Table 8.8 shows the results of the multiple linear regression analysis used to formulate D_{SL}. The performance of this measure is fair, as its correlation with SL is 0.73, which corresponds to an explanation of 53 percent of the variability in the subjective quality SL. In all probability, this poor performance is due to the difficulty of modeling the diverse mix of distortions which excite SL. This was, nevertheless, the best objective measure found for this parametric quality.

8.4.3 SF: Signal Fluttering, Bubbling

This subjective quality quantifies the degree to which the speech signal has a fluttering or bubbling quality. Table 8.9 lists those distortions which excite the SF subjective scale. The dominant class of distortions in this table is that of pole distortions. The controlled pole distortions explicitly alter the original speech pole locations, while the parametric coder distortions based on an all-pole vocal tract model distort the speech pole locations in a more complex way through modeling errors and parameter quantization. Two prominent exceptions are the first and fourth table entries: the interrupted distortions. These are understandably perceived as fluttering because their interruptions are periodic. The presence of these interrupted distortions in Table 8.9 suggests that it is the periodic quality of the controlled and coder pole distortions which correlate most highly with subjective fluttering and bubbling.

Figure 8.7 illustrates an interesting split in the characterization of distortions on the SF scale. The histogram has a concentrated main

Table 8.8: Summary statistics for SL objective measure.

Multiple R	0.7301	Std. Error of Estimate	3.5471
Multiple R-Squared	0.5330		

Prediction Statistics

Multiple R	0.7217	Std. Error of Prediction	3.5772
Multiple R-Squared	0.5209		

Analysis of Variance

Source of Variation	Sum of Squares	Degrees of Freedom	Mean Square	F Ratio	Area in Tail
Regression	15034.63	8	1879.32	149.37	0.0000
Residual	13173.04	1047	12.58		

Objective Variable	Regression Coefficient	t-statistic	Two-tail Significance	Contribution to R-Square
Intercept	83.1344	309.29	0.000	
$D_1(1)$	-0.0512	-1.48	0.139	0.0010
$D_1(2)$	0.7568	8.29	0.000	0.0307
$D_1(3)$	0.4732	4.92	0.000	0.0108
$D_1(4)$	0.4952	6.74	0.000	0.0203
$D_1(5)$	0.0158	4.67	0.000	0.0097
$D_1(6)$	-0.0057	-2.58	0.011	0.0029
$D_2(1)$	-0.0205	-1.55	0.121	0.0011
$D_2(2)$	-0.0570	-4.36	0.000	0.0085

lobe with a highly skewed left tail, with the main lobe corresponding to a nearly vertical mass of points in the scatterplot at an SF score of 85. These are distortions whose composite acceptability does vary, but not because of SF degradations. On the other hand, there is a clearly separate mass of points extending diagonally down and to the left which do vary on the SF scale with corresponding decrease in values on the CA scale.

Irregularities in the transition of speech formants are a possible cause of perceived fluttering. In order to measure these irregularities, one needs a means of determining formant frequency and bandwidth for a given speech frame. The vocal tract system function, as derived from LPC analysis is one vehicle for formant estimation. However when tried, it was found that the frame length of 30 ms which is commonly chosen for LPC analysis is too long for observing formant fluttering.

Table 8.9: Distortions which most prominently excite subjective quality SF, listed in order of decreasing significance.

SF fluttering bubbling

DISTORTION	MAX	MIN	RANGE
interrupted, period = 1024	85.50	50.90	34.60
LPC	85.30	51.10	34.20
PD 400 - 800, frequency	85.80	52.60	33.20
interrupted, period = 300	80.90	48.70	32.20
VEV 13	84.90	60.70	24.20
VEV 7	83.30	60.30	23.00
PD 1300 - 1900, frequency	87.90	66.30	21.60
PD 400 - 800, radial	83.60	63.90	19.70
APC	86.60	67.30	19.30
BD 400 - 800	83.20	64.40	18.80
ECHO	88.60	70.80	17.80
BD 2600 - 3400	79.80	62.90	16.90
BD 100 - 3500	80.40	63.90	16.50
PD 800 - 1300, frequency	84.10	68.10	16.00
PD 000 - 400, radial	88.60	72.90	15.70
PD 1300 - 1900, radial	88.60	73.10	15.50
PD 2600 - 3400, radial	87.50	72.60	14.90
center clipping	85.60	72.10	13.50

Therefore the critical band estimates of the short-time speech spectrum were used in this measure. They work well because the spectral estimates are for 10 ms frames, and transient phenomena such as fluttering or bubbling can be observed. Fluttering was measured as the frame to frame spectral distance of the distorted speech normalized by the frame to frame spectral distance of the reference speech. As might be expected in hindsight, this measure performed better when the speech frames were restricted to being vocalic. In vowel segments formant transitions are naturally smooth, and erratic departures from smoothness are quite perceivable. The objective measure for a given frequency band is computed as

$$D(p,s,d,k) = (1/M)\left[\sum_{fv}\left|\log_{10}\frac{(V(m,s,d,k)/V(m-1,s,d,k)}{(V(m,s,\phi,k)/V(m-1,s,\phi,k)}\right|^{p}\right]^{1/p} \tag{8.11}$$

where $V(m,s,d,k)$ is the critical band estimate of the speech spectrum

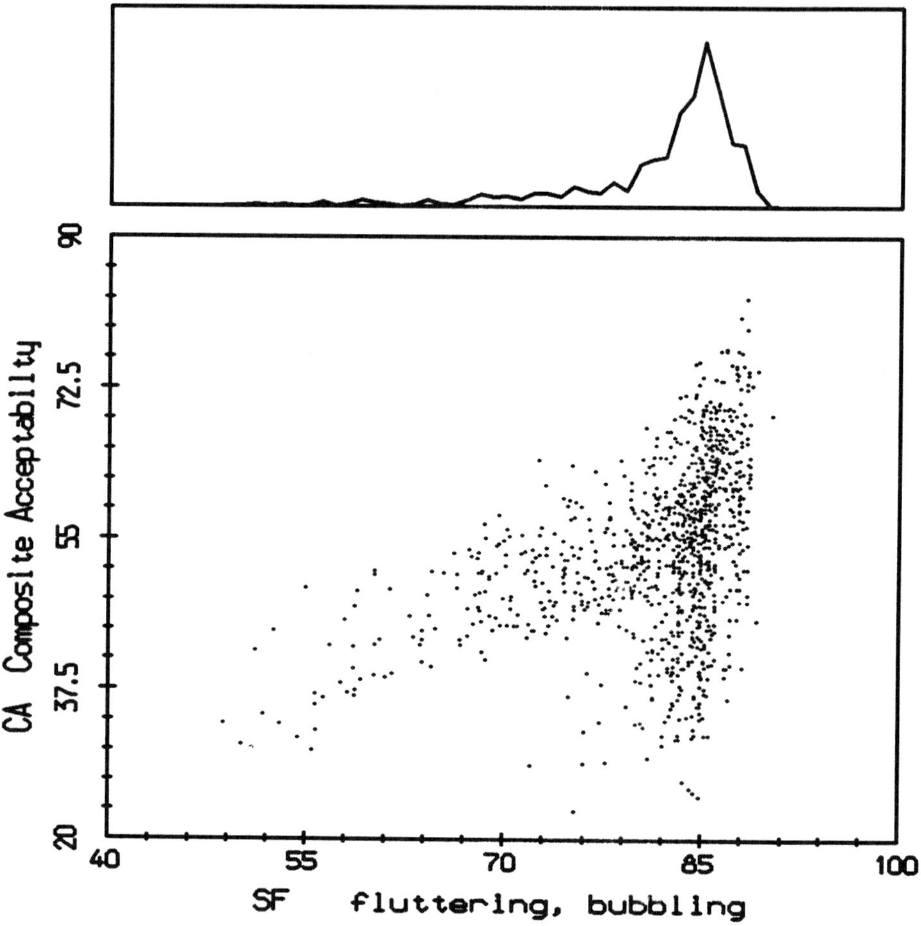

Figure 8.7: Scatterplot and histogram for subjective quality SF.

in frame m. The sum is over all speech frames which are identified as vocalic. Segmenting the speech into vowel and non-vocalic frames was done by hand. The objective measure for SF is formulated as

$$D_{SF} = b_0 + b_1 D(p, s, d, 25) \tag{8.12}$$

where only the highest band was found to contribute significantly to the regression model. This is rather surprising since SF was hypothesized as formant fluttering and there are no significant formants in the

Table 8.10: Summary statistics for SF objective measure.

Multiple R	0.5321	Std. Error of Estimate	5.7662
Multiple R-Squared	0.2831		

Prediction Statistics			
Multiple R	0.5286	Std. Error of Prediction	5.7756
Multiple R-Squared	0.2794		

Analysis of Variance

Source of Variation	Sum of Squares	Degrees of Freedom	Mean Square	F Ratio	Area in Tail
Regression	13831.75	1	13831.75	416.28	0.0000
Residual	35044.87	1054	33.24		

Objective Variable	Regression Coefficient	t-statistic	Two-tail Significance	Contribution to R-Square
Intercept	84.6805	357.88	0.000	
D(1)	-4.3302	-20.40	0.000	0.2831

highest frequency band. Perhaps the highest band provides an indication of fluttering which is occurring in the lower bands where the high frequencies allow for more prominent transition effects.

Table 8.10 summarizes the performance of D_{SF}, which is generally poor. Its correlation to SF is only 0.53, accounting for only 28 percent of the variability of SF. Setting p equal to 1.0 produced the best measure for SF.

8.4.4 BN: Background Hissing, Rushing

This scale specifies the extent to which the background of the distorted signal has a hissing or rushing quality. Table 8.11 lists those distortions which most excite the BN subjective scale. This scale is in contrast to the ones discussed thus far in that a very homogeneous set of controlled distortions produce this subjective quality, namely additive noise distortions. The middle frequency narrowband additive noise distortions have the greatest perceptual impact, with the broadband additive noise being perceived as almost equal in degree of distortion. At the bottom of the table is quantization distortion. This is not an anomaly, since, for medium to fine quantization levels, the quantization noise is nearly

Table 8.11: Distortions which most prominently excite subjective quality BN, listed in order of decreasing significance.

BN hissing, rushing

DISTORTION	MAX	MIN	RANGE
800 - 1300 Hz noise	80.40	49.30	31.10
broadband additive noise	83.40	54.00	29.40
400 - 800 Hz noise	79.10	50.40	28.70
0 - 400 Hz noise	85.80	66.40	19.40
1300 - 1900 Hz noise	82.10	69.60	12.50
2600 - 3400 Hz noise	87.20	74.80	12.40
1900 - 2600 Hz noise	84.00	72.80	11.20
quantization	85.30	75.60	9.70

uncorrelated with the signal and is perceived as a background process.

Figure 8.8 illustrates a dichotomy on the BN scale similar to that found on the SF scale. The scatter plot clearly shows two separate groups of distortions. The first group, the larger, all have about the same value for BN regardless of the value of CA. In the second group, BN and CA are approximately linearly related.

The set of the distortions which excite the BN subjective scale indicate that a function which measures additive noise would be an appropriate objective measure for this scale. The objective measure used is that of Eq. (8.9), but here it is used alone to estimate BN. The measure $D_2(s,d,k)$ is transformed into $D_2'(s,d,k)$ in order to consolidate the number of bands. The transformation is as follows:

$D_2'(s,d,j)$ Band No.	$D_2(s,d,k)$ Band Nos.
1	$1-5$
2	$6-16$

Note that bands 17 through 25 were not used in this measure. The final objective measure used to estimate BN was then

$$D_{BN}(s,d) = b_0 + \sum_{j=1}^{2} b_j D_2'(s,d,j) \qquad (8.13)$$

The performance of this measure is extremely good. The objective measure results are shown in Table 8.12. The primary reason for

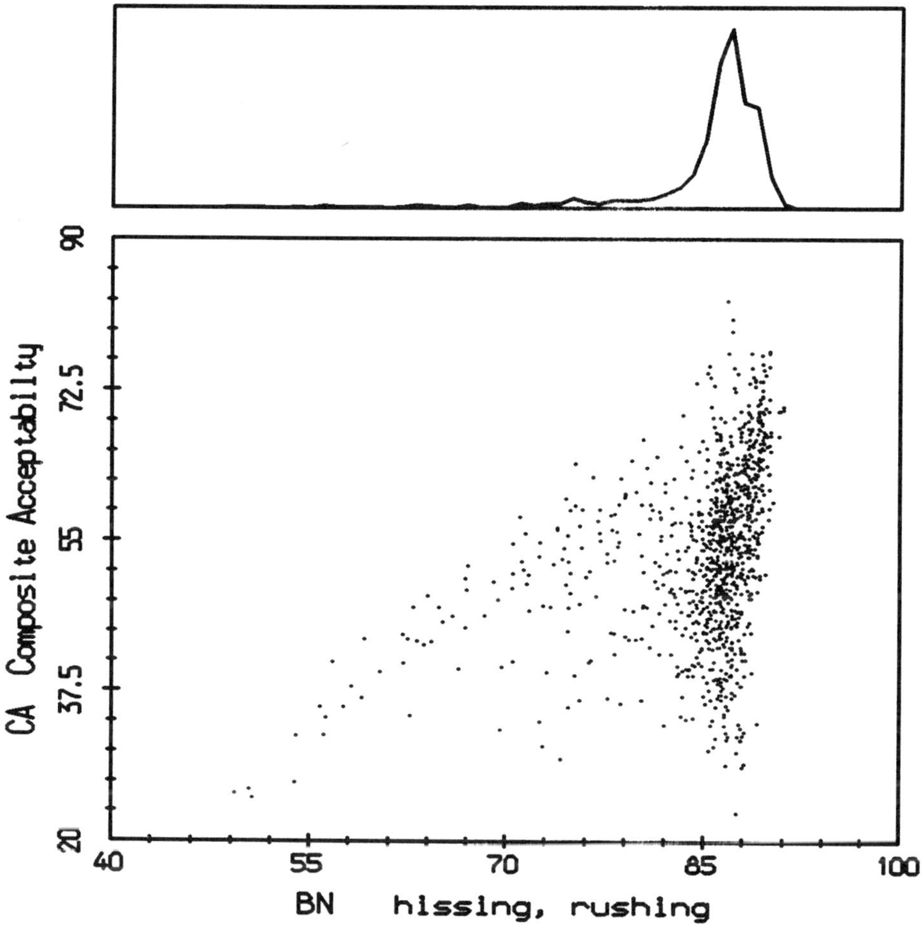

Figure 8.8: Scatterplot and histogram for subjective quality BN.

such good performance, correlation of 0.91, is that all distortions which excite BN are very similar and hence can be modeled well as a group. Another reason is that there are relatively few distortions which excite BN, as can be seen from the narrow central lobe and the low left tail of Fig. 8.6. This means that the regression model need only account for the variability of these few distortions, and can approximate the quality scores of the other distortions with a constant. Of all parametric objective measures studied, this measure was by far the most successful.

Table 8.12: Summary statistics for BN objective measure.

Multiple R	0.9141	Std. Error of Estimate	2.3142
Multiple R-Squared	0.8356		

Prediction Statistics

Multiple R	0.9122	Std. Error of Prediction	2.3341
Multiple R-Squared	0.8321		

Analysis of Variance

Source of Variation	Sum of Squares	Degrees of Freedom	Mean Square	F Ratio	Area in Tail
Regression	28631.68	3	9543.89	1782.14	0.0000
Residual	5633.78	1052	5.35		

Objective Variable	Regression Coefficient	t-statistic	Two-tail Significance	Contribution to R-Square
Intercept	87.1909	1093.33	0.000	
$D_1(1)$	0.2183	11.34	0.000	0.0201
$D_1(2)$	-0.4144	-38.99	0.000	0.2375
$D_1(2)$	0.2303	2.49	0.013	0.0001

8.4.5 BF: Background Chirping, Bubbling

This subjective quality quantifies the degree to which the speech background has a chirping or bubbling quality. Table 8.13 lists those distortions which excite the BF subjective scale. This scale is very similar to SF, or signal fluttering and bubbling, with a few important differences. First, although interrupted does not excite BF, it was at the top of the list for SF (see Table 8.9). This is understandable since an interruption of the speech waveform is a distortion entirely associated with the speech signal and produces no spurious or uncorrelated background distortion. The second difference is that high frequency narrowband noise distortions excite the BF scale, where they did not excite SF. These distortions are most likely perceived as chirping background distortions. The rest of the distortions listed in Table 8.13 are for the most part the same distortions associated with SF, listed in Table 8.9. Therefore an objective measure for estimating BF should be similar to

Table 8.13: Distortions which Most prominently excite subjective quality BF, listed in order of decreasing significance.

BF chirping, bubbling

DISTORTION	MAX	MIN	RANGE
PD 1300 - 1900, radial	85.70	54.40	31.30
PD 800 - 1300, frequency	86.40	57.20	29.20
LPC	85.10	56.00	29.10
PD 1900 - 2600, radial	85.10	57.90	27.20
PD 400 - 800, radial	85.30	59.20	26.10
PD 400 - 800, frequency	85.60	59.60	26.00
PD 000 - 400, radial	85.20	59.60	25.60
PD 800 - 1300, radial	87.50	65.90	21.60
PD 1300 - 1900, frequency	86.90	66.30	20.60
VEV 7	77.40	59.00	18.40
VEV 13	76.20	59.90	16.30
PD 2600 - 3400, frequency	86.70	70.70	16.00
PD 2600 - 3400, radial	90.00	74.60	15.40
APC	84.40	69.10	15.30
PD 1900 - 2600, frequency	87.50	72.30	15.20
PD 2600 - 3400, frequency	87.50	73.80	13.70
BD 2600 - 3400	83.60	70.80	12.80
2600 - 3400 Hz noise	85.10	72.70	12.40
BD 100 - 3500	81.80	69.80	12.00
1900 - 2600 Hz noise	83.60	71.60	12.00
BD 400 - 800	83.50	71.80	11.70
1300 1900 Hz noise	86.40	74.70	11.70
BD 1300 - 1900	83.00	71.90	11.10
quantization	85.10	74.10	11.00
BD 800 - 1300	81.60	70.80	10.80

a measure for SF. Referring back to the multidimensional scaling of the subjective data base, Fig. 8.1, one can see that SF and BF are rather close together, confirming the fact that the two quality scales detect perceptually similar distortions.

However, the scatterplot for BF versus CA, Fig. 8.9, shows a scatter of points which is more like that for SD than SF. That is, BF has a rather broad distribution of scores as does SD, and the scatter does not exhibit as strong a separation into two groups as does SF.

This research was unable to identify good objective measures for BF. The measure with the highest correlation to BF was the same

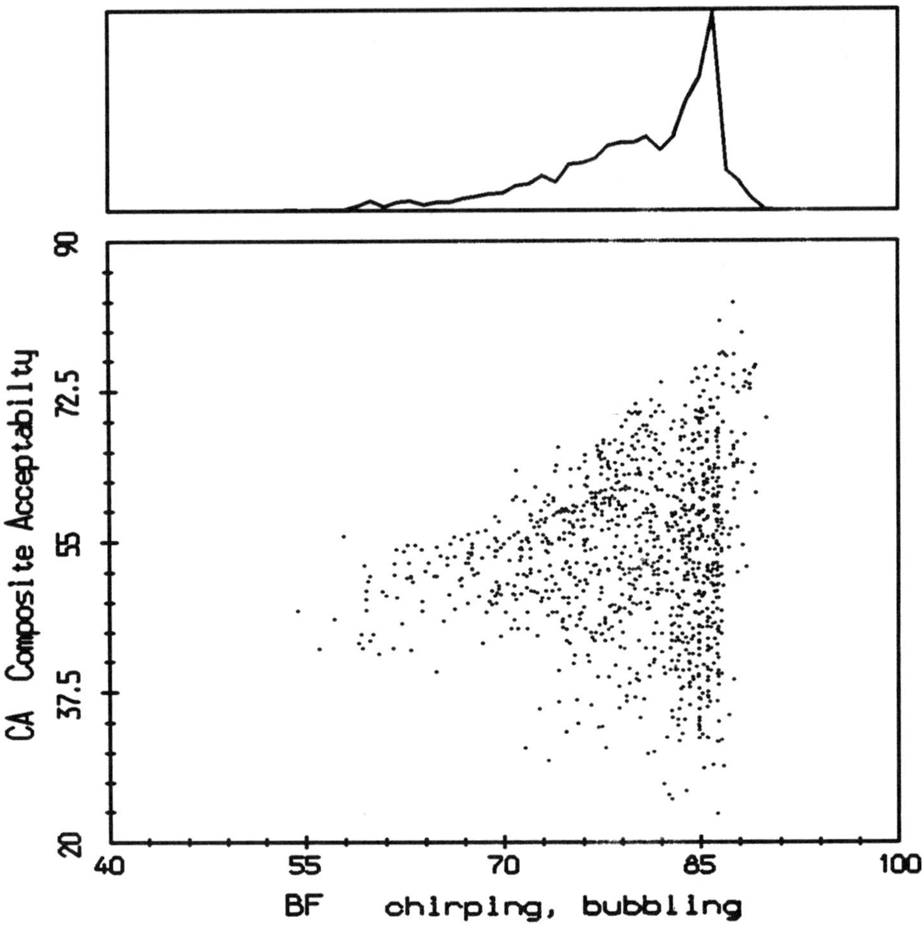

Figure 8.9: Scatterplot and histogram for subjective quality BF.

measure used to estimate SF. This is not surprising, since the same insights into objective measures for SF largely hold true for those of BF. However an important difference is that BF is a background distortion, and distortions which excite this scale are, in general, not correlated with the signal. Because of this difference, the performance of D_{BF} was quite poor, as listed in Table 8.14, with a correlation to subjective quality of only 0.44, accounting for only 19 percent of the variability of BF. Since the regression model is inaccurate, no regression coefficients

Table 8.14: Summary statistics for BF objective measure.

Multiple R	0.4387	Std. Error of Estimate	5.9607
Multiple R-Squared	0.1925		

Table 8.15: Distortions which most prominently excite subjective quality SI, listed in order of decreasing significance.

SI irregular, interrupted

DISTORTION	MAX	MIN	RANGE
interrupted, period = 1024	87.90	38.40	49.50
ADM	91.00	49.60	41.40
2600 - 3400 Hz noise	87.10	62.50	24.60
ADPCM	85.00	60.50	24.50
center clipping	87.60	63.90	23.70
interrupted, period = 300	86.80	66.20	20.60
APCM	85.20	65.10	20.10
ECHO	89.90	76.20	13.70
PD 800 - 1300, frequency	89.50	76.50	13.00
PD 1900 - 2600, radial	89.90	77.80	12.10
PD 000 - 400, radial	89.60	79.10	10.50
PD 1900 - 2600, frequency	89.20	79.30	9.90
1900 - 2600 Hz noise	87.10	78.70	8.40

are presented here.

8.4.6 SI: Signal Irregular, Interrupted

This parametric quality scale describes the degree to which the speech signal is irregular and interrupted. Table 8.15 lists distortions which excite this subjective scale. The most prominent distortion is the slow periodic interruption, with the fast periodic interruption falling in the middle of the table. These two distortions certainly produce perceptually interrupted speech. It is difficult to find an objective quality which is common to the remainder of the distortions which excite SI. They most likely excite the subjective quality irregular, rather than interrupted. The remaining distortions are not totally disjoint, however. Both APCM and ADPCM, the two highest bands of the narrowband additive noise and several pole distortions also excite SI.

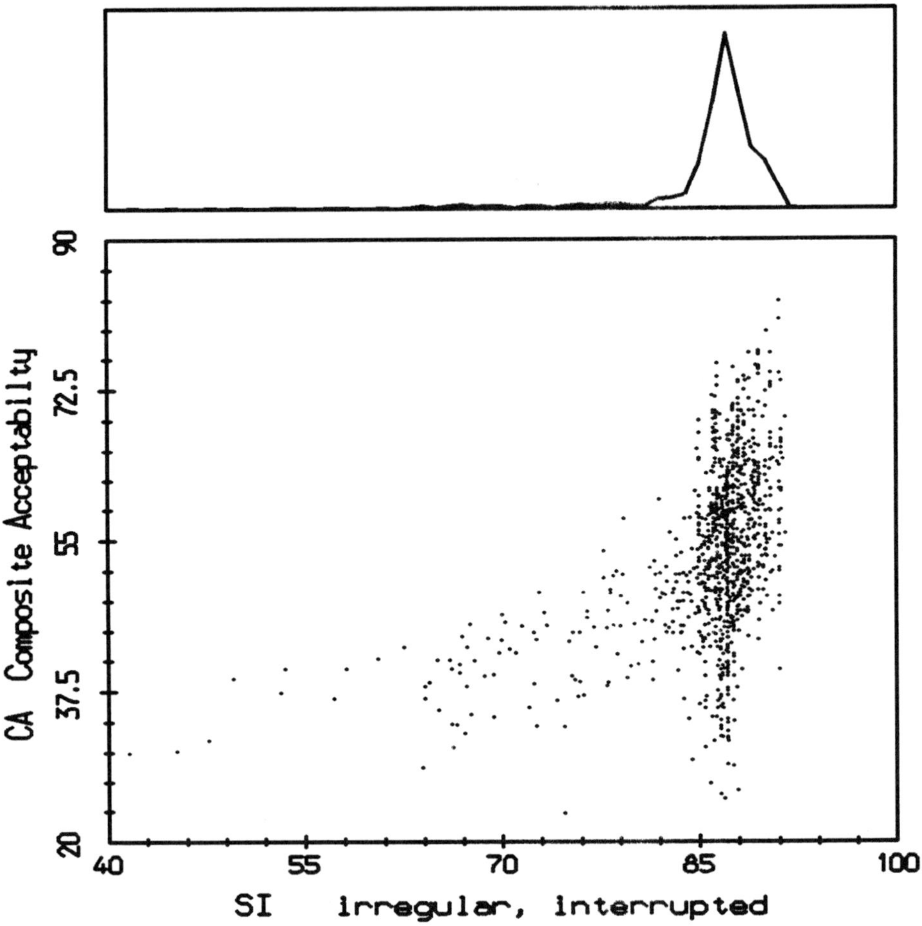

Figure 8.10: Scatterplot and histogram for subjective quality SI.

Figure 8.10 shows the histogram for SI and the scatterplot of SI versus CA. Most distortions do not excite SI, as indicated by the very compact main lobe in the histogram, centered at an SI score of 90, and by the corresponding dense vertical mass of points in the scatterplot. The few systems which do excite the SI scale lie to the left of this mass in a scatter configuration very much like that of subjective quality SL.

Since interrupted is the most important aspect of the SI scale and since this quality is easy to model objectively, the measure used for estimating SI was designed to respond only to interruptions of the speech waveform. In particular the average number of consecutive frames for which the distorted speech energy was either below a specified threshold or above the threshold is measured as a gauge of interruption. The measure is best expressed using intermediate variables as follows:

$$RATIO(n,s,d) = \frac{\log_{10}\left\{[(1/FL)\sum_m x_n(m,s,d)^2]^{1/2}\right\}}{\log_{10}\left\{[(1/FL)\sum_m x_n(m,s,\phi)^2]^{1/2}\right\}} \tag{8.14}$$

$$ON(s,d) = \text{Average run length of frames for which } RATIO(n,s,d) > TH \tag{8.15}$$

$$OFF(s,d) = \text{Average run length of frames for which } RATIO(n,s,d) < TH \tag{8.16}$$

$$D(s,d,1) = OFF(s,d) \tag{8.17}$$

$$D(s,d,2) = \left[\frac{OFF(s,d)}{ON(s,d)+OFF(s,d)}\right] \tag{8.18}$$

$$D(s,d,3) = D(s,d,1)^2 \tag{8.19}$$

$$D(s,d,4) = D(s,d,2)^2 \tag{8.20}$$

$$D_{SI}(s,d) = b_0 + \sum_{j=1}^{4} b_i D(s,d,j) \tag{8.21}$$

Parameters FL and TH can be varied as desired to alter the measure. Parameter FL is the number of samples in a frame of speech and parameter TH specifies the threshold between objectively interrupted and non-interrupted speech. Setting TH equal to 0.2 and FL equal to 16 produced the best measure. In the formula specifying $RATIO$, m is the index of the speech samples comprising frame n. The objective measure variables are specified in Eqs. (8.17) through (8.20). Note that the last two objective variables are simply the first two objective variables squared. Therefore the final objective measure specified in Eq. (8.21) is actually a multiple linear and polynomial regression equation.

The results of a regression analysis to find the best estimate, D_{SI}, of quality SI are shown in Table 8.16. The measure performed well,

Table 8.16: Summary statistics for SI objective measure.

Multiple R	0.8533	Std. Error of Estimate	2.3771
Multiple R-Squared	0.7282		

Prediction Statistics

Multiple R	0.8441	Std. Error of Prediction	2.4382
Multiple R-Squared	0.7126		

Analysis of Variance

Source of Variation	Sum of Squares	Degrees of Freedom	Mean Square	F Ratio	Area in Tail
Regression	14738.41	4	3684.60	652.05	0.0000
Residual	5509.53	975	5.65		

Objective Variable	Regression Coefficient	t-statistic	Two-tail Significance	Contribution to R-Square
Intercept	87.2668	1086.30	0.000	
D(1)	0.3522	5.48	0.000	0.0084
D(2)	-71.4805	-16.83	0.000	0.0790
D(3)	-0.0372	-16.33	0.000	0.0743
D(4)	73.1667	9.55	0.000	0.0254

as measured by a multiple R of 0.85, with the qualification that the ADPCM, APCM and ECHO distortions were not included in the analysis. The decision to exclude this data, 72 points in all, was not based on statistical criteria alone, and in this respect is a departure from the general approach to objective measure design, in that the exclusion of an entire distortion instead of only the statistical outliers in that distortion were made.

Because the ECHO distortion is very different from other distortions, it was excluded in the hopes of improving the performance of the measure. ADPCM and APCM were excluded because these distortions were not being modeled well by this objective measure, as indicated by residual analysis. Since the latter two systems were associated with large estimation errors, leaving them out greatly improved the measure's correlation with the SI scale. As mentioned previously, these two coder distortions produce a subjectively irregular distortion, which suggests why they are modeled poorly by a measure of periodic interruption.

Table 8.17: Distortions which most prominently excite subjective quality SH, listed in order of decreasing significance.

SH distant, thin

DISTORTION	MAX	MIN	RANGE
highpass filtering	84.70	54.20	30.50
bandpass filtering	85.00	60.60	24.40
0 - 400 Hz noise	86.90	75.40	11.50
CVSD	90.30	79.30	11.00
lowpass filtering	87.90	78.00	9.90
peak clipping	87.10	79.60	7.50

8.4.7 SH: Signal Distant, Thin

This last subjective quality measures the degree to which the distorted speech sounds distant or thin. The distortions which most dramatically excite this parametric quality scale are all bandlimiting distortions, specifically highpass and bandpass distortions. These two distortions are ordered one and two in Table 8.17. Since for the higher frequency bands, the bandpass filtering is very similar to highpass filtering, it is reasonable that these two distortions are grouped together. This indicates that highpass filtering is the most important objective correlate to speech being perceived as distant and thin. Two seemingly out of place distortions found in Table 8.17 are CVSD and lowpass filtering. It can be seen, however, that CVSD does in fact produce a bandlimiting distortion which slightly decreases the energy of speech in a broad band centered at approximately 2000 Hz. The only feature CVSD and lowpass filtering have in common is that they both diminish speech energy in middle frequency range, although lowpass filtering eliminates virtually all out-of-band energy. A possible conclusion is that like SL, SH is correlated to a decrease in midband speech energy. Other distortions which excite SH are peak clipping and the lowest band of narrowband additive noise. Peak clipping smears energy across the entire spectrum which is perceived primarily as high frequency distortion due to the low level of speech energy at high frequencies. Therefore these two distortions produce noise at opposite ends of the spectrum. This effect may be correlated to the quality distant and thin.

Figure 8.11 shows the scatterplot of SH versus CA. Of all the

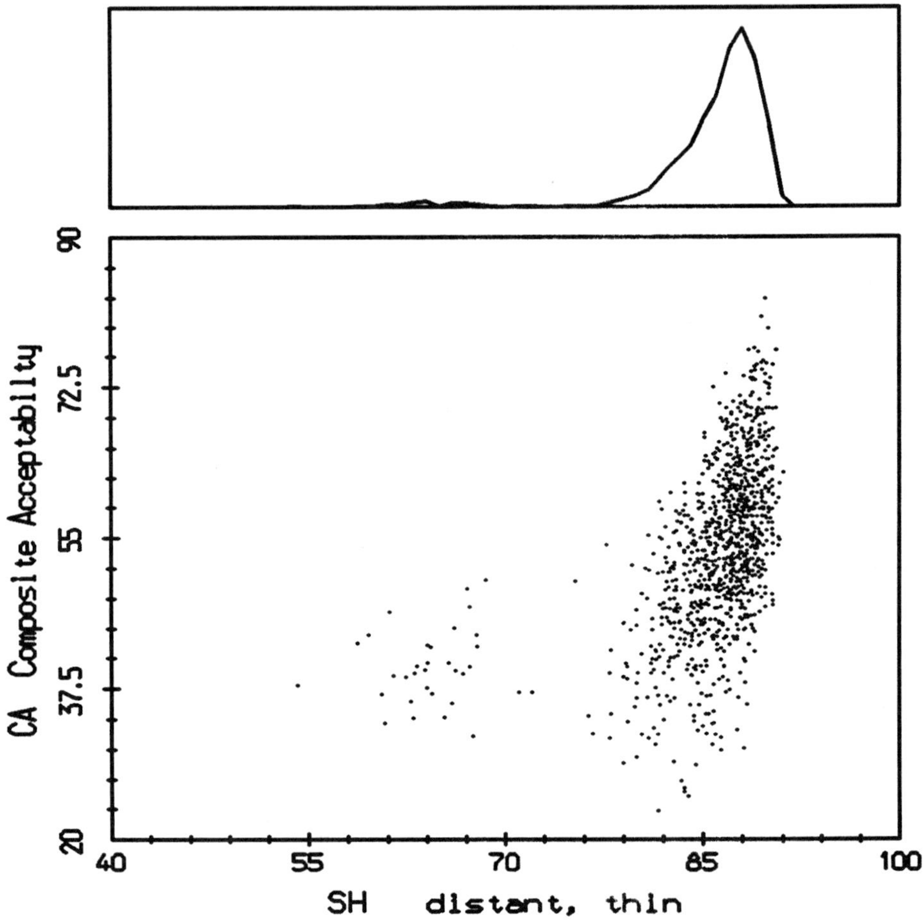

Figure 8.11: Scatterplot and histogram for subjective quality SH.

parametric scales, SH has the most distinct clustering in this two-dimensional space. This clearly indicates two classes of distortion: those which do not excite the SH scale (indicated by the mass of points centered at the SH value of 90), and those which do, indicated by the distinct cluster of points centered at the SH value of 60. As one would expect, the smaller cluster consists of highpass and bandpass bandlimiting distortions.

The objective measure used to estimate SH concentrated on the principal objective feature of SH, which is highpass filtering. The objective measure is a spectral distance measure, and has the form

$$D(p,s,d,k) = \left\{ \log_{10} \left[\left| \frac{V(\cdot,s,d,k)}{V(\cdot,s,\phi,k)} \right|^p \right] \right\}^{1/p} \tag{8.22}$$

where $V(\cdot,s,d,k)$ and $V(\cdot,s,\phi,k)$ are the average critical band spectra for the distorted and reference waveforms, respectively. Setting p, the power in the norm, equal to 2 produced the best measure. The number of bands in this objective measure was reduced to four to produce D', as follows:

$D'(s,d,k)$ Band No.	$D(s,d,k)$ Band Nos.
1	1 - 5
2	6 - 10
3	11 - 15
4	16 - 25

The total number of independent variables used in this objective measure was six: four spectral distance variables for the four frequency bands and two stress variables, one for the functional form lowpass, represented as $D_1'(s,d,5)$, and one for bandpass, represented as $D_1'(s,d,6)$. The remaining stress variables did not significantly contribute to the regression model. The objective variables are combined in a regression equation for estimating SH as follows:

$$D_{SH}(s,d) = b_0 + \sum_{j=1}^{6} b_j D'(s,d,j) \tag{8.23}$$

Table 8.18 shows the results of this analysis. Performance for this measure was significantly better than for the somewhat similar measure for estimating SL. For this measure a correlation of 0.85 was obtained. This is primarily due to the fact that the distortions which produce most of the variance in SH, highpass and bandpass filtering, are relatively homogeneous and can therefore be effectively modeled.

8.5 Summary

Seven parametric objective measures were presented in the previous section. The performance of these measures, summarized in Table 8.19,

Table 8.18: Summary statistics for SH objective measure.

Multiple R	0.8540	Std. Error of Estimate	2.4543
Multiple R-Squared	0.7293		

Prediction Statistics

Multiple R	0.8489	Std. Error of Prediction	2.4853
Multiple R-Squared	0.7206		

Analysis of Variance

Source of Variation	Sum of Squares	Degrees of Freedom	Mean Square	F Ratio	Area in Tail
Regression	17024.22	6	2837.36	471.05	0.0000
Residual	6318.65	1049	6.02		

Objective Variable	Regression Coefficient	t-statistic	Two-tail Significance	Contribution to R-Square
Intercept	7.6093	3.60	0.000	
D(1)	0.5852	10.65	0.000	0.0292
D(2)	0.4948	7.08	0.000	0.0129
D(3)	0.1894	4.35	0.000	0.0049
D(4)	-0.1230	-6.70	0.000	0.0116
D(5)	0.0270	8.71	0.000	0.0196
D(6)	0.0176	7.14	0.000	0.0132

ranges from very good (a correlation of 0.91 for BN) to poor (a correlation of 0.44 for BF). On the whole these results are good, though they are even more significant because the objective measures estimate subjective quality over the entire distorted data base (with the exception of D_{SI}), which indicates that these objective measures are applicable to a broad range of speech distortions.

The regression models associated with each parametric objective measure were carefully constructed so that all of the regression weights had an intuitively satisfying physical interpretation. The assignment of a meaning to the regression coefficients is a check on the appropriateness of the regression model. More complex models with relatively meaningless regression weights have been avoided. Even though such models are able to provide improved performance, it is suspected that they do so by accounting for artifactual structure in the data and do not provide improved modeling of the subjective speech perception process.

Table 8.19: Summary of the performance of parametric objective measures.

Subjective Quality Scale	Correlation of Objective to Subjective
SF	0.53
SH	0.85
SD	0.65
SL	0.73
SI	0.85
BN	0.91
BF	0.44

In some cases the parametric objective measure results may have utility by themselves. For example, a low score on the BN objective measure may indicate excessive additive noise distortion in the speech system, while a low score on the SF objective measure may indicate insufficient quantization levels in the vocal tract parameters of an LPC speech coder. In general, the parametric measures yield specific information which may be extremely useful in diagnosing the cause of voice quality degradation in a communication system.

The key to designing a good measure of composite acceptability is to measure all significant perceptual aspects of acceptability in the model. This point was illustrated in the multidimensional scaling of Fig. 8.1. This chapter has presented objective measures which are able to track subjective parametric quality, though with varying accuracy. Even though some of the parametric measures have marginal performance in estimating their associated subjective quality, all that is needed for producing improved measures of speech acceptability is parametric objective quality measures which measure different aspects of speech acceptability. Measures are needed which in some sense span the perceptual space of subjective composite acceptability. With this in mind, the real measure of the adequacy of the parametric objective measures is their ability to produce accurate estimates of composite acceptability when combined into a composite measure of speech acceptability. This will be discussed in the next chapter.

References

[8.1] R. E. Barlow, *Statistical Inference Under Order Restrictions.* John Wiley & Sons, Limited, 1972.

Chapter 9

Composite Objective Measures

In this chapter composite objective measures for estimating acceptability in the Diagnostic Acceptability Measure will be discussed. Two sets of measures are presented, one based on the basic objective measures described in Chap. 5, and the other based on the parametric objective measures described in Chap. 8. For both sets, the regression model used can be specified as

$$\hat{y}_i = b_0 + \sum_{j=1}^{k} b_j x_{ij} \tag{9.1}$$

with x_{ij} being an objective variable, b_j being a regression coefficient, and $\hat{y}_i$ being estimated composite acceptability. The index over all objective variables is j, and the index over observations is i. The observation index is the index over all distortion systems for all speakers, and can alternatively be written in terms of s, the speaker index, and d, the distortion system index.

9.1 Composite Basic Measures

This section presents the results of a preliminary study on the viability of composite measures for speech quality assessment. The goal of this work was to determine if significant gains in performance could be achieved by combining a number of individual measures. This research led directly to the more sophisticated parametric composite measures discussed in the next section. Because of the preliminary character of this research, the composite objective measures presented here should be considered as indicators of potential performance rather than as viable measures in their own right.

Composite measures were formed as linear combinations of either the simple or the frequency variant measures presented in Chap. 5. These composite measures were constructed from objective measures which had relatively high correlation with subjective estimates, and which, at the same time, seemed to be measuring different characteristics of the distorted speech signal.

For addressing this last point, a test was performed to assess the degree to which different basic measures contained the same information. A large number of composite measures, each containing six independent variables (six basic objective measures), were tested to determine to what extent the coefficient of correlation between estimated and actual composite acceptability could be improved by combining various types of objective measures. In this way it could be determined, for example, if all spectral distance measures contain similar information despite different formulations and parameterizations, or if the log area ratio measure contains information different from a spectral distance measure. Considering the large number of basic measures (including their parameterizations) presented in Chap. 5, the composite measures could by no means test all combinations of measures. However, a reasonable sampling of combinations is presented for the distortion groups WFC and WBD, which illustrates several general points. (See Table 5.1 for an explanation of WFC and WBD.)

The results of the test are presented in Table 9.1. In the top part of the table, objective measures are denoted by number, with the a key at the bottom of the table. An asterisk next to the line number indicates that the difference between the estimated correlation coefficients of the single and composite measures is significant at the 0.05 level (as computed from Eq. 4.64), while a plus indicates a difference at the 0.10 level. Each of these results will be discussed. First, line 1 suggests that all log spectral distance measures contain similar information. This is indicated by the fact that the best composite measure (composed of six different log spectral distance measures) performs only slightly better

Table 9.1: Results of mutual information test.

		Measures	Maximum Single Result		Maximum Composite Result		Distor. Subset
			$\hat{\rho}$	$\hat{\sigma}_e$	$\hat{\rho}$	$\hat{\sigma}_e$	
1.		1	.63	7.0	.64	6.9	WBD
2.	+	1,2,3	.63	7.0	.67	6.7	WBD
3.		1,2,3,15	.72	6.2	.74	6.0	WBD
4.		4,5	.77	5.7	.78	5.7	WFC
5.		13,14	.93	3.3	.93	3.3	WFC
6.	+	10,11	.65	6.8	.69	6.6	WBD
7.	+	1,11	.65	6.8	.69	6.6	WBD
8.		11,12	.63	6.8	.67	6.7	WBD
9.	*	1,2,3,6,7,8, 9,10,11,12	.63	6.8	.69	6.6	WBD
10.	*	8,9,10,11	.65	6.8	.70	6.5	WBD
11.	*	6,7,10,11	.65	6.8	.70	6.6	WBD

1. Log Spectral Dist.
2. Linear Spectral Dist.
3. Nonlinear Spectral Dist.
4. Signal-to-noise ratio
5. Segmental SNR
6. Linear Feedback Dist.
7. Log Feedback Dist.
8. Linear PARCOR Dist.
9. Log PARCOR Dist.
10. Linear Area Ratio Dist.
11. Log Area Ratio Dist.
12. Energy Ratio
13. Frequency Variant SNR
14. Frequency Variant Seg. SNR
15. Frequency Variant Spectral Dist.

than the best single measure. This is true to a slightly lesser extent for all forms of spectral distance measures (line 2). There is little increase in performance even if frequency variant spectral distance measures are also included in this group (line 3).

There is little to be gained from combining standard and segmental SNR or standard and segmental frequency variant SNR, as shown in lines 4 and 5. However, combinations of linear and log area ratio offer some improvement (line 6), and hence contain different information to some extent. Likewise, the best log spectral distance measure combined with the best LPC-parameter measure, the log area ratio, provides a small gain in performance (line 7). But log area ratio measures combined with energy ratio measures show no significant increase in performance (line 8). This underscores the point that different measures do not necessarily provide different information.

When six members from the entire set of LPC-parameter measures combined with the entire set of spectral distance measures is used to form a composite measure, a statistically significant improvement is obtained (line 9). This illustrates a general phenomenon: more improvement is obtained by combining a good measure with a bad measure of vastly different type than is obtained by combining two or more similar measures. The measures of lines 10 and 11 also provide examples of this phenomenon.

The study of composite measures is an overwhelming task if all possible combinations of a great number of basic measures are to be considered. Therefore, the method for choosing members of a composite measure is based on educated guesswork, experimental evidence, and intuition. In this respect the study was not exhaustive, but nevertheless was sufficient to indicate some of the improvements which are possible when using composite measures. In all, twelve measures were investigated, with the best of these described in Table 9.2, which lists its performance for predicting CA, TSQ, and TBQ over the set of all distortions and the set of WBD distortions. Although the ability of this measure to predict CA across all distortions is very good, a qualification must be made. The measure is based on multiple polynomial regression analysis, in which 3rd order polynomial models were used for each of the measures listed in Table 9.2, and all polynomial models were combined to estimate composite acceptability. Therefore, thirty-four parameters were estimated in the regression analysis (the frequency variant spectral distance measure had six bands, each fit with a 3rd order model). The reason that this result is not a strong one is that the validity and appropriateness of this model was not investigated, either by using traditional regression analysis diagnostics, or by cross validation of the model on a split data set. To the extent that this measure models artifacts in the data, it can be expected to have significantly poorer performance for different data set.

9.2 Composite Parametric Measures

This section presents the results of a study of composite measures based on parametric objective measures. The basic technique used in this research was to combine the individual parametric objective measures

Table 9.2: Performance of composite measure.

Measures included in Composite
1. Log Area Ratio [$\alpha = 0,\ p = 1$]
2. Frequency Variant Log Spectral Distance Measure [$\alpha = 0,\ \gamma = 1.0,\ p = 2$]
3. PARCOR [$\alpha = 0,\ p = 1$]
4. Linear Spectral Distance Measure [$\delta = 1,\ \alpha = 2,\ \gamma = 0,\ p = 2$]
5. Energy Ratio [$\alpha = 0,\ \delta = .25$]
6. Feedback [$\alpha = 0,\ p = 1$]

	CA		TSQ		TBQ	
	$\hat{\rho}$	$\hat{\sigma}_e$	$\hat{\rho}$	$\hat{\sigma}_e$	$\hat{\rho}$	$\hat{\sigma}_e$
ALL	.84	4.6	.82	4.9	.38	6.2
WBD	.86	4.2	.86	4.6	.48	6.0

designed in Chap. 8 into composite measures for the estimation of composite acceptability (CA). In this research, considerable effort was expended in order to simplify and validate the composite measures. Therefore, in contrast to the previous section, the performance predictions of the composite measurements presented in this section are quite well founded.

The composite measures presented in this section are formed in the same manner as the parametric objective measures discussed in Chap. 8. Regression analysis is used, but with the exception that composite acceptability is the dependent variable and that all of the parametric objective variables are included as the independent variables. Therefore the regression weights for each of the parametric objective measures (presented in Chap. 8) are discarded and new weights are computed so that the importance of each objective variable can be determined relative to all others included in the analysis.

9.2.1 Measure with Thirty-Five Variables

The first composite measure includes all of the parametric objective variables discussed in the previous chapter, with little regard for their mutual information. In this respect, there may be too many independent variables, with some of the variables highly correlated. A difference in the formulation of the composite objective measure is that objective variables based on a critical band spectral representation have a uniform frequency band reduction as follows:

$D'(s,d,j)$ Band No.	$D(s,d,k)$ Band Nos.
1	1 – 5
2	6 – 10
3	11 – 15
4	16 – 20
5	21 – 25

where $D'(s,d,k)$ is the new objective variable formed by summing together the original objective variables associated with the bands listed in the right-hand column, and k is the band index. The single exception is the objective variables for BN which were combined as follows:

$D'(s,d,j)$ Band No.	$D(s,d,k)$ Band Nos.
1	1 – 5
2	6 – 15
3	16 – 25

When the lowpass, highpass, bandpass, and band reject monotonic regression stresses are included for the spectral distances associated with the objective measures of SL and SH, there are a total of thirty-five objective variables in the regression analysis. These can be specified as follows:

Number of Variables	Parametric Quality	Specified by Equation
5	SF	8.11
9	SH	8.22
5	SD	8.7
9	SL	8.8
4	SI	8.17 – 8.20
3	BN	8.9

where the column labeled "parametric quality" indicates in which parametric objective measure the objective variables are used.

Table 9.3: Results of regression using twenty-one independent variables.

Multiple R	0.8135	Standard Error of Estimate	5.8255
Multiple R-Squared	0.6618		

The results of combining the preceding thirty-five objective variables in an multiple linear regression analysis is shown in Table 9.3. The correlation of estimated to actual composite acceptability is quite good (0.81) with the regression model explaining slightly more than 65 percent of the variability of subjective composite acceptability. However, the correlation matrix for independent variables has large off-diagonal elements (greater than 0.75), indicating excessive correlation between some of the variables in the model.

9.2.2 Measure with Twenty-One Variables

A second regression analysis was done with the goal of reducing the magnitude of the off diagonal terms in the correlation matrix while still retaining the level of performance in the objective measure. This was done by combining some of the objective variables. Table 9.4 lists the objective variables included in the regression model. There are now twenty-one objective variables in the model, with the reduction from thirty-five accomplished by either dropping a variable or altering the way critical bands are combined to form a new objective variable.

One additional objective variable was added to the parametric variables of Chap. 8, the weighted slope spectral distance measure. This measure is described in Sec. 2.2.9 and in more detail in Sec. 6.3.3. For this measure the speech spectrum was split into the critical bands listed in Table 6.1, K_{max} was set equal to 40, and K_{locmax} set equal to 1000. The weighted slope spectral distance measure was found to give quite good estimates of composite acceptability over the speech quality data base. It was included in the composite objective measure of acceptability as a direct measure of speech acceptability and in this respect complements the parametric objective measures.

The results of the analysis are shown in Tables 9.5 and 9.6. The performance is quite good, with a multiple R of 0.82, so that 67 percent of the variability of subjective acceptability is explained by the regres-

Table 9.4: Listing of twenty-one objective variables.

Parametric Quality	Objective Variable Index	Critical Band Nos.	Equation
SF	1	1 - 5	(8.11)
	2	6 - 10	
	3	11 - 15	
	4	16 - 20	
	5	21 - 25	
SH	1	11 - 15	(8.22)
	2	16 - 20	
	3	21 - 25	
	4	Stress (Lowpass)	
	5	Stress (Highpass)	
SD	1	6 - 10	(8.4)
	2	11 - 20	
SL	1	11 - 25	(8.8)
	2	Stress (Lowpass)	
	3	Stress (Highpass)	
SI	1		(8.17)
	2		(8.18)
	3		(8.19)
	4		(8.20)
BN	1	6 - 15	(8.9)
CA (WSSD)	1	1 - 25	(6.12)

sion model. The jackknife validation method (see Sec. 4.4.2) results in a prediction correlation of 0.81, a negligible decrease in performance. A more stringent test of model performance is obtained by the cross-validation procedure, where the data base is randomly split in half to yield training and prediction data sets. The regression coefficients are computed based on the training data and, using these coefficients, the estimates are computed for the prediction data. A cross validation test for this model results in multiple R for the training analysis of 0.82 and a prediction correlation of 0.77. The analysis correlation is nearly identical to that obtained with the full data set. Though the prediction

Table 9.5: Results of Regression using twenty-one independent variables, (a).

Multiple R	0.8206	Std. Error of Estimate	5.6859
Multiple R-Squared	0.6734		

Jackknife Prediction Statistics:

Multiple R	0.8106	Std. Error of Prediction	5.7659
Multiple R-Squared	0.6571		

Cross-Validation Prediction Statistics:

Multiple R	0.7701	Std. Error of Prediction	6.0369
Multiple R-Squared	0.5931		

Analysis of Variance

Source of Variation	Sum of Squares	Degrees of Freedom	Mean Square	F Ratio	Area in Tail
Regression	68924.33	21	3282.11	101.52	0.0000
Residual	33428.91	1034	32.32		

correlation of 0.77 is statistically different from 0.82 at a 0.05 level of significance, it is still able to account for more that 59 percent of the variability of subjective acceptability when predicting speech acceptability and is only a modest decrease in performance.

The correlation matrix, shown in Fig. 9.1, has relatively few large off diagonal terms, as is desired. This indicates that no superfluous objective variables are included in the model. The appropriateness of the included variables can also be judged by their regression coefficients. Variables which are highly correlated tend to have coefficients of opposite sign. Physically this is illogical, since highly correlated variables usually measure similar information and should therefore have regression weights with at least the same sign.

Table 9.6 lists the regression coefficients for the objective variables in the model. Prior to the regression analysis all independent variables were normalized to zero mean and unit variance. This results in the intercept of the model being nearly equal to the mean of the subjective acceptability ratings. Since most of the parametric objective variables measure speech *distortion*, most all of the regression weights are negative. This means that large distortions result in lower acceptability

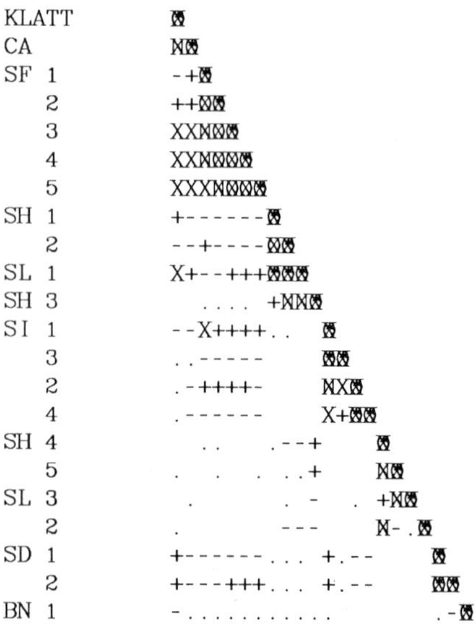

The Absolute Values of the Matrix Entries Have Been Printed Above in Shaded Form According to the Following Scheme:

		LESS THAN OR EQUAL TO	.118
.	.118	TO AND INCLUDING	.236
-	.236	TO AND INCLUDING	.355
+	.355	TO AND INCLUDING	.473
X	.473	TO AND INCLUDING	.591
M	.591	TO AND INCLUDING	.709
⊠	.709	TO AND INCLUDING	.827
▩		GREATER THAN	.827

Figure 9.1: Twenty-one variable correlation matrix in shaded form.

Table 9.6: Results of regression using twenty-one independent variables, (b).

Objective Variable		Regression Coefficient	t-statistic	Two-tail Signif.	Contribution to R-Square
INTERCEPT		53.2388	272.13	.000	
WSSD		-2.9891	-8.36	.000	.02208
BN	1	-1.5523	-6.02	.000	.01143
SI	1	-.5817	-1.11	.269	.00039
	2	-3.8538	-5.68	.000	.01020
	3	.0864	1.55	.120	.00076
	4	.3494	4.14	.000	.00541
SH	1	-1.4625	-3.48	.001	.00382
	2	-.9539	-2.11	.035	.00140
	3	-3.5327	-10.14	.000	.03249
	4	.2261	.77	.443	.00019
	5	-1.5746	-4.95	.000	.00773
SL	1	2.1125	6.93	.000	.01517
	2	-1.0216	-4.44	.000	.00624
	3	1.1061	4.78	.000	.00722
SF	1	1.6316	4.78	.000	.00722
	2	-.0553	-.11	.909	.00000
	3	-2.2809	-5.58	.000	.00984
	4	-1.2635	-3.09	.002	.00302
	5	-.9784	-2.54	.011	.00204
SD	1	-4.4717	-5.99	.000	.01133
	2	2.4899	6.15	.000	.01195

estimates, as would be expected. This does not hold for $D_{SI}(3)$ and $D_{SI}(4)$ which are the squares of $D_{SI}(1)$ and $D_{SI}(2)$. Since these variables are in a sense correction factors they do not have negative weight. Two other exceptions to this general trend are $D_{SF}(1)$ and $D_{SD}(2)$, both of which have positive weight. Although there is no ready physical interpretation for these weights, they are nevertheless statistically significant and contribute to the model.

Figure 9.2 is a scatterplot of the residual, or error in the estimate, versus estimated composite acceptability. It shows no structure in the pattern of the residuals, indicating that the regression model is appropriate. Furthermore, the spread of the residuals is nearly constant over the range of CA which confirms the important assumption of constant error variance. Figure 9.3 is a normal probability plot for the regression model. It is quite linear, indicating that the residuals are very nearly normally distributed. This of course means that parametric statistical tests of the type described in Chap. 4 are valid for this data.

Figure 9.4 shows a plot of y versus $\hat{y}$, or actual subjective acceptability versus estimated acceptability. The numerals in the plot represent both a position in the y, $\hat{y}$ space and the number of observations which fall at that position. The scatter about the regression line is rather large, though the linear trend relating $\hat{y}$ to y is very clear. The points at the low end of the plot represent poor quality systems, some of which are also very influential on the regression analysis. The distortion systems which are at the upper or lower edges of the y versus $\hat{y}$ scatterplot are outliers, since they are furthest from the regression line. If they are eliminated from the analysis, the result is more reliable coefficient estimates and a better performing measure.

Many of the distortion systems in the data base are quite severe, and are not typical of medium to good quality communication systems. Not surprisingly, these systems are often the statistical outliers of the objective measure of composite acceptability. Table 9.7 lists all the distortion systems which are associated with a prediction error greater than two standard deviations of the error distribution for the model of Table 9.5. The four speakers in the data base are referred to as CH, LL, RS and QQ. Of those distortions in Table 9.7, the additive noise distortions are at very low signal-to-noise levels and the interruption distortions are at 50 to 75 percent signal duty cycle, both of which are very severe distortions. Most of the other systems listed in the table are also the most severe of the six levels in each distortion. Echo is a exception, in which the most severe spectral distortion, that caused by a small time delay, corresponds to the perceptually least severe distortion. Therefore small echo delays receive a much lower acceptability rating from the objective measure than from the subjective measure. In the ADM and CVSD coder systems both the least and most severe levels of distortion appear in the table. Closer examination reveals that the objective measure estimates consistently lower speech acceptability for the least severe distortion and consistently higher acceptability for the most severe distortion.

If those systems whose error is greater than two standard deviations are removed, the multiple R of the regression increases to 0.85 as

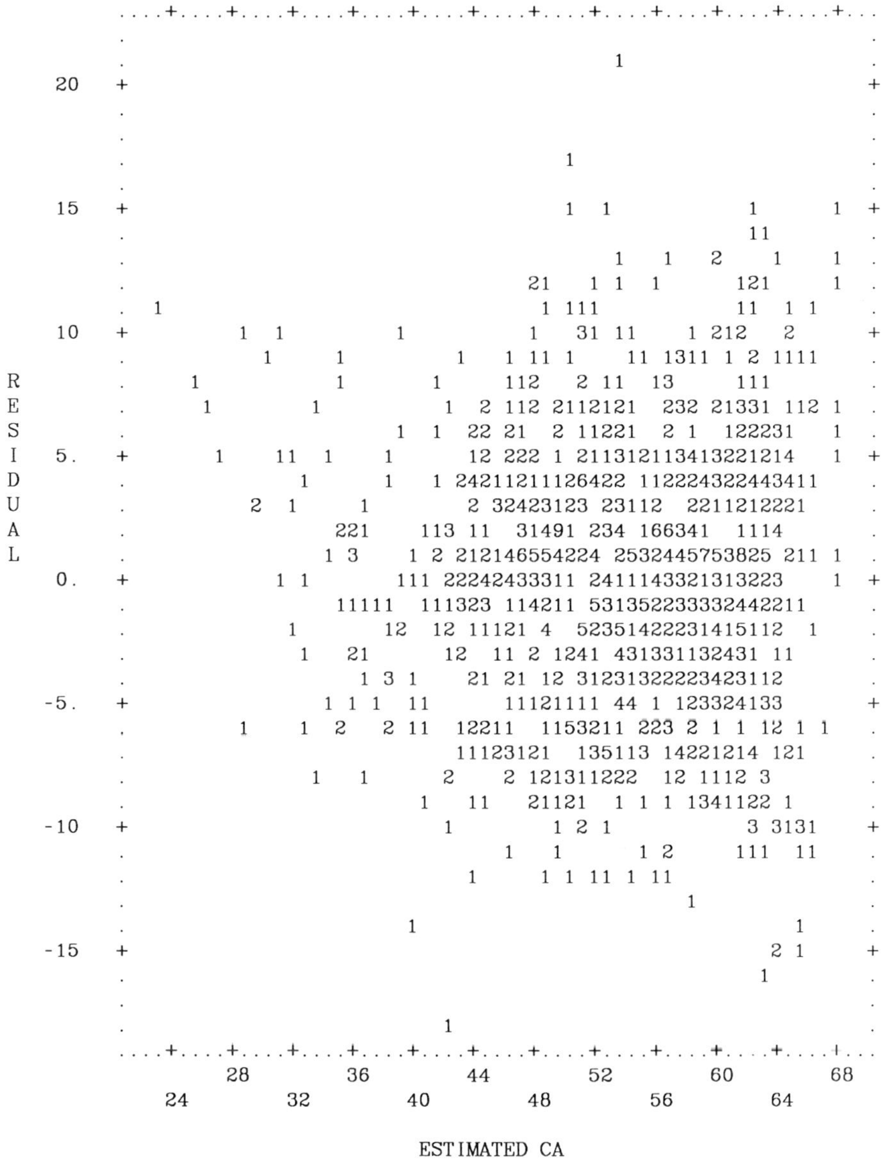

Figure 9.2: Residual versus estimated composite acceptability.

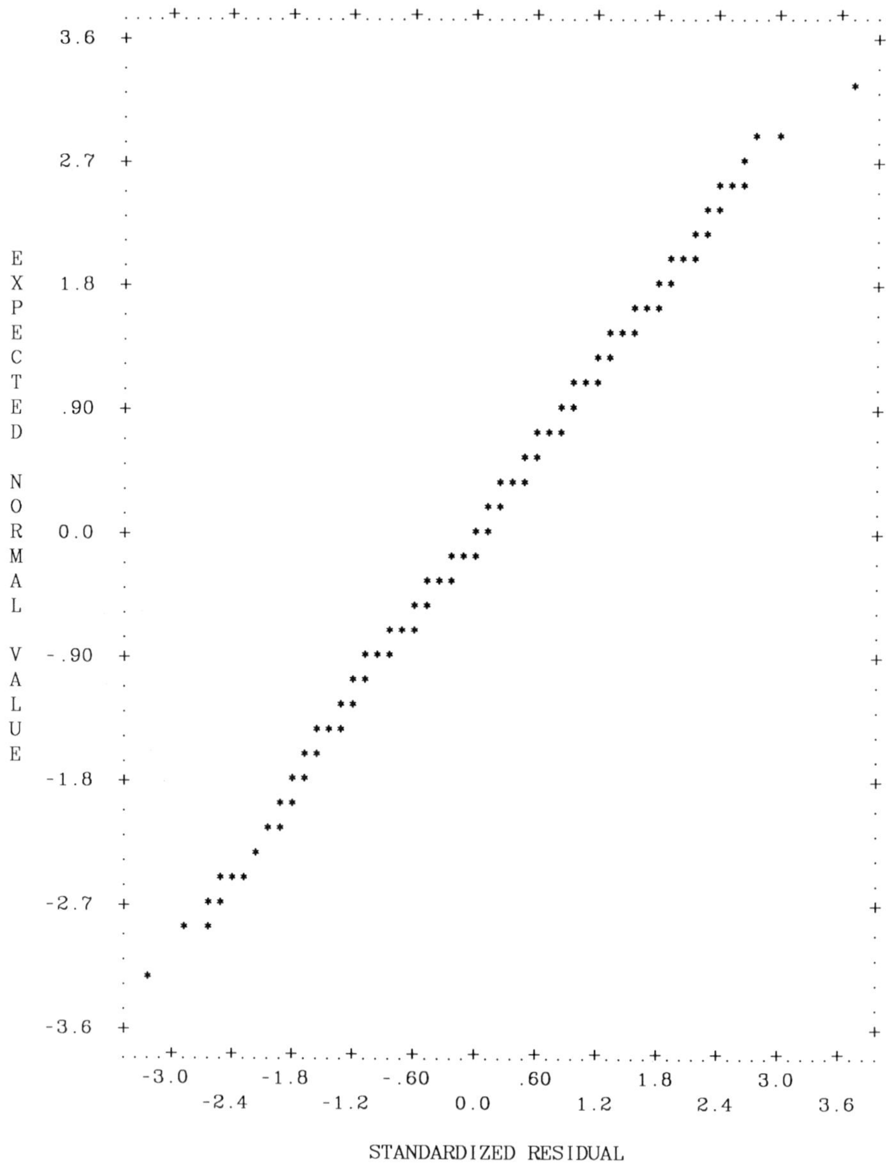

Figure 9.3: Normal probability plot.

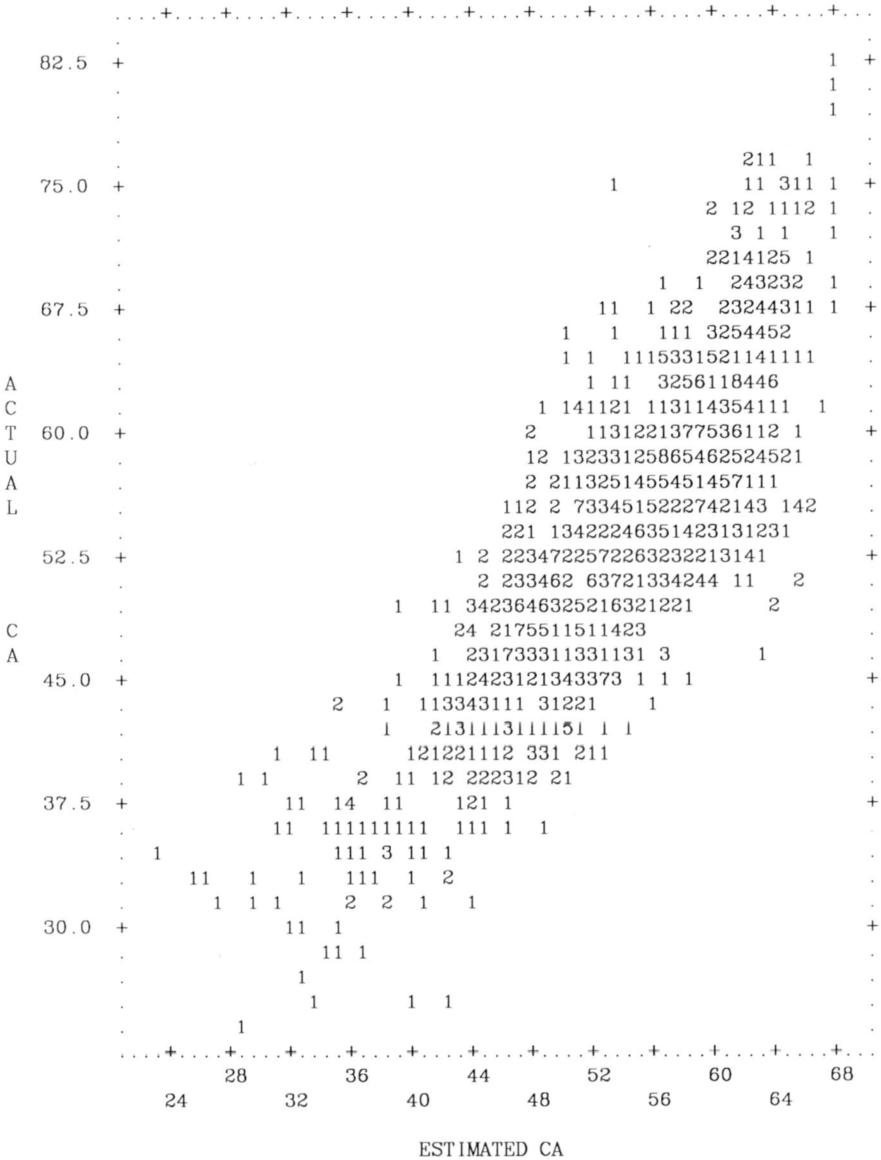

Figure 9.4: Actual versus estimated composite acceptability.

Table 9.7: List of statistical outliers.

Obs. No.	Spkr	Distortion	Obs. No.	Spkr	Distortion
18	RS	SNR 30DB	598	QQ	Pole Distor.
37	RS	L.P. 400 Hz	599	QQ	Pole Distor.
111	RS	Center Clip	629	CH	Pole Distor.
116	QQ	Center Clip	661	RS	Pole Distor.
117	QQ	Center Clip	667	QQ	Pole Distor.
138	RS	Interruption	668	QQ	Pole Distor.
153	LL	Interruption	801	LL	Pole Distor.
154	LL	Interruption	802	LL	Pole Distor.
199	LL	Narrowband Noise	803	LL	Pole Distor.
217	CH	Narrowband Noise	804	LL	Pole Distor.
223	LL	Narrowband Noise	817	CH	Echo
339	CH	ADPCM	829	RS	Echo
342	CH	ADPCM	852	LL	Pole Distor.
427	QQ	Narrowband Freq. Noise	874	LL	ADM
428	QQ	Narrowband Freq. Noise	875	LL	ADM
438	CH	Narrowband Freq. Noise	877	RS	ADM
492	LL	Pole Distor.	878	RS	ADM
493	RS	Pole Distor.	880	RS	ADM
541	RS	Pole Distor.	882	RS	ADM
547	QQ	Pole Distor.	901	RS	CVSD
557	CH	Pole Distor.	907	QQ	CVSD
565	RS	Pole Distor.	912	QQ	CVSD
589	RS	Pole Distor.			

Speaker	Number of Outliers
CH	7
LL	12
RS	15
QQ	11
Total Outliers	45

shown in Table 9.8. This is a better representation of the performance of the objective measure of acceptability, since most very poor quality systems are excluded from the analysis. A correlation of 0.85 accounts for more than 70 percent of the variability of subjective composite acceptability. However, it is important to note that for this data 0.85 is not statistically different from 0.82 at a 0.05 level of significance.

Although Table 9.7 indicates that many of the outliers are un-

Table 9.8: Performance of composite objective measures.

Data Included	Correlation to CA
All data	0.8201
All data, outliers removed	0.8461
One speaker	0.8661
One speaker, outliers removed	0.8778

usual from the point of view of their severity of distortion (or lack of distortion), from a purely statistical point of view 45 points of greater than 2σ deviation is nearly what would be expected in a normal distribution. In particular, out of 1056 observations, one would expect 48 such outliers, which is very close to the 45 obtained in the data.

If the data is restricted to be from only one speaker rather than four, the performance of the resulting regression model increases to a correlation of 0.87, as listed in Table 9.8, which is statistically different from 0.82 at a five percent level. This result suggests that the objective measures are sensitive to physical differences between talkers, while the Diagnostic Acceptability Measure, through careful control procedures, is not.

A much smaller increase in correlation results if the the same outliers are removed as before and the analysis again restricted to just one speaker. In this case the multiple R is 0.88. Obviously a larger decrease in performance results from the objective measure's difficulty in modeling differences between speakers as compared to its difficulty in modeling outlying data.

9.2.3 Dimensionality of Parametric Measures

In regression analysis the correlation matrix of the independent variables gives some indication of the dimensionality of the data, which in this case are the variables associated with the parametric objective measures. However, a more appropriate method for assessing the dimensionality of a set of data is factor analysis. Principal components analysis, one of the simplest forms of factor analysis, was used to analyze the objective measure data.

Factor analysis postulates a model of the form

$$x_{ji} = \sum_{k=1}^{p} a_{jk} F_{ki} + e_{ji} \tag{9.2}$$

where x are the data, F are the *factor scores*, a are the *factor loadings*, i is the observation index, j is the data variable index, k is the factor index, and e is the model error. The goal of factor analysis, as used here, is to summarize the information contained in many data variables into only a few *factors*, while sustaining negligible loss of information. Geometrically, Eq. (9.2) can be interpreted as projecting the data (from j space) onto a lower k-dimensional factor space, in which the factor loadings form the basis of the factor space and the factor scores are the coordinates which locate the data in the factor space. Ideally, the elements of the factor loadings (a_{jk}) are either very large or very small, so that each data variable is associated with a minimum number of factors. This can be accomplished by rotation of the factor loadings to form a new, but equivalent, basis.

In the principal components method of analysis, the sample covariance matrix is computed for all variables in the analysis. The matrix is normalized to have determinant equal to 1.0 and eigenvalues and unit eigenvectors are extracted. Each eigenvalue greater than 1.0 indicates a *factor*, with the square root of the eigenvalue multiplied by the associated eigenvector the *factor loading*. The variance explained by each factor is equal to the eigenvalue associated with that factor. In order to limit the dimensionality of the factor space, eigenvectors with eigenvalues less than 1.0 are interpreted as aspects of the data which are not significant. Factor loadings are rotated, and factors and loading elements within each factor are sorted in order of decreasing significance. In the case of the data presented here, loading components less than 0.25 are set equal to zero to emphasize the most significant factors for the various measures. Factor analysis was performed on the same thirty-five variables used in the regression analysis of section 9.2.1. The analysis identified nine factors which were able to account for 85 percent of the variance in the data space, as shown in Table 9.9, a number which agrees well with the ten dimensions represented by the DAM's parametric scales. Though not shown in the table, augmentation by five factors accounted for an additional 10 percent of the variance.

Table 9.9: Percent of variance explained by factors.

Factor	Variance Explained	Cumulative Proportion of Variance in Data Space	in Factor Space
1	8.6899	.2483	.2921
2	6.1328	.4235	.4983
3	3.7912	.5318	.6257
4	2.9885	.6172	.7262
5	2.6051	.6916	.8137
6	1.7332	.7412	.8720
7	1.5413	.7852	.9238
8	1.2120	.8198	.9645
9	1.0547	.8500	1.0000

The rotated factor loadings are shown in Table 9.10. Each factor is very nearly composed of one, and only one, of the parametric objective measures, indicating that these measures are statistically orthogonal. The exception is the parametric objective measures associated with SH and SL, which are grouped together in factor 1. This is to be expected, since the same objective measure, long-time spectral distance, was used for both of these parametric measures.

9.2.4 Adequacy of the Parametric Objective Measures

A final issue concerning the regression model for composite acceptability is whether or not the objective variables are measuring *all* of the information contained in the parametric subjective measures of the DAM. A way to evaluate this is to observe the increase in model performance, as measured by R-squared, when the following model variable substitution is performed: replace the set of parametric objective quality variables (x_{ij}'s) associated with a parametric subjective scale with the actual subjective quality evaluation for that scale for all distortions in the data base, and then re-run the regression analysis with the new variables. A small increase in the performance of the new model indicates that the information measured by the parametric subjective measures is also measured by the objective variables.

Table 9.11 presents this information. The first few lines of the

Table 9.10: Sorted and rotated factor loadings.

Measure	Factor								
	1	2	3	4	5	6	7	8	9
SL 2	.96	0	0	0	0	0	0	0	0
SH 2	.95	0	0	0	0	0	0	0	0
SL 1	.93	0	0	0	0	0	0	0	0
SH 1	.90	0	0	0	0	0	0	0	0
SL 3	.83	0	0	0	0	0	0	0	0
SH 3	.82	0	0	0	0	0	0	0	0
SL 4	.66	0	-.26	.34	0	0	0	0	.44
SH 4	.59	0	0	.39	0	0	0	0	.42
SD 2	0	.97	0	0	0	0	0	0	0
SD 4	0	.97	0	0	0	0	0	0	0
SD 3	0	.97	0	0	0	0	0	0	0
SD 1	0	.95	0	0	0	0	0	0	0
SD 5	0	.94	0	0	0	0	0	0	0
SF 3	0	0	.86	0	0	0	0	0	0
SF 2	0	0	.85	0	0	0	0	0	0
SF 4	0	0	.85	0	0	0	0	0	0
SF 5	0	.31	.77	0	0	0	0	0	-.33
SF 1	0	0	.76	0	.31	0	0	0	0
SH 6	0	0	0	.86	0	0	0	0	0
SL 6	0	0	0	.78	0	0	0	0	0
SH 7	0	0	0	-.70	0	0	.46	.30	.25
SH 5	.36	.26	0	.59	0	.28	0	0	.30
SI 1	0	0	.28	0	.84	0	0	0	0
SI 2	0	0	0	0	.83	0	0	0	0
SI 3	0	0	0	0	.81	0	0	0	0
SI 4	0	0	0	0	.78	0	0	0	0
BN 1	0	0	0	0	0	.93	0	0	0
BN 2	0	0	0	0	0	.91	0	0	0
BN 3	0	0	0	0	0	.88	0	0	0
SL 8	0	0	0	0	0	0	.88	0	0
SL 7	0	0	0	-.32	0	0	.78	0	0
SH 8	0	0	0	0	0	-.30	.59	0	.48
SH 9	0	0	0	0	0	0	0	.84	0
SL 9	0	0	0	0	0	0	0	.70	0
SL 5	.51	0	0	.48	0	0	0	0	.53

Table 9.11: Performance of Objective versus Subjective Parametric Measures in Composite Measure.

Composite Objective Measure	Multiple R	R^2	
21 Variable Measure	0.81	0.66	
Substituted Subjective Measure	**Multiple R**	**R^2**	**Increase in Multiple R^2**
SF	0.81	0.66	0.00
SH	0.84	0.70	0.04
SD	0.86	0.74	0.08
SL	0.85	0.72	0.06
SI	0.83	0.69	0.03
BN	0.82	0.68	0.02
BF	0.82	0.67	0.01

table show the multiple R and multiple R-squared for the "baseline" composite objective measure of Table 9.3. In the remainder of the table this model was altered by replacing the set of parametric *objective* variables associated with a particular parametric quality with the corresponding parametric *subjective* measure of the DAM. At any instance only a single parametric measure was substituted into the model. The single exception to this is BF, in which the objective variables for BF are the same as those for SF. In this case the subjective measure BF was added to the model without deleting any parametric objective variables.

The significant statistic in Table 9.11 is the increase in multiple R-squared when a subjective parametric measure is substituted into the composite model. The increase is measured relative to the baseline, or fully objective, composite model. This shows that the greatest disparity between subjective and objective measures is for the parametric qualities of SD and SL and, to a lesser extent SH. Therefore the greatest gains in performance of the composite objective measure can be had by reassessing the objective variables for SD and SL so as to match them more closely to their subjective measures. On the other hand, the objective variables for SF appear to capture all of the information provided by the SF parametric subjective measure. Another encouraging point is that objective variables matched to BF would contribute little additional information to the composite model. Therefore not having an objective measure specifically for BF does not appreciably diminish the performance of the composite model.

9.3 Discussion

The very good correlation between the composite objective measures presented in this chapter and subjective acceptability, as summarized in Table 9.8, is clear evidence that the parametric objective measures have led to significantly improved measures of speech acceptability. The hypothesis was that all significant aspects of subjective acceptability had to be included in a composite objective measure of acceptability. The parametric quality scales of the DAM provided means of developing a set of broad and perceptually distinct parametric objective measures, which could then be incorporated in to a large multiple linear regression model for estimating speech acceptability. The performance of these measures, though good in their own right, was validated by their ability to produce very good estimates of composite acceptability in the composite regression model.

The multidimensional scaling of Fig. 9.5 summarizes the relationship between the DAM subjective quality scales and both the parametric and the composite objective measures. Although the stress in this scaling is 20 percent, it still yields an interpretation very closely aligned with the objective measure correlation figures. In the scaling the white dots are the subjective measures, with each identified by the appropriate mnemonic (see Table 3.1). A line is drawn between the subjective measure and the corresponding objective measure, where the black dots represent the objective measures. The most significant aspect of the scaling is that the parametric objective measures are dispersed throughout the realization space. Though some parametric measures are not closely correlated to their respective subjective measure, the dispersion indicates that all the parametric objective measures are measuring different aspects of speech quality. Finally, the most useful objective measure, composite acceptability, is extremely close to subjective composite acceptability which graphically indicates the very good performance of this objective measure.

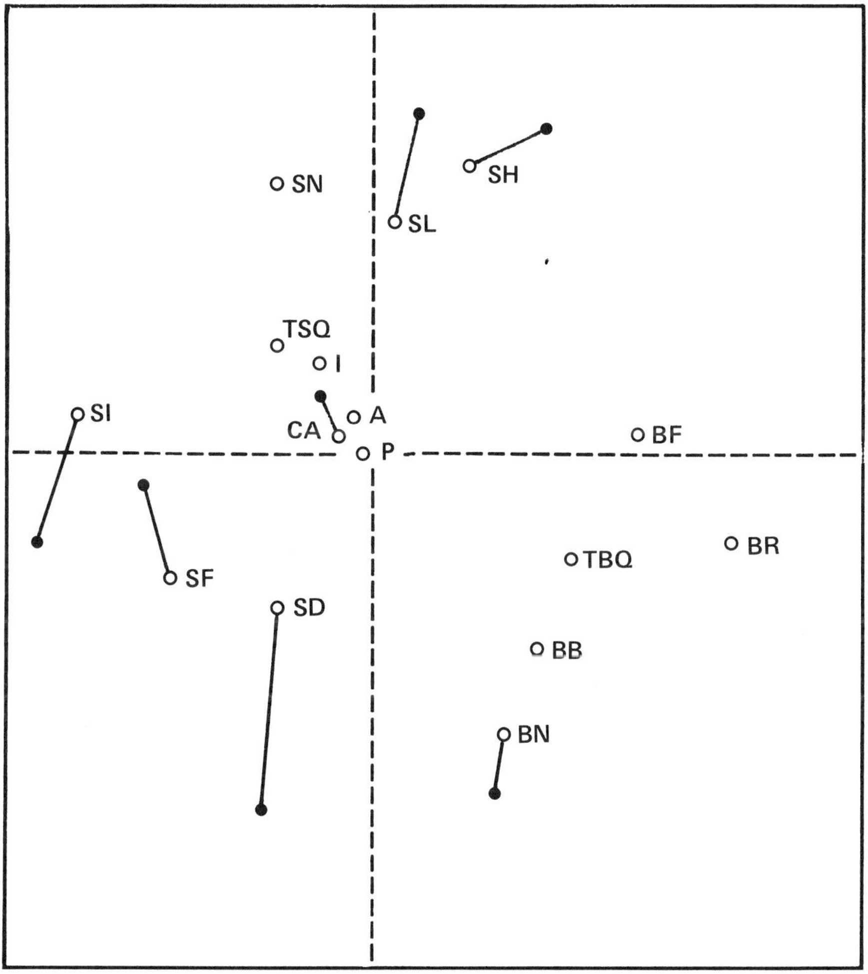

Figure 9.5: Multidimensional Scaling of Subjective and Objective Quality Measures.

Index